Lightning Physics and Lightning Protection

Lightning Physics and Lightning Protection

E M Bazelyan

and

Yu P Raizer

CRC Press
Taylor & Francis Group
Boca Raton London New York

CRC Press is an imprint of the
Taylor & Francis Group, an **informa** business

CRC Press
Taylor & Francis Group
6000 Broken Sound Parkway NW, Suite 300
Boca Raton, FL 33487-2742

First issued in paperback 2019

© 2000 by Taylor & Francis Group, LLC
CRC Press is an imprint of Taylor & Francis Group, an Informa business

ISBN-13: 978-0-367-39904-7

British Library Cataloguing-in-Publication Data

A catalogue record for this book is available from the British Library.

Library of Congress Cataloging-in-Publication Data are available

Publisher: Nicki Dennis
Commissioning Editor: John Navas
Production Editor: Simon Laurenson
Production Control: Sarah Plenty
Cover Design: Victoria Le Billon
Marketing Executive: Colin Fenton

Typeset in 10/12pt Times by Academic + Technical, Bristol

Visit the Taylor & Francis Web site at
http://www.taylorandfrancis.com

and the CRC Press Web site at
http://www.crcpress.com

Contents

Preface

Today, we know sufficiently much about lightning to feel free from the mystic fears of primitive people. We have learned to create protection technologies and to make power transmission lines, skyscrapers, ships, aircraft, and spacecraft less vulnerable to lightning. Yes, the danger is getting less but it still exists! With every step of the technical progress, lightning arms itself with a new weapon to continue the war by its own rules against the self-confident engineer. As we improve our machines and stuff them with electronics in an attempt to replace human beings, lightning acts in an ever refined manner. It takes us by surprise where we do not expect it, making us feel helpless again for some time.

We do not intend to present in this book a set of universal lightning protection rules. Such a task would be as futile as advertising a universal antibiotic lethal to every harmful microbe. The world is changeable, and today's panacea often becomes a useless pill even before the advertising sheet fades. Technical progress has so far failed to take lightning unawares. Improvement and miniaturization of devices increase our concern about the refined destructive behaviour of lightning, but no prophet is able to foresee all of its destructive effects.

We do not plan to discuss in detail all available information on lightning. There are already some excellent books providing all sort of reference data, among them the two volumes of *Lightning* edited by R H Golde and *Lightning Discharge* by M Uman. Our aim is different. We think it important to give the reader some clear, up-to-date physical concepts of lightning development, which cannot be found in the books referred to. These will serve as a basis for the researcher and engineer to judge the properties of this tremendous gas discharge phenomenon. Then we shall discuss the nature of various hazardous manifestations of lightning, focusing on the physical mechanisms of interaction between lightning and an affected construction. The results of this consideration will further be used to estimate the effectiveness of conventional protective measures and to predict technical means for their improvement. We give, wherever possible, technical advice and recommendations. Our main goal, however, is to help the reader to make his own predictions by providing information on the whole arsenal of potentionally hazardous effects of lightning on a particular construction. We have often witnessed situations when an engineer was trying hard to 'impose' this or that protective device on an operating experimental structure which resisted his unnatural efforts. Ideally, the designer must be able to foresee all details of the relationship between lightning and the construction being designed. It is only in this case that lightning protection can become functionally effective and the protective device can be made compatible with the construction elements.

If an engineer is determined to follow this approach, both expedient and well-grounded, he will find this book useful. It is a natural extension of our previous book *Spark Discharge*, published by CRC Press in 1997, which dealt with streamer–leader breakdown of long gas gaps. The streamer–leader process is part of any lightning discharge when a plasma spark closes a gigantic air gap. Although the destructive effect of lightning is primarily due to the return stroke which follows the leader, it is the leader that makes the discharge channel susceptible to it. This is why we give an overview of the streamer–leader process, focusing on extremal estimations and presenting some new ideas. We hope that the second chapter will prove informative even for those familiar with our book of 1997.

Some results of the lightning investigation run in the Krzhizhanovsky Power Institute are used in the book. The authors would like to thank Dr B N Gorin and Dr A V Shkilev who kindly allowed us to use the originals of lightning photographs. We are also grateful to L N Smirnova for translation of this book.

Chapter 1

Introduction: lightning, its destructive effects and protection

If you want to observe lightning, the best thing to do is to visit a special lightning laboratory. Such laboratories exist in all parts of the globe except the Antarctic. But you can save on the travel if you just climb onto the roof of your own house to give a good field of vision. Better, fetch your camera. Even an ordinary picture can show details the unaided human eye often misses. You might as well sit back in your favourite armchair, having pulled it up to a window, preferably one overlooking an open space. The camera can be fixed on the window sill. There is nothing else to do but wait for a stormy night.

There is enough time for the preparations to be made because the storm will be approaching slowly. At first, the air will grow still, and it will get much darker than it normally is on a summer night. The cloud is not yet visible, but its approach can be anticipated from the soundless flashes at the horizon. They gradually pull closer, and the brightest of them can already be heard as delayed and yet amiable roaring. This may go on for a long time. It may seem that the cloud has stopped still or turned away, but suddenly the sky is ripped open by a fire blade. This is accompanied by a deafening crash, quite different from a cannon shot because it takes a much longer time. The first lightning discharge is followed by many others falling out of the ripped cloud. Some strike the ground while others keep on crossing the sky, competing with the first discharge in beauty and spark length. This is the right time to start observations: remove the camera shutter and try to take a few pictures.

1.1 Types of lightning discharge

The above recommendation to remove the camera shutter should be taken literally. Much information on lightning has been obtained from photographs taken with a preliminarily opened objective lens. It is important, however, that

1

Figure 1.1. A static photograph of a lightning stroke at the Ostankino Television Tower in Moscow.

no other bright light source should be present within the vision field of the camera lens. The film can then be exposed for many minutes until a spark finds its way into the frame. After this, the lens should be closed with the shutter and the camera should be set ready for another shot. Experience has shown that at least one third of pictures taken during a good night thunderstorm prove successful.

All lightning discharges can be classified, even without photography, into two groups – intercloud discharges and ground strikes. The frequency of the former is two or three times higher than that of the latter. An intercloud spark is never a straight line, but rather has numerous bends and branchings. Normally, the spark channel is as long as several kilometres, sometimes dozens of kilometres.

The length of a lightning spark that strikes the ground can be defined more exactly. The average cloud altitude in Europe is close to three kilometres. Spark channels have about the same average length. Of course, this parameter is statistically variable, because a discharge from a charged cloud centre may start at any altitude up to 10 km and because of a large number of spark bends. The latter are observable even with the unaided eye. In a photograph, they may look strikingly fanciful (figure 1.1). A photograph can show another important feature inaccessible to the naked eye – the main bright spark reaching the ground has numerous branches which have stopped their development at various altitudes. A single branch may have a length comparable with that of the principal spark channel (figure 1.2).

Branches can be conveniently used to define the direction of lightning propagation. Like a tree, a lightning spark branches in the direction of

Figure 1.2. A photograph of a descending lightning with numerous branches.

growth. In addition to descending sparks outgrowing from a cloud toward the ground, there are also ascending sparks starting from a ground construction and developing up to a cloud (figure 1.3). Their direction of growth is well indicated by branches diverging upward.

In a flat country, an ascending spark can arise only from a skyscraper or a tower of at least 100–200 m high, and the number of ascending sparks grows with the building height. For example, over 90% of all sparks that strike the 530-m high Ostankino Television Tower in Moscow are of the ascending type [1]. A similar value was reported for the 410-m high Empire State Building in New York City [2]. Buildings of such a height can be said to fire lightning sparks up at clouds rather than to be attacked by them. In mountainous regions, ascending sparks have been observed for much lower buildings. As an illustration, we can cite reports of storm observations made on the San Salvatore Mount in Switzerland [3]. The receiving tower there was only 70 m high but most of the discharges affecting it were of the ascending type.

Skyscrapers and television towers are, however, quite scarce on the Earth. So the researcher has a natural desire to construct, in the right place and for a short time, a spark-generating tower of his own. For this, a small probe pulling up a thin grounded wire is launched towards a

Figure 1.3. A photograph of an ascending lightning.

storm cloud [4]. When the probe rises to 200–300 m above the earth, an ascending spark is induced from it. A discharge artificially induced in the atmosphere is often referred to as triggered lightning. To raise the chances for a successful experiment, the electric field induced by the storm charges at the ground surface are measured prior to the launch. The probe is triggered when the field strength becomes close to 200 V/cm, which provides spark ignition in 60–70% of launches [5].

The value 200 V/cm is two orders of magnitude smaller than the threshold value of $E = 30$ kV/cm, at which a short air gap with a uniform field is broken down under normal atmospheric conditions. Clearly, no spark ignition would be possible without the local field enhancement by electric charges induced on the probe and the wire. Below, we shall discuss the triggered discharge mechanism in more detail.

A field detector on the Earth's surface (it might as well be placed on the window of your own room) can easily determine the polarity of the charge transported by a lightning spark to the ground. The polarity of the spark is defined by that of the charge. About 90% of descending sparks occurring in Europe during summer storms carry a negative charge, so these are known as negative descending sparks. The other descending sparks are positive. The proportion of positive sparks has been found to be somewhat larger in tropical and subtropical regions, especially in winter, when it may be as large as 50%.

There is no special name for lightning sparks generated by aircraft during flights, when they are entirely insulated from the ground. Such discharges arise fairly frequently. A modern aircraft experiences at least one lightning stroke every 3000 flight hours. Almost half of the strokes start from the aircraft itself, not from a cloud. This often happens in heap rather than clouds carrying a relatively small electric charge. The reason for a discharge from a large ground-insulated object is principally the same as from a grounded object and is due to the electric field enhancement by surface polarization charge. This issue will be discussed after the analysis of ascending sparks in section 4.2.

1.2 Lightning discharge components

An observer can notice a lightning spark flicker which, sometimes, may become quite distinct. Even the first cinematographers knew that the human eye could distinguish between two events only if they occurred with a time interval longer than 0.1 s. Since lightning flicker is observable, the pause between two current impulses must be longer than 0.1 s.

A current-free pause can be measured quite accurately by exposing a moving film to a lightning discharge. With up-to-date lenses and photographic materials, one can obtain a good 1 mm resolution of the film. In order to displace an image by 1 mm over a time period of 0.1 s, the film speed must be about 1 cm/s. It can be achieved by manually moving the film keeping the camera lens open (alas, an electrically driven camera is unsuitable for this). Then, with some luck, one may get a picture like the one in figure 1.4. The spark flashes up and dims out several times. Unless the pause is too long, a new flash follows the previous trajectory; otherwise, the spark takes a partially or totally new path.

A lightning spark with several flashes is known as a multicomponent spark. One may suggest that the channel of the first component formed in unperturbed air differs in its basic characteristics from the subsequent channels, if they take exactly the same path through the ionized and heated air. The formation of subsequent components is considered in sections 4.7 and 4.8. Note only that multicomponent sparks are usually negative, both ascending and descending. The average number of components is close to three, while the maximum number may be as large as thirty. Generally, the average duration of a lightning flash is 0.2 s and the maximum duration is 1–1.5 s [6], so it is not surprising that the eye can sometimes distinguish between individual components. Positive sparks normally contain only one component.

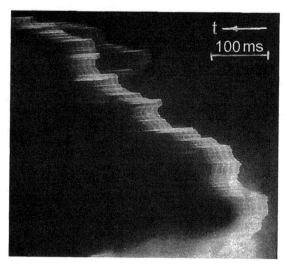

Figure 1.4. The image of a multicomponent lightning in a slowly moving film.

1.3 Basic stages of a lightning spark

The affinity of lightning to a spark discharge was demonstrated by Benjamin Franklin as far back as the 18th century. Historically, basic spark elements were first identified in lightning, and only much later were they observed in laboratory sparks. This is easy to understand if one recalls that a lightning spark has a much greater length and takes a longer time to develop, so that its optical registration does not require the use of sophisticated equipment with a high space and time resolution. The first streak photographs of lightning, taken in the 1930s by a simple camera with a mechanically rotated film (Boys camera), are still impressive [7]. They show the principal stages of the lightning process – the leader stage and the return stroke.

The leader stage represents the initiation and growth of a conductive plasma channel – a leader – between a cloud and the earth or between two clouds. The leader arises in a region where the electric field is strong enough to ionize the air by electron impact. However, it mostly propagates through a region in which the external field induced by the cloud charge does not exceed several hundreds of volts per centimetre. In spite of this it does propagate, which means that there is an intensive ionization occurring in its tip region, changing the neutral air to a highly conductive plasma. This becomes possible because the leader carries its own strong electric field induced by the space charge concentrated at the leader tip and transported together with it. A rough analogue of the leader field is that of a metallic needle connected with a thin wire to a high voltage supply. If the needle is sharp enough, the electric field in the vicinity of its tip will be very strong even at a relatively low voltage. Imagine now that the needle is falling down on to the earth, pulling the wire behind it. The strong field region, in which the air molecules become ionized, will move down together with the needle.

A lightning spark has no wire at its disposal. The function of a conductor connecting the leader tip to the starting point of the discharge is performed by the leader plasma channel. It takes a fairly long time for a leader to develop – up to 0.01 s, which is eternity in the time scale of fast processes involving an electric impulse discharge. During this period of time, the leader plasma must be maintained highly conductive, and this may become possible only if the gas is heated up to an electric arc temperature, i.e. above 5000–6000 K. The problem of the channel energy balance necessary for the heating and compensation for losses is a key one in leader theory. It is discussed in chapters 2 and 4, as applied to various kinds of lightning discharge.

A leader is an indispensable element of any spark. The initial and all subsequent components of a flash begin with a leader process. Although its mechanism may vary with the spark polarization, propagation direction and the serial number of the component, the process remains essentially the same. This is the formation of a highly conductive plasma channel due to the local enhancement of the electric field in the leader tip region.

A return stroke is produced at the moment of contact of a leader with the ground or a grounded object. Most often, this is an indirect contact: a counterpropagating leader, commonly termed a counterleader, may start from an object to meet the first leader channel. The moment of their contact initiates a return stroke. During the travel from the cloud to the ground, the lightning leader tip carries a high potential comparable with that of the cloud at the spark start, the potential difference being equal to the voltage drop in the leader channel. After the contact, the tip receives the ground potential and its charge flows down to the earth. The same thing happens with the other parts of the channel possessing a high potential. This 'unloading' process occurs via a charge neutralization wave propagating from the earth up through the channel. The wave velocity is comparable with the velocity of light and is about 10^8 m/s. A high current flows along the channel from the wave front towards the earth, carrying away the charge of the unloading channel sites. The current amplitude depends on the initial potential distribution along the channel and is, on average, about 30 kA, reaching 200–250 kA for powerful lightning sparks. The transport of such a high current is accompanied by an intense energy release. Due to this, the channel gas is rapidly heated and begins to expand, producing a shock wave. A peal of thunder is one of its manifestations.

The return stroke is the most powerful stage of a lightning discharge characterized by a fast current change. The current rise can exceed 10^{11} A/s, producing a powerful electromagnetic radiation affecting the performance of radio and TV sets. This effect is still appreciable at a distance of several dozens of kilometres from the lightning discharge.

Current impulses of a return stroke accompany all components of a descending spark. This means that the leader of every component charges the channel as it moves down to the earth, but some of the charge becomes neutralized and redistributed at the return stroke stage. Prolonged peals of thunder result from the overlap of sound waves generated by the current impulses from all subsequent spark components.

An ascending spark is somewhat different. The leader of the first component starts at a point of zero potential. As the channel travels up, the tip potential changes gradually until the leader development ceases somewhere deep in the cloud. There is no fast charge variation during this process; as a result, the first component has no return stroke. However, all subsequent spark components starting from the cloud do develop return strokes and behave exactly in the same way as a descending spark.

Of special interest is the return stroke of an intercloud discharge. Its existence is indicated by peals of thunder as loud as those of descending sparks. Clearly, an intercloud leader is generated in a charged region of a storm cloud, or in a storm cell, and travels towards an oppositely charged region. The charged region of a cloud should not be thought of as a conductive body, something like a plate of a high voltage capacitor. Cloud charges are distributed throughout a space with a radius of hundreds of metres and are localized on water droplets and ice crystals, known as hydrometeorites, having no contact with one another. The formation of a return stroke implies that the leader comes in contact with a highly conductive body of an electrical capacitance comparable with, or even larger than, that of the leader. It appears that the role of such a body in an intercloud discharge is played by a concurrent spark coming in contact with the first one.

Measurements made at the earth surface have shown that the current impulse amplitude of a return stroke decreases, on average, by half for about 10^{-4} s. This parameter variation is very large – about an order of magnitude around the average value. Current impulses of positively charged sparks are usually longer than those of negatively charged ones, and the impulses of the first components last longer than those of the subsequent ones.

A return stroke may be followed by a slightly varying current of about 100 A, which may persist in the spark channel for some fractions of a second. At this final stage of continuous current, the spark channel remains electrically conductive with the temperature approximately the same as in an arc discharge. The continuous current stage may follow any lightning component, including the first component of an ascending spark which has no return stroke. This stage may be sporadically accompanied by current overshoots with an amplitude up to 1 kA and a duration of about 10^{-3} s each. Then the spark light intensity becomes much higher, producing what is generally termed as M-components.

1.4　Continuous and stepwise leaders

This introductory chapter contains no theory, and this makes the discussion of leader details a very complicated task. So we shall mention only its principal features which can be registered by a continuously moving film. Continuous streak photographs show lightning development in time. One needs, however, a certain skill and experience to be able to interpret them adequately. Suppose a small light source moves perpendicularly to the earth at a constant velocity. It may be a luminant bomb descending with a parachute. A film moving horizontally, i.e. in the transverse direction, at a constant speed will show a sloping line (figure 1.5(*a*)). Given the film speed (the display rate), one can easily calculate the light source velocity from the line slope. A uniformly propagating vertical channel will leave on a film a sloping wedge (figure 1.5(*b*)) rather than a line. From its slope, too, one can find the channel velocity, or its propagation rate. The higher the rate of the process in question, the higher must be the display rate in streak photography. The highest display rates can be obtained using an electron–optical converter, in which an image is converted to an electron beam scanned across the screen by an electric field. A conventional photo-camera registers the displayed electronic image from the screen onto an immobile film. Electron–optical converters have provided much information on long sparks, but their application in lightning observations has been limited. The main results here have been obtained using mechanical streak cameras. We described this technique and analysed streak pictures in our book on long sparks [8].

　　Figure 1.6(*a*) shows the leader of an ascending lightning spark going up from the top of a grounded tower in the electric field of a negatively charged cloud cell. The leader carries a positive space charge and, therefore, it should be referred to as a positive leader. One can clearly see the bright trace of the channel tip, which looks like a nearly continuous line. This kind of leader is known in literature as a continuous leader. The changing trace slope suggests that the leader velocity changes during its propagation. These changes are, however, quite smooth, not interrupting the tip travel up to the cloud.

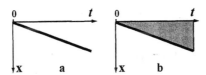

Figure 1.5. The analysis of an image of a vertically descending light source in a horizontally moving film (image display in streak photography): (*a*) point source, (*b*) elongating channel.

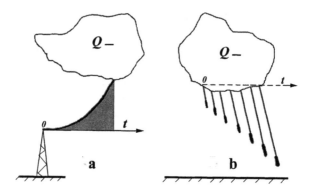

Figure 1.6. A schematic streak picture of a positive ascending (*a*) and a negative descending (*b*) lightning leader.

An essentially different behaviour is exhibited by the leader shown in figure 1.6(*b*). The channel grows in a stepwise manner, covering several dozens of metres in each step. Hence, this kind of leader is termed as a stepwise leader. The new step in the photograph is especially bright; its appearance makes the whole channel behind it also a little brighter. The step length varies between 10 and 200 m with an average of 30 m. The time lapse between two steps is 30–90 µs [9]. The stepwise pattern is characteristic of negatively charged leaders. Positive leaders, both ascending and descending, usually grow in a continuous manner. When averaged over the total time of development, the velocity of stepwise and continuous leaders prove nearly the same, 10^5–10^6 m/s, with an average of about 3×10^5 m/s.

If the leader of the next component moves along the hot track of the first one, it always develops continuously. The new process, termed a dart leader, differs from the first one exclusively in a high leader velocity, about $(1-4) \times 10^7$ m/s. It does not change much along its trajectory from the cloud to the earth. Streak photographs clearly show the bright head of a dart leader, while the channel light intensity is much lower. If the next component takes its own path, its leader behaves in the same way as that of the first component, i.e. it develops more slowly and often in a stepwise pattern.

Dart leaders have not had a fair share of attention from researchers. There is neither theory nor laboratory analogue of this type of gas discharge. Still, it is a most fascinating form of discharge developing record high leader velocities. The contact of a dart leader with the earth produces the fastest current rise, which can reach its amplitude maximum within 10^{-7} s. This is the source of record strong electromagnetic fields which exert one of the most hazardous effects on modern equipment. An attempt at a theoretical treatment of the dart leader will be made in section 4.8.

1.5 Lightning stroke frequency

1.5.1 Strokes at terrestrial objects

Experience shows that lightning most frequently strikes high objects, especially those dominating over an area. In a flat country, it is primarily attracted by high single objects like masts, towers, etc. In mountains, even low buildings may be affected if they are located on a high hill or on the top of a mountain. Common sense suggests that it is easier for an electrical discharge, such as lightning, to bridge the shortest gap to the highest object in the locality. In Europe, for example, a 30 m mast experiences, on the average, 0.1 lightning stroke per year (or 1 stroke per 10 years), whereas a single 100 m construction attracts 10 times more lightnings. On closer inspection, the strong dependence of stroke frequency on the construction height does not look trivial. The average altitude of the descending discharge origin is about 3 km, so a 100 m height makes up only 3% of the distance between the lightning cloud and the earth. Random bendings make the total lightning path much longer. One has to suggest, therefore, that the near-terrestrial stage of lightning behaviour involves some specific processes which predetermine its path here. These processes lead to the attraction of a descending leader by high objects. We shall discuss the attraction mechanism in chapter 5.

    Scientific observations of lightning show that there is an approximately quadratic dependence of the stroke frequency N_1 on the height h of lumped objects (their height is larger than the other dimensions). Extended objects of length l, such as power transmission lines, show a different dependence, $N_1 \sim hl$. This suggests the existence of an equivalent radius of lightning attraction, $R_{eq} \sim h$. All lightnings displaced from an object horizontally at a distance $r \leqslant R_{eq}$ are attracted by it, the others missing the object. This primitive pattern of lightning attraction generally leads to a correct result. For estimations, one can use $R_{eq} \approx 3h$ and borrow the stroke frequency per unit unperturbed area per unit time, n_1, from meteorological observations. The latter are used to make up lightning intensity charts. For example, the lightning intensity in Europe is $n_1 < 1$ per $1 \, \mathrm{km}^2$ per year for the tundra, 2–5 for flat areas, and up to 10 for some mountainous regions such as the Caucasus. A tower of $h = 100$ m is characterized by $R_{eq} = 0.3$ km with $N_1 = n_1 \pi R_{eq}^2 \approx 1$ stroke per year at the average value of $n_1 \approx 3.5 \, \mathrm{km}^{-2} \, \mathrm{year}^{-1}$. This estimation is meaningful for a flat country and only for not very high objects, $h < 150$ m, which do not generate ascending lightnings.

1.5.2 Human hazard

It has long been proved that Galvani was wrong suggesting a special 'animal electricity'. A human being is, to lightning, just another sticking object, like a tree or a pole, only much shorter. The lightning attraction radius for humans is as small as 5–6 m and the attraction area is less than 10^{-4} km^2. If a man had stopped alone in the middle of a large field two thousand years ago, he might have expected to attract a direct lightning stroke only by the end of the third, coming millennium. In actual reality, however, the number of lightning victims is large, and direct strokes have nothing to do with this. It is known from experience that one should not stay in a forest or hide under a high tree in an open space during a thunderstorm. A tree is about 10 times higher than a man, and a lightning strikes it 100 times more frequently. When under the tree crown, a man has a real chance to be within the zone of the lightning current spread, which is hazardous.

After a lightning strikes the tree top, its current I_M runs down along its stem and roots to spread through the soil. The root network acts as a natural grounding electrode. The current induces in the soil an electric field $E = \rho j$, where ρ is the soil resistivity and j is the current density. Suppose the current spreads through the soil strictly symmetrically. Then the equipotentials will represent hemispheres with the diagonal plane on the earth's surface. The current density at distance r from the tree stem is $j = I_M/(2\pi r^2)$ the field is $I_M \rho/(2\pi r^2)$ and the potential difference between close points r and $r + \Delta r$ is equal to $\Delta U = (I_M \rho/2\pi)[r^{-1} - (r + \Delta r)^{-1}] \approx E(r)\Delta r$. If a person is standing, with his side to the tree, at distance $r \approx 1$ m from the tree stem centre and the distance between his feet is $\Delta r \approx 0.3$ m, the voltage difference on the soil with resistivity $\rho = 200\,\Omega/$m will be $\Delta U \approx 220$ kV for a moderate lightning of $I_M = 30$ kA. This voltage is applied to the shoe soles and, after a nearly inevitable and fast breakdown, to the person's body. There is no doubt that the person will suffer or, more likely, will be killed – the applied voltage is too high. Note that this voltage is proportional to Δr. This means that it is more dangerous to stand with one's feet widely apart than with one's feet pressed tightly together. It is still more dangerous to lie down along the radius from the tree, because the distance between the extreme points contacting the soil becomes equal to the person's height. It would be much safer to stand still on one foot, like a stork. But it is, of course, easier to give advice than to follow it. Incidentally, lightning strikes large animals more frequently than humans, also because the distance between their limbs is larger.

If you have a cottage equipped with a lightning protector, take care that no people could approach the grounding rod during a thunderstorm. The situation here is similar to the one just described.

1.6 Lightning hazards

1.6.1 A direct lightning stroke

In the case of a direct lightning stroke, the current flows through the conducting elements of the affected object, with the hot channel contacting the construction element which has received the stroke.

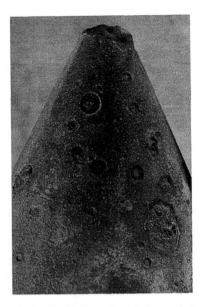

Figure 1.7. Traces of lightning strokes at the steel tip of the Ostankino Television Tower in Moscow.

Thermal effects of lightning are most hazardous at the site of contact of a high temperature channel with combustible materials. This often leads to a fire which becomes most probable when the continuous current stage has a long duration. A return stroke is unlikely to cause a fire even in the case of a powerful lightning discharge, because the strong shock wave produced blows off the flames and combustion products. In combustible dielectric materials a lightning stroke contacts on its way may first be broken down by the strong electric field of the leader tip and then, in the return stroke and continuous current stages, they may be melted through at the site of contact with the hot spark. A burn-through or a burn-off often occurs at the point where the spark contacts a metallic surface several millimetres thick. The holes and burn-offs are usually of the same size. The photograph in figure 1.7 demonstrates the traces of numerous lightning strokes on the steel tip of the Ostankino Television Tower. Slight faults cannot disturb the mechanical strength of a massive metallic construction. Normally, the hazards of burn-offs and fuses are associated with the melted metal in-flow into an object which may contain inflammable and explosive materials or gas mixtures. Incidentally, not only is a burn-through of a metallic wall dangerous but also the local overheating when the temperature of the inner metal surface may go up to 700–1000°C. Unfortunately, the surface often acts as a lighter.

Thermal damage of conductors, through which lightning current flows, occurs fairly rarely. It is characteristic of miniature antennas and various detectors mounted on the outer construction surfaces. The probability of emergency increases if lightning current encounters bolted or riveted joints. The electric contact thus formed always has an elevated contact resistance which may cause a local overheating. This results in the metal release and rivet loosening, disturbing the mechanical strength of the joint. Mobile joints (hinges, ball bearings, etc.) are subject to a similar damage. The site of a sliding contact becomes locally overheated to produce cavities which hamper the motion of mobile parts. In extreme conditions, they may become welded.

Electrodynamic effects of lightning current rarely become hazardous. Mechanical stress arising in electrically loaded and closely spaced metallic structures or in a single structure with an abruptly changing direction of the current is not appreciable and lasts less than 100 ms (it is the attenuation time of a current impulse). However, lightning current has been repeatedly observed to narrow down thin metallic pipes, to change the tilt of rods and to strain thin surfaces. Such effects are not vitally dangerous in themselves but, under certain conditions, may lead to an emergency. As an illustration, imagine the situation when the lightning-affected pipe is part of an aircraft speed control. What will happen if the crew take the readings for granted and do not receive corrections from a ground air traffic controller?

Electrohydraulic effects of lightning are much more hazardous than those discussed above. Modern machines have parts made from a variety of composite materials. These may include, along with plastics, superthin metallic films (both outer and inner), nearly as thin metallic meshes, and miniature conductors monolithic with a dielectric wall. Under the action of lightning current, these metallic parts evaporate, the arising arcs contacting the plastic making it decompose and evaporate. A shock wave appears which splits and bloats the composite wall. A similar effect arises when a lightning spark partially penetrates through a narrow slit between vaporizable plastic walls (most plastics possess gas-generating properties). No one questions a great future of composite materials, but their peaceful coexistence with lightning is still a challenge to the engineer.

Direct stroke overvoltage represents a hazardous rise of voltage when a lightning current impulse propagates across the construction elements. We shall analyse this very dangerous effect of lightning with reference to a power transmission line, because engineers first encountered the phenomenon of overvoltage in such lines. Moreover, the problem of electric insulation for a transmission line can be stated most clearly. Figure 1.8 shows schematically a metallic tower with a ground rod (the grounding resistance is R_g) and a high voltage wire suspended by an insulator string. Above the wire, there may be a lightning conductor attached right to the tower. It stretches along all the line and is to trap lightning sparks aimed

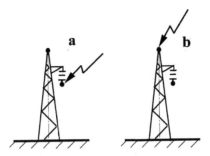

Figure 1.8. Lightning current as an overvoltage source on a power transmission line during a stroke at the power wire (*a*) and a grounded tower (*b*).

at the line wires. A rigorous solution to this problem is given later in this book. Here, we should only like to explain the nature of overvoltages phenomenologically.

Suppose at first that the lightning conductor has proved unreliable, and a discharge has struck the wire (figure 1.8(*a*)). At the point of the stroke, the current will branch to produce two identical waves of the amplitude $I_M/2$, where I_M is the lightning current amplitude. The two waves will run towards the ends of the line with a velocity nearly equal to vacuum light velocity, $c = 3 \times 10^8$ m/s. Until the end-reflected waves return, the wire potential relative to the ground will rise to $U_M = I_M Z/2$. The wave resistance $Z = (L_1/C_1)^{1/2}$ in this expression is defined by the inductance L_1 and the capacitance C_1 per unit wire length; it varies slightly, between 250 and 350 Ω, with the height and the wire radius. With this wave resistance, the average lightning current with an amplitude $I_M = 30$ kA will raise the wire potential up to $U_M = 3750$–5250 kV. The tower potential will practically remain unchanged and equal to zero, so the insulation overvoltage will be close to the calculated value of U_M. This will be clear if we compare U_M with the operating line voltage which does not exceed 1000 kV even in high power lines but normally is 250–500 kV.

In reality, the distance to the line ends l is as large as many dozens of kilometres. The time it takes the reflected wave of the opposite sign cutting down the overvoltage to arrive back at the stroke point is $\Delta t = 2l/c$, or many hundreds of microseconds. This time is much longer than the strong current duration in the return stroke (100 μs). For this reason, reflected waves, which become strongly attenuated, do not normally have enough time to interfere with the process so that the overvoltage acts as long as a lightning current impulse. Practically, any lightning stroke at a wire represents a real hazard: the insulation will be broken down to produce short-circuiting. The power line in that case must be disconnected.

Suppose now that lightning has struck a tower. More often, this is actually not a tower but rather an overhead grounded wire connected to it.

The lightning current will flow down the metallic tower to the ground electrodes to be dissipated in the earth. Let us take point A at the height of the insulator string connection. Due to the lightning current $i(t)$, the potential at this point, φ_A, will differ from the zero potential of the earth by the voltage drop in the grounding resistance R_g and in the tower inductance L_s between the tower base up to the point A:

$$\varphi = R_g i + L_s \frac{di}{dt}. \tag{1.1}$$

However, the power wire potential will practically remain the same (in this qualitative description, we ignore all inductances between the power wires, tower and grounded wire). The power wire potential is due to the operating voltage source of the power line: $\varphi_w = U_{op}$. Then, the insulator string voltage will be

$$U = \varphi_w - \varphi_A = U_{op} - R_g i - L_s \frac{di}{dt}. \tag{1.2}$$

Note that the lightning current and operating voltage may have different polarities. As a result, the overvoltage U may prove to be the sum of the three terms in equation (1.2).

The inductance component of the overvoltage, $L_s\, di/dt$, has a short lifetime: it acts about as long as the lightning current rises. For a current impulse with an average amplitude I_M 30 kA and an average rise time $t_f = 5\,\mu s$, the inductance voltage at L_s 50 μH will be about 300 kV. The resistance component U_M at a typical grounding resistance $R_g = 10\,\Omega$ will have about the same value but will act an order of magnitude longer, i.e., as long as the lightning current flows. For this reason, this component makes the principal contribution to the insulation flashover.

The emergency situation just described is not as bad as a direct stroke at a power wire when the same lightning current can induce an order of magnitude higher voltage. The insulation of a ultrahigh voltage line can withstand short overvoltages up to 1000–1500 kV and seldom suffers from lightning strokes at a tower or a lightning protection wire. To produce a harmful effect, the lightning current must be 3–5 times the average value. Lightning strokes of this power do not occur frequently, making up less than 1% of all strokes. Quite different is the effect of a direct stroke at a power network with an operating voltage of 35 kV and lower. The insulation system will suffer equally from a stroke at a power wire or a tower. It is no use protecting such a line with grounded wire.

Insulation flashover due to the tower potential rise is referred to as reverse flashover. This name does not imply the definite direction of the discharge development but only indicates the direction from which the potential rises, i.e., the grounded end of the insulator string rather than the power wire.

The above illustration of overvoltages on the line transmission insulation demonstrates, to some extent, a variety of mechanisms of direct lightning current effects. In actual reality, such mechanisms are much more diverse. It is important to remember that in modern technologies, overvoltages are not always measured in hundreds of kilovolts, as for high-voltage transmission lines. Short voltage rises of only 100–10 V may be hazardous to microelectronic devices. Of special interest in this connection are situations when lightning current flows across solid metallic jackets with electric circuits inside. These problems are discussed in chapter 6.

1.6.2 Induced overvoltage

Induced overvoltage is the most common and dangerous effect of lightning on electric circuits of modern technical equipment. This effect is brought about by electromagnetic induction. The current flowing through the lightning spark and the metallic structures of an affected object generates an alternating magnetic field which can induce an induction emf in any of the circuits in question. The procedure of estimating induced overvoltages is quite simple. If $B_{av}(t)$ is the magnetic induction averaged over the circuit cross section S, the induction emf is expressed as

$$E_{emf} \approx S \frac{dB_{av}}{dt}. \tag{1.3}$$

When the length of the current conductor inducing the magnetic field is much longer than the distance to the circuit, r_c, and when the width of the circuit normal to the magnetic field is much smaller than r_c, we have

$$E_{emf} \approx \frac{\mu_0 S}{2\pi r_c} \frac{di}{dt} \tag{1.4}$$

where $\mu_0 = 4\pi \times 10^{-7}$ H/m is vacuum magnetic permeability. The order of magnitude of the induction emf amplitude is defined as

$$E_{emf} \approx \frac{\mu_0 S}{2\pi r_c} A_{max} \tag{1.5}$$

where A_{max} is the maximum rate of the current impulse rise equal to 10^{11} A/s for the subsequent components of a powerful lightning flash. A circuit of area $S = 1$ m^2 located at a distance $r_c = 10$ m from a lightning current conductor may become the site of induced overvoltage with an amplitude up to 20 kV.

This value is only an arbitrary guideline, because induced overvoltage may vary with the circuit area, its orientation and distance from the lightning current. Circuits with an area of hundreds and thousands of square metres may be created by large industrial metallic constructions and power transmission lines. The distance between the circuit and the current flow may also vary greatly. For such diverse parameters of a system, the problem will be more complicated in the case of fast current variations along the spark and in time. It cannot then be approached as a quasi-stationary problem, but one must take into account the law of current wave propagation along the spark channel and the finite velocity of the electromagnetic field in the space between the channel and the circuit. Solutions to such problems are illustrated in chapter 6.

There is another class of situations associated with electromagnetic induction in screened volumes. Of special interest is the situation when the lightning current i flows across a solid metallic casing and the circuit in question is inside it. Unless the casing is circular, an emf-inducing magnetic field gradually appears inside the casing. It is remarkable that the time variation of the emf is not at all defined by di/dt. The magnetic field going through the circuit is affected more by the relatively slow current redistribution along the casing perimeter than by the time variation of the lightning current. The problem of pulse induction in aircraft inner circuits or in screened multiwire cables ultimately reduces to the problem above. Some approaches to its solution will be considered in chapter 6.

1.6.3 Electrostatic induction

Benjamin Franklin felt the effect of electrostatic induction when he raised his finger up to a lifted wire during a thunderstorm. The electric field of a storm cloud had polarized the wire by separating its electric charges. The strong electric field of the polarization charge had broken down the air gap between the thin wire end and the explorer's finger and carried the charge through his body to the earth.

Electrostatic induction induces a charge in any grounded conductor or a metallic object. Suppose it is a vertical metallic rod of length l located in an external vertical field E_0. When insulated from the earth, the rod would take the potential of the space at its centre, $\varphi_c = E_0 l/2$, which follows from the symmetry consideration. The grounded rod potential is zero; hence, the external field potential is compensated by the charge q_i induced by this field on the rod. The charge can be estimated from the rod capacitance C_r as $q_i = C_r \varphi_c = E_0 l C_r/2$.

The production of the charge q_i implies the existence of current through a ground electrode of the object. This is a low current, because it takes several seconds for the cloud charge, creating the field E_0, to be formed. As much time, Δt, is necessary for the charge $-q_i$ to flow down into the earth, leaving behind the induced charge q_i on the conductor. If the field E_0 is largely created by the leader charge of a close lightning discharge, the exposure time of the induced charge reduces to $\Delta t \approx 10^{-3} - 10^{-2}$ s. But in this case, too, the current through the ground electrode is low. For example, at $E_0 \approx 1$ kV/cm characteristic of close discharges, $l = 10$ m, and $C_r = 100$ pF,† the average current is

---

† Approximately, $C_r \approx 2\pi\varepsilon_0 l/\ln h/r$, where r is the rod radius, h is an average distance between the rod and the earth ($h = l/2$), $\varepsilon_0 = 8.85 \times 10^{-12}$ F/m is the vacuum dielectric permittivity. At $l = 10$ m and $r = 2$ cm, $C_r \approx 100$ pF.

$i_i \approx q_i/\Delta t = E_0 l C_r/2\Delta t \approx 0.5\text{--}0.05\,\text{mA}$. Even if the grounding resistance is as high as $R_g \approx 10\,\text{k}\Omega$ (we deal here with a damaged ground electrode when the connection with the grounding circuit is made across the high contact resistance of the break), the induced charge current will change the rod potential relative to the earth by the value $\Delta\varphi = i_i R_g = 5\text{--}0.5\,\text{V}$. This potential rise can be ignored in any situation.

When a lightning channel reaches the earth and the process of leader charge neutralization begins in the return stroke, the field at the earth, E_0, rapidly drops to zero, eliminating the charge q_i. The same charge now flows back through the grounding resistance for a much shorter time, about $1\,\mu\text{s}$, and the current i_i increases to about $0.5\,\text{A}$. At the same resistance $R_g \approx 10\,\text{k}\Omega$ the voltage will rise to $5\,\text{kV}$. In practice, it may rise even higher, producing a spark breakdown at the site of poor contact. The breakdown may become very dangerous if there are explosive gas mixtures nearby, since the spark energy is sufficiently high to set a fire.

There is another mechanism of igniting sparks in an induced charge field, which may be hazardous even in the case of perfect grounding of a metallic construction. Suppose that a grounded rod of length l and radius r is in the leader electric field E_0 of a nearby lightning discharge. The charge induced on the rod will enhance the field at its top approximately by a factor of l/r. With $l \gg r$, this is sufficient to excite a weak counterpropagating leader process. Of course, if this leader is only about $10\,\text{cm}$ long, it will have no effect on the lightning trajectory. Its energy is, however, large enough to ignite an inflammable gas mixture, if there is any in the vicinity, since the channel temperature is close to $5000\,\text{K}$ and its lifetime is as long as that of a lightning leader.

1.6.4 High potential infection

This unsuitable term has long been used in Russian literature on lightning protection. It means that the surface and underground service lines, which get into a construction to be protected, may introduce in it a potential different from the zero potential of the construction metalwork connected to earth connection. This may become possible if a service line is not linked to the grounding of the construction but connected or passes close to the earth connection of another construction loaded by lightning current during a stroke (figure 1.9). This may also be a natural earth connection formed at the moment of lightning contact with the earth due to an intense ionization in it. If the introduced potential is high, it causes a spark breakdown between the service line and a nearby metallic structure of the object, whose potential is zero owing to the earth connection. The scenario of the emergency that follows has been described above.

To avoid sparking induced by high potential infection, all metallic service lines subject to explosion zooms are linked to the earth connection of the construction. All metalwork potentials are equalized. The connection,

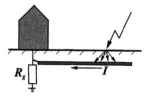

Figure 1.9. Schematic input of high potential from remote lightning strokes.

however, becomes loaded by additional current, which finds its way there from a remote lightning stroke, using the service line as a conductor. When the earth connection resistance is low and the service line goes through the ground with a high resistivity so that the current leakage through the side surface is not large, nearly all of the lightning current arrives at the connection from the stroke site. This situation appears to be somewhat similar to a direct lightning stroke. Sometimes, special measures must be taken to restrict the infection current. A detailed treatment of the problem of current and potential infections will be offered in chapter 6.

1.6.5 Current inrush from a spark creeping along the earth's surface

This phenomenon is familiar to all communications men who have to repair communications cables damaged by lightning. The damaged site can be detected easily, because it is indicated by a furrow in the ground extending far away from the stroke site. A furrow may be as long as several dozens of metres, or 100–200 m in a high resistivity ground. Such a long gap can be bridged by a spark because of the electric field created by the spark current spreading out through the ground. The mechanism of spark formation along a conducting surface differs from that of a 'classical' leader propagating through air. A creeping spark can develop in low fields and have a very high velocity.

Underground cables are not the only objects suffering from creeping spark current. Similarly, it can find its way to underground service lines and to the earth connections of constructions well equipped by lightning protectors. But a protector palisade cannot stop lightning. When the conventional way from the earth surface is blocked, it breaks through from beneath, making a bypass in the ground. Lightning thus behaves very much like a clever general in ancient times, who ordered his soldiers to make a secret underground passageway instead of attacking openly the impregnable castle walls. It is reasonable to suggest that the contact of a creeping spark with combustible materials is as frequent a cause of a fire as a direct lightning stroke.

The details of the creeping discharge mechanism have been unknown until quite recently. They are analysed in chapter 6.

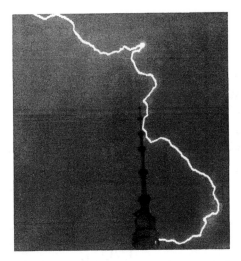

Figure 1.10. This lightning has missed the teletower tip by over 200 m.

1.6.6 Are lightning protectors reliable?

Lightning protectors are believed to be reliable, since their design has changed but little over two and a half centuries. Nevertheless, the photograph in figure 1.10 makes one question this judgement: the lightning struck the Ostankino Television Tower 200 m below its top, i.e., the Tower could not protect itself. This is not an exception to the rule. Most descending discharges missed the Tower top more or less closely, contrary to what had been expected. This is a serious argument against the vulgar explanation of the major principle of protector operation that lightning takes a shortcut at the final stage of its travel to the earth. There are also other arguments, perhaps not as obvious but still convincing.

Breakdown voltage spread is registered in long gaps even under strictly identical conditions. The breakdown probability Ψ varies with the pulse amplitude of test voltage U (figure 1.11). Deviations from the 50% probability voltage, $U_{50\%}$, are appreciable and may be 10–15% either way. Curve 2 in figure 1.11 shows the probability function $\Psi(U)$ for a shorter gap. In certain voltage ranges, both curves promise breakdown probabilities remarkably different from zero. This means that if two different gaps are tested simultaneously, there is a chance that any of them (the smaller and the larger gap) will be bridged. In general, this situation is similar to that arising when a lightning discharge is choosing a point to strike at. It does not always take the shortest way to a protector but, instead, may follow a longer path in order to attack the protected object.

For solving the lightning path problem, one has to treat a multielectrode system consisting of several elementary gaps. For lightning, all elementary

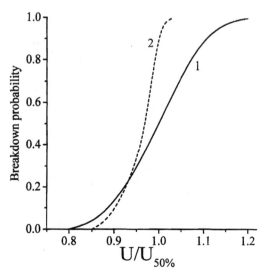

Figure 1.11. Distributions of breakdown voltages in air gaps of various lengths with a sharply non-uniform electric field.

gaps have a common high voltage electrode (the leader that has descended to a certain altitude), while the zero potential electrodes are formed by the earth's surface with grounded objects and protectors distributed on it. The problem of protector effectiveness thus reduces to the calculation of breakdown probabilities for the elementary gaps in a multielectrode system. The general formulation of this problem is very complex, since the spark development in the elementary gap in real conditions cannot be taken to be independent. The discharge processes affect one another by redistributing their electric fields, which eliminates straightforward use of statistical relations describing independent processes.

We cannot say that the spark discharge theory for a multielectrode system has been brought to any stage of completion. But what has been done, theoretically and experimentally, allows the formulation of certain concepts of the lightning orientation mechanism and the development of engineering approaches to estimate the effectiveness of protectors of various heights (see chapter 5).

Investigation of multielectrode systems is also important from another point of view: we must find ways of affecting lightning actively. It would be reasonable to leave the discussion of this issue for specialized chapters of this book, but they will, however, attract the attention of professionals only, or of those intending to become professionals. It is not professionals but amateurs who, most often, try to invent lightning protectors. They have at their disposal a complete set of up-to-date means: lasers, plasma jets, corona-forming electrodes for cloud charge exchange, radioactive sources, high voltage generators stimulating counterpropagating leaders, etc. That lightning management has a future has been confirmed by laboratory studies on sparks of multimetre length. These experiments and their implications will be analysed below, so there is no point in discussing them here. Still, it is hard to resist the temptation to make some preliminary comments addressed to those who like to invent lightning protection measures.

When explaining the leader mechanism at the beginning of this chapter, we noted that the leader tip carries a strong electric field sufficient for an intense air ionization. It is very difficult to act on this field directly, because it would be necessary to create charged regions close by, whose charge density and amount would be comparable with those in the immediate vicinity of the tip. Pre-ionization of the air by radioactive sources is of little use because of the low air conductivity after radiative treatment. The initial electron density behind the ionization wave front in the leader process is higher than 10^{12} cm^{-3}, and in a 'mature' leader it is at least an order of magnitude higher. These and even much lower densities are inaccessible to radiation at a distance of dozens of metres from the radiation source which must present no danger to life. The same is true of a gradual charge accumulation due to a slow corona formation between special electrodes. Besides, one cannot predict the polarity of a particular spark to decide which charge is to be pumped into the atmosphere.

Quite another thing is plasma generation. In principle, we could create a plasma channel comparable with the lightning rod height, thus increasing its length. A high power laser could, in principle, be used as a plasma source. It is clear that it should be a pulse source and the plasma produced should have a short lifetime. It must be generated exactly at the right moment, when a lightning leader is approaching the dangerous region near the object to be protected. This is a new problem associated with synchronization of the laser operation and lightning development, giving a new turn to the task of lightning protection, which does not at all become easier.

Finally, we should always bear in mind that most lightning discharges are multicomponent. In about half of them, the subsequent components do not follow the path of the first component. In fact, these are new discharges which would require individual handling. To prepare a laser light source for a new operation cycle for a fraction of a second is possible but difficult technically. The cost of such protection is anticipated to be close to that of gold.

It is not our intention to intimidate lightning protection inventors. We just want to warn them against excessive enthusiasm.

1.7   Lightning as a power supply

The question of whether lightning could serve as a power supply cannot be answered positively, no matter how much we wish it to be one. Some authors of science fiction books force, quite inconsiderately, their characters to harness lightning in order to use its electric power. Even without this service, lightning has done much for people by stimulating their thought. The energy of a lightning flash is not very high. The voltage between a cloud and the earth can hardly exceed 100 MV even in a very powerful storm, the transported charge is less than 100 C, and maximum energy release is 10^{10} J. This is equivalent to one ton of trinitrotoluene or 2–4 ordinary airborne bombs. A family cottage consumes more power for heating, illumination, and other needs over a year. Actually, only a small portion of the lightning power is accessible to utilization, while most of it is dissipated in the atmosphere.

Normally, a person lives through 40–50 thunder storm hours during a year. All storms send to the earth an average of 4–5 lightning sparks per square kilometre of its surface providing a power of less than 1 kW/km^2 per year. In a country of 500×400 km^2, this is about 200 MW, which is a very small value compared with the electrical power produced by an industrial country. Just imagine the immense net which would be necessary for trapping lightning discharges in order to collect such a meagre power! Other natural power sources, such as wind, geothermal waters, and tides, are infinitely more powerful than lightning, but they are still not utilized much. Clearly, we should not even raise the problem of lightning power resources.

1.8   To those intending to read on

There will be no more popularized stories about lightning in this book. Nor shall we mention ball lightning here. The next chapter will contain a thorough analysis of available data and theoretical treatments of the long spark, because we believe that without these preliminaries the lightning mechanism may not become clear to the reader. Nature has eagerly employed standard solutions to its problems, so lightning is quite likely to represent the limiting case of the long spark. It would be useful for readers to familiarize themselves with our previous book *Spark Discharge*, because it is totally concerned with this phenomenon. But even without it, they will be able to find here basic information on long sparks. We have tried to describe their general mechanisms and to give predictions as to their extension to air gaps of extremal length. Even for this reason alone, the next chapter is not a summary of the previous book. Lightning is as complicated a phenomenon as the long spark and is definitely more diverse. It is a multicomponent process. Since its subsequent components sometimes take the path of an earlier component, we must consider the effects of temperature and residual conductivity in the spark channel on the behaviour of new ionization waves.

Even a simple model should not treat a kilometre spark in terms of electrical circuits with lump parameters. A lightning spark is a distributed system. The time for which the electric field perturbation spreads along the sparks is comparable with the duration of some of its fast stages. The allowance for the delay can, in some cases, change the whole picture radically. This requires new approaches to lightning treatments. Experimental data and theoretical ideas concerning the lightning leader and return stroke are discussed together. First, there are not many of them. On the other hand, we have tried to point out ideological relationships between experiment and theory and to offer a more or less consistent physical description.

Spark discharges in a multi-electrode system are the subject of a special chapter. We present available data and analyse possible mechanisms of lightning orientation. This is, probably, the most debatable part of the book. Field studies of lightning orientation are very difficult to carry out primarily because constructions of even 100–200 m high are affected by descending discharges only once or twice a year. The observer must have exceptional patience and substantial support to be able to reveal statistical regularities in lightning trajectories. From field observations, one usually borrows the statistics of lightning strokes at objects of various height and, sometimes, the statistics of strokes at protected objects, such as power transmission lines with overhead grounding wire connections. This material, however, is too scarce to build a theory. For this reason, one has to refer to laboratory experiments on long sparks generated in 10–15 m gaps. No one has ever proved (or will ever do so) the geometrical similarity of sparks; therefore, experimental data can be extended to lightning only qualitatively. Nevertheless theoretical treatments must be brought to conclusion when one develops recommendations on particular protector designs. We analyse the reliability of engineering designs, wherever possible.

The last chapter of the book discusses lightning hazards and protection not only in terms of applications. Even the classical theory of atmospheric overvoltages in power transmission lines required the solution of complicated electrophysical problems. Thorough theoretical treatments are necessary for the analysis of lightning current effects on internal circuits of engineering constructions with metallic casings, on underground cables, aircraft, etc. The range of problems to be considered is not limited to electromagnetic field theory. We shall also discuss gas discharge mechanisms of a spark creeping along a conducting surface, the excitation of leader channels in air with the composition and thermodynamic characteristics locally changed by hot gas outbursts, and the lightning orientation under the action of the superhigh operating voltage of an object. These theoretical considerations will not screen our practical recommendations concerning effective lightning protection and the application of particular types of protectors.

References

[1] Bazelyan E M, Gorin B N and Levitov V I 1978 *Physical and Engineering Fundamentals of Lightning Protection* (Leningrad: Gidrometeoizdat) p 223 (in Russian)
[2] McEachron K 1938 *Electr. Engin.* **57** 493
[3] Berger K and Vogrlsanger E 1966 *Bull. SEV* **57** No 13 1
[4] Newman M M, Stahmann J R, Robb J D, Lewis E A *et al* 1967 *J. Geophys. Res.* **72** 4761
[5] Uman M A 1987 *The Lightning Discharge* (New York: Academic Press) p 377
[6] Berger K, Anderson R B and Kroninger H 1975 *Electra* **41** 23
[7] Schonland B, Malan D and Collens H 1935 *Proc. Roy. Soc. London Ser A* **152** 595
[8] Bazelyan E M and Raizer Yu P 1997 *Spark Discharge* (Boca Raton: CRC Press) p 294
[9] Schonland B 1956 *The Lightning Discharge. Handbuch der Physik* **22** (Berlin: Springer) p 576

Chapter 2

The streamer–leader process in a long spark

This chapter will deal with the spark discharge in a long air gap. We have already mentioned in chapter 1 that this material should not be ignored by the reader. But for the long spark, specialists would know much less about lightning. Today, high voltage laboratories are able to produce and study long sparks of several tens and even hundreds of metres long [1–3]. Many of the long spark parameters and properties lie close to the lower boundary of respective lightning values. Most effects observable in a lightning discharge were, sooner or later, reproduced in the laboratory. One exception is a multicomponent discharge, but the obstacles lie in the technology rather than in the nature of the phenomenon. It would be very costly to instal and synchronize several high voltage power generators, making them discharge consecutively into the same air gap.

As for the fine structure of gas-discharge elements, long spark researchers are far ahead of lightning observers. This could not be otherwise, since a laboratory discharge can be reproduced as often as necessary, by starting the generator at the right moment, within a microsecond fraction accuracy, and strictly timing the switching of all fast response detectors. But with lightning, the situation is different. It strikes every square kilometre of the earth's surface in Europe approximately 2 to 4 times a year. So, even such a high construction as the Ostankino Television Tower (540 m) is struck by lightning only 25–30 times a year. Of these, only 2–3 discharges are descending, while the others go up to a cloud. Normally, lightning observations have to be made from afar, so that many details of the process are lost. The gaps in the study of its fine structure must, of necessity, be filled in laboratory conditions.

The long spark theory is far from being completed, and there is no adequate computer model of the process. Still, there has lately been some progress, primarily owing to laboratory investigations. It would be unwise to discard these data and not to try to use them for the description of lightning. In this chapter, we shall outline our conception of the basic phenomena in a long spark. We shall present some newer data and ideas which emerged after the book [4] on the long spark had been published. We should like to emphasize again that many details of the spark physics are still far from being clear.

2.1 What a lightning researcher should know about a long spark

The key point is how a spark channel develops in a weak electric field, by 1–2 orders of magnitude lower than what is necessary to increase the electron density in air ($E_i \approx 30\,\text{kV/cm}$ under normal conditions). Naturally, we speak of a discharge in a sharply non-uniform field. Near an electrode with a small curvature radius (suppose this is a spherical anode of radius $r_a \approx 1$–$10\,\text{cm}$), the field is $E_a(r_a) \equiv E_a > E_i$ at the voltage $U \approx 50$–$500\,\text{kV}$. This is the site of initiation of a discharge channel. At a distance $r = 10r_a$ from the electrode centre, the channel tip enters the outer gap region, where the initial value of $E = E_a(r_a/r)^2$ is one hundredth of that on the electrode. This weak field is incapable of supporting ionization. Nevertheless, the channel moves on, changing the neutral gas to a well-ionized plasma.

There is no other reasonable explanation of this fact except for a local enhancement of the electric field at the tip of the developing channel. The enhancement is due to the action of the channel's own charge. Indeed, a conductive channel having a contact with the anode tends to be charged as much as its potential U_a relative to the grounded cathode. Current arises in the channel, which transports the positive electric charge from the anode (more exactly, from the high voltage source, to which the anode is connected). (In reality, electrons moving through the channel toward the anode expose low mobility positive ions.) Such would be exactly the mechanism of charging a metallic rod if it could be pulled out of the anode like a telescopic antenna. Then the strongest field region would move through the gap together with the rod tip. We can say that a strong electric field wave is propagating through a gap, in which ionization occurs and produces a new portion of the plasma channel. We can also name it as an ionization wave, and this term is commonly used.

The wave mechanism of spark formation was suggested as far back as the 1930s by L Loeb, J Meek, and H Raether. The channel thus formed was termed a streamer (figure 2.1). Experiments showed that the streamer velocity could be as high as $10^7\,\text{m/s}$. In lightning, this velocity is demonstrated by the dart leader of a subsequent component. Even the mere fact that these velocities are comparable justifies our interest in the streamer mechanism. It is important to know what determines the streamer velocity and how it changes with the tip potential. For this, we have to analyse

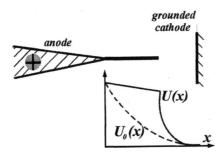

Figure 2.1. A schematic cathode-directed streamer: $U_0(x)$, external field potential; $U(x)$, potential along the conductive streamer axis.

processes taking place in the streamer tip region where ionization occurs. It is necessary to find out how the processes of charged particle production are related to electron motion in the electric field, due to which the charged region travels through the gap like the crest of a sea wave.

The specific nature of spark breakdown is not restricted to the ionization wave front, because its crucial parameter is the channel tip potential U_t. Its value may be much smaller than the potential U_a of the electrode, from which the streamer has started, since the channel conductivity is always finite and the voltage drops across it. Therefore, the analysis of streamer propagation for a large distance will require a knowledge of the electron density behind the wave front and the current along the channel in order to eventually calculate the electric field in the travelling streamer and to derive from it the voltage drop on the channel. Incidentally, the field and the current preset the power losses in the channel. It will become clear below how important this parameter is for spark theory.

The streamer creates a fairly dense plasma. Without this, it would be unable to transport an appreciable charge into the gap. A quantitative description of the ionization wave propagation provides the initial electron density in the channel and defines its initial radius. Behind the wave front, the streamer continues to live its own life. A streamer channel may expand, through ionization, in the radial electric field of its intrinsic charge, provided that the latter grows. The cross section of the current flow then becomes larger. The channel continuously loses the majority current carriers – electrons. The rates of electron attachment to electronegative particles and electron–ion recombination strongly affect the fate of the discharge as a whole. If the air through which a streamer propagates is cold and the power input into the channel is unable to increase its temperature considerably (by several thousands of degrees), the process of electron loss is very fast, since the attachment alone limits the electron average lifetime to 10^{-7} s. This is a very small value not only at the scale of lightning but also of a laboratory spark, whose development in a long gap takes 10^{-4}–10^{-3} s. One must be able to analyse kinetic processes in the channel behind the ionization wave front. Without the knowledge of their parameters, one will be unable to define the conditions, in which a streamer breakdown in air will be possible.

Here and below, we shall mean by a breakdown the bridging of a gap by a channel which, like an electric arc, is described by a falling current–voltage characteristic. The channel current is then limited mostly by the resistance of the high voltage source. Such a situation in technology is usually called short circuiting.

Current rise without an increase of the gap voltage inevitably suggests a considerable heating of the gas in the channel. Due to thermal expansion, the molecular density N decreases, thereby increasing the reduced electric field E/N and the ionization rate constant (see [4]). Another consequence of the heating is a change in the channel gas composition because of a partial dissociation of O_2, N_2 and H_2O molecules and the formation of easily ionizable NO molecules. The significance of many reactions of charged particle production and loss changes. The importance of electron attachment decreases, because negative ions produced in a hot gas rapidly disintegrate to set free the captured electrons. The electron–ion recombination rate becomes lower. But of greater importance is associative ionization involving O and N atoms. The reaction is accelerated with temperature rise but it does not depend directly on the electric field. This creates prerequisites for a falling current–voltage characteristic.

Clearly, a researcher dealing with long sparks and lightning cannot avoid considering the energy balance in the discharge channel, which determines the gas temperature. It is here that the final result is most likely to depend on the scale of the phenomenon and the initial conditions. In the laboratory, a streamer crossing a long gap seldom causes a breakdown directly. A streamer propagating through cold air remains cold. It will be shown below that the specific energy input into the gas is too small to heat it. Even during its flight, the old, long-living portions of a streamer lose most of their free electrons. In actual fact, it is not a plasma channel but rather its nonconductive trace which crosses a gap. The researcher must possess special skills to be able to produce an actual streamer breakdown of a cold air gap in laboratory conditions.

The situation with lightning may be different. Most lightnings are multi-component structures. With the next voltage pulse, the ionization wave often propagates through the still hot channel of the previous component. It is not cold air but quite a different gas with a more favourable chemical composition and kinetic properties. Surrounded by cold air, the hot tract shows some features of a discharge in a tube with a fixed radius and, hence, with a more concentrated energy release. It seems that the mechanism of the phenomenon known as a dart leader is directly related to streamer breakdown. One should

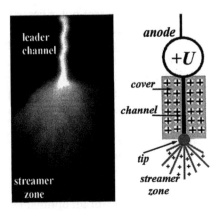

Figure 2.2. A photograph and a scheme of a positive leader.

be ready to give a quantitative description or make a computer simulation of this process.

Long gaps of cold air are broken down by the leader mechanism. During the leader process, a hot plasma channel (5000–10 000 K) is travelling through the gap. Numerous streamers start at high frequency from the leader tip, as from a high voltage electrode, and form a kind of fan. They fill up a volume of several cubic metres in front of the tip (figure 2.2).

This region is known as the streamer zone of a leader, or leader corona, by analogy with a streamer corona that may arise from a high voltage electrode in laboratory conditions. The total current of the streamers supplies with energy the leader channel common to the streamers, heating it up.

The streamer zone is filled up with charges of streamers that are being formed and those that have died. As the leader propagates, the zone travels through the gap together with its tip, so that the leader channel enters a space filled with a space charge, 'pulling' it over like a stocking. A charged leader cover is thus formed which holds most of the charge (figure 2.2). It is this charge that changes the electric field in the space around a developing spark and lightning. It is neutralized on contact of the leader channel with the earth, creating a powerful current impulse characteristic of the return stroke of a spark. Thus, we can follow a chain of interrelated events, which unites the simplest element of a spark (streamer) with the leader process possessing a complex structure and behaviour.

All details of the leader development directly follow from the properties of a streamer zone. In lightning, it is entirely inaccessible to observation because of the relatively small size and low luminosity. Today, there is no other way but to study long sparks in laboratory conditions and to extrapolate the results obtained to extremely long gaps. This primarily concerns a stepwise negative leader, whose streamer zone has an exclusively complex structure. It contains streamers of different polarities, starting not only from the leader tip but also from the space in front of them.

The leader channel of a very long spark, let alone of lightning, is its longest element. An appreciable part of the voltage applied to the gap may drop on this element. This is why one should know the time variation of the channel conductivity. The channel properties mostly depend on the current flowing through a given channel cross section. If the current is known, it is not particularly important whether it belongs to a long spark or lightning. The parameter that changes is the time during which one observes this process: for lightning, it is one or two orders of magnitude longer than for a spark. By analysing the self-consistent process of leader current production in the streamer zone and its effects on plasma heating and conversion in the channel, one can derive the conditions for an optimal leader development in a gap of a given length. There are reasons to believe that these conditions are realized in lightning when it develops in an extremely weak field. Nature always strives for perfection, not because it is animated but because optimal conditions most often lead to the highest probability of an event.

To conclude this section, long spark theory is of value in its own right to specialists in lightning protection. Lightning current is the cause of the most common type of overvoltage in electric circuits. The amplitude of lightning overvoltages reaches the megavolt level. In order to design a lightning-resistant circuit, one must be able to estimate breakdown voltages in air gaps of various lengths and configurations. This can be done only with a clear understanding of the long spark mechanism.

2.2   A long streamer

2.2.1   The streamer tip as an ionization wave

Let us consider a well developed 'classical' streamer, which has started from a high voltage anode and is travelling towards a grounded cathode. The main ionization process occurs in the strong field region near the streamer tip. We shall focus on this region. The front portion of a streamer is shown schematically in figure 2.3 together with a qualitative axial distribution of the longitudinal field E, electron density n_e, and a difference between the densities of positive ions and electrons, or the density of the space charge $\rho = e(n_+ - n_e)$ (the time is too short for negative ions to be formed).

The strong field near the tip is created mostly by its own charge. In front of the tip where the space charge is small, the field decreases approximately as $E = E_m(r_m/r)^2$, which is characteristic of a charged sphere of radius r_m. Here, E_m is the maximum streamer field at the tip front point. In fact, the radius at which the field is maximum should be termed the tip radius r_m. It approximately coincides with the initial radius of the cylindrical channel extending behind the tip. The front portion of a conventionally hemispherical tip

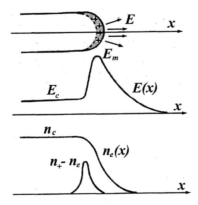

Figure 2.3. Schematic representation of the front portion of a cathode-directed streamer and qualitative distributions of the electron density n_e, the density difference $n_+ - n_e$ (space charge), and longitudinal field E along the axis.

should be called the ionization wave front. The streamer tip charge is primarily concentrated in the region behind the wave front. The field there becomes low, dropping to a value E_c in the channel, small as compared with E_m. The lines of force going radially away from the tip in front of it become straight lines inside the tip and align axially along the streamer channel.

Let us mentally subdivide the continuous process of streamer development into stages. The strong field region in front of the tip is the site of ionization of air molecules by electron impact. The initial seed electrons necessary for this process are generated by the streamer in advance. Their production is due to the emission of quanta, accompanying the ionization process because of electronic excitation of molecules. In our case, highly excited N_2 molecules are active so that the quanta emitted by them ionize the O_2 molecules, whose ionization potential is lower than that of N_2. The radiation is actively absorbed, but its intensity is high enough to provide an initial electron density n_0 of about 10^5–10^6 cm^{-3} at a distance of 0.1–0.2 cm from the tip. Each of these electrons gains energy from the strong field, giving rise to an electron avalanche. Since the number of avalanches developing simultaneously is very large, they fill up the space in front of the streamer tip to form a new plasma region. Owing to the electron outflow towards the channel body, the positive space charge of the plasma becomes exposed. Simultaneously, electrons that have advanced from the ahead region neutralize the positive charge of the 'old' tip which turns to a new channel portion, thereby elongating the streamer.

The gas in the wave front region must be highly ionized for the electron–ion separation to produce an appreciable charge capable of creating a strong ionizing field in front of the newly formed tip. For this reason, the region of concentrated tip charge is somewhat shifted towards the channel body relative to the intensive ionization site (figure 2.3). Normally, the electric field is pushed out of a good plasma conductor, and the space charge (if the conductor is charged) quickly concentrates near its surface as a 'surface' charge. The plasma of a fast streamer ('fast' in the sense that will be specified below) possesses a fairly high conductivity, and these properties apply to such a streamer. Therefore, the region of strong field and space charge in the tip looks like a thin layer, as is shown in figure 2.3.

If the streamer length is $l \gg r_m$, its velocity and the tip parameters change little during the time the tip travels a distance of its several radii. This means that, depending on the time t and the axial coordinate x, all parameters are the functions of the type $E(x, t) = E(x - V_s t)$, and what is shown in figure 2.3 moves to the right as a whole, without noticeable distortions. The picture changes only as the streamer velocity changes relatively slowly. This kind of process represents a wave, in this case a wave of strong field and ionization. The external parameter determining the wave characteristics (its velocity V_s, maximum field E_m, tip radius r_m, electron density behind the wave n_e) is the tip potential U_t. It is indeed an external characteristic of the tip, although it partly depends on the properties of the wave itself. The potential U_t is equal to the anode potential U_a minus the voltage drop on the streamer channel. The channel properties, however, are initially determined by the ionization wave parameters, so that the problem of streamer development is, strictly speaking, just one problem. Still, it can be approximately subdivided into two parts: the ionization wave problem and the problems of voltage drop and current in the channel. Both parts will be related by the dependencies of $V_s(U_t)$ and current i_l at the channel front on velocity V_s.

2.2.2 Evaluation of streamer parameters

The formulas to be derived in this and subsequent sections of this chapter do not claim high accuracy. The streamer and leader problems are very complex, and a rigorous solution can be obtained only by numerical computation. But a simplified analytical treatment may also be useful because it provides an understanding of basic laws and relations among the process parameters. In other words, one is able to get a general idea of the physics of the phenomenon under study and to estimate the order of values of its characteristics.

Let us consider a fast streamer, whose velocity is much higher than the electron drift velocity in the wave. Streamers are fast in many situations of practical interest. The calculation of electron production can ignore the slight drift of electrons from a given site in space for the short time the ionization wave passes through it. In this case, the ionization kinetics along the streamer axis is described by the following simple equations:

$$\frac{\partial n_e}{\partial t} = \nu_i n_e, \qquad \frac{n_e}{n_0} = \exp \int \nu_i \, dt = \exp \int \nu_i \frac{dx}{V_s} \qquad (2.1)$$

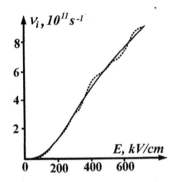

Figure 2.4. Ionization frequency of air molecules by electron impact under normal conditions (from the data on ionization coefficient α and electron drift velocity V_e in [11]).

where $\nu_i = \nu_i(E)$ is the frequency of electron ionization of molecules. Its time integral has been transformed to the integral over the coordinate x along the wave axis, according to the equality $dx = V_s\,dt$ corresponding to the coordinate system moving together with the wave. Due to the sharp increase of the ionization frequency with field (figure 2.4), the region where the field is not much less than its maximum makes the largest contribution to the electron production. This region in the wave is of the same order of magnitude as the tip radius (figure 2.3). So we can write the approximate expressions for the integral (2.1) and streamer velocity:

$$V_s \approx \frac{\nu_{im} r_m}{\ln\left(n_c/n_0\right)}, \qquad \nu_{im} \equiv \nu_i(E_m). \qquad (2.2)$$

This type of formula was first suggested by Loeb [5] and has been used since that time, in this or modified form, in all streamer theories [6–10]. The velocity of a fast streamer is weakly related to the initial n_0 and final n_c electron densities and is determined only by the maximum field E_m and the tip radius r_m.

The quantities E_m and r_m which determine V_s are not independent. They are interrelated by the tip potential U_t. For an isolated conductive sphere with a uniformly distributed surface charge Q', we have $U = r_m E_m = Q'/4\pi\varepsilon_0 r_m$, where $\varepsilon_0 = (36\pi \times 10^{11})^{-1} \approx 8.85 \times 10^{-12}\,\mathrm{F/m}$ is vacuum permittivity. A streamer looks more like a cylinder with a hemispherical rounded end (see figure 2.3). We can show [4] that in a long perfect conductor of this shape, one half of the potential at the hemisphere centre is created by charges concentrated on the hemisphere surface and the other half by those on the cylinder surface, so that the tip charge is $Q = 2\pi\varepsilon_0 r_m U_t$. The field at the tip front point is, to good approximation, only one half of that in an isolated sphere with the same potential, or

$$U_t = 2E_m r_m. \tag{2.3}$$

The tip charge moves because of the electron drift under the field action. The electron density in the wave plasma and the respective plasma conductivity must provide the charge transport with the same velocity as that of the wave. This permits estimation of the plasma density in the streamer just behind its tip. With the same assumptions as those in (2.2), the electron density in the strong field region on the streamer axis increases as $n_e \approx n_0 \exp(\nu_{im} t)$ for the time $\Delta t \approx r_m / V_s$. During this period of time, the electron density rises to its final value $n_c \approx n_0 \exp(\nu_{im} \Delta t)$, and the electron drift towards the channel with velocity $V_e = \mu_e E_m$ (where μ_e is electron mobility taken, for simplicity, to be constant) exposes the charge which creates the field E_m in the region of the new streamer tip.

The electron charge that flows through a unit cross section normal to the axis in the wave front region over time Δt is

$$q = e\mu_e E_m n_0 \int_0^{\Delta t} \exp(\nu_{im} t)\, dt = \frac{\mu_e E_m n_c}{\nu_{im}}. \tag{2.4}$$

It leaves behind it a positive charge of the same surface density q. It is this charge that creates the field E_m. We shall see soon that the effective thickness of a positively charged layer is $\Delta x \ll r_m$. In electrostatics, the field of such a layer at the conductor surface is equal to $E_m \approx q/\varepsilon_0$ (this equality is absolutely exact for the surface charge of a perfect conductor). By substituting q from (2.4), we get an estimate for the density behind the ionization wave:

$$n_c \approx \varepsilon_0 \nu_{im} / e\mu_e.$$

The plasma density n_c is not related directly to the streamer velocity and is essentially determined by the maximum field value which defines the ionization frequency.

Let us make sure that the tip charge is indeed concentrated in the thin layer. The effective time for the charge to be formed in the layer approximately is

$$\Delta t_e \approx \int_0^{\Delta t} (\Delta t - t) \exp(\nu_{im} t)\, dt \Big/ \int_0^{\Delta t} \exp(\nu_{im} t)\, dt = \nu_{im}^{-1}.$$

The space charge layer of thickness Δx moves to a new site at velocity $\Delta x / \Delta t_e$ which is equal to the streamer velocity V_s, since the ionization wave moves as a whole. Hence, using (2.2), we obtain

$$\Delta x = V_s \Delta t_e \approx V_s / \nu_{im} \approx r_m / \ln(n_c / n_0). \tag{2.5}$$

The final plasma density is many orders of magnitude larger than the initial density n_0, so that the logarithm in (2.5) is a large value. Therefore, we have $\Delta x \ll r_m$.

The formulas derived here claim for nothing more than an illustration of functional relations among streamer parameters. Numerical factors allowing the transition from an order of magnitude to a specific value have been deliberately ignored. This is justified because we have simplified all initial conditions and the derivation procedure in order to reveal the physics of the phenomenon in question. The significance of a formula will increase if it is 'equipped' with numerical factors, even approximate ones. Since we know the origin of these factors, we can partly judge about the theory validity and meaningfully compare the analytical results with computations and measurements.

A more rigorous treatment of a fast ionization wave, using one-dimensional equations consistent with the streamer physical model [4, 9], yields the expressions

$$V_s = \frac{\nu_{im} r_m}{(2k-1)\ln(n_m/n_0)}, \qquad n_c = \frac{\varepsilon_0 \nu_{im}}{k e \mu_e}, \qquad \frac{n_c}{n_m} = \ln\frac{n_m}{n_0} \qquad (2.6)$$

where k is the power index from the approximate formula $\nu_i(E) \sim E^k$ and n_m is the electron density in the wave front at the point of maximum field (the density is an order of magnitude smaller than the maximum achievable density n_c). In the field range characteristic of an air streamer, $k \approx 2.5$.

The issue of the streamer tip radius or maximum field represents the most complicated and least convincing point in streamer theory. It is likely that their values are established under the action of a self-regulation mechanism related to the proportionality $V_s \approx \nu_i(E_m)$ and to the rapidly increasing (at first) and then slowly growing dependence of ν_i on E (figure 2.4). If, at constant tip potential, the tip radius turns out to be too small and the field E_m respectively too high, corresponding to the slow growth of $\nu_i(E)$, the channel front end will not only move forward quickly but it will expand as fast under the action of a strong transverse field. The value of r_m will rise while that of E_m, according to (2.3), will fall.

Suppose, on the contrary, that the radius r_m is too large and the field E_m is too low, corresponding to a rapid growth of $\nu_i(E)$. Any slight plasma protrusion in the tip front will locally enhance the field. The ionization rate will greatly increase there, and the protrusion will run forward as a channel of a smaller radius. Some qualitative considerations of this kind were suggested in an old work of Cravath and Loeb [12], but these authors discussed a lightning channel obeying other mechanisms, because lightning develops via the leader process. These ideas were used in [6, 7] to formulate an approximate streamer theory. A semi-qualitative criterion was suggested for choosing the maximum field feasible in the streamer tip. According to [6, 7], E_m corresponds to the saturation point or bending in the function $\nu_i(E)$. This criterion was refined in [13] by establishing a quantitative relation of E_m to the slope of the $\nu_i(E)$ function and to the charge and normal field distributions over the streamer tip surface. It was shown that the field E_m at the tip front point is established such that the normal field on its lateral surface corresponds to the point of transition from the rapid to the slow growth of the $\nu_i(E)$ function (figure 2.4).

The streamer's choice of maximum field was manifested during a numerical simulation of short fast streamers within the framework of a complete two-dimensional formulation [14–16]. The mechanism of automatic establishment of E_m was demonstrated in [13] in the calculation of long streamer development with arbitary initial conditions and considerably simplified equations (see also [4 Suppl.]). No one has been able yet to calculate a long streamer within a complete two-dimensional model. We can conclude from these results that the field at the front point of an air streamer propagating at atmospheric pressure and room temperature seems to be $E_m \approx 150$–$170\,kV/cm$. The tip radius varies with the tip potential, approximately satisfying (2.3).

We shall give a numerical example as an illustration. At $E_m = 170\,kV/cm$ in air ($\nu_{im} \approx 1.1 \times 10^{11}\,s^{-1}$, $\mu_e \approx 270\,cm^2/V\,s$) and $r_m = 0.1\,cm$ (corresponding to $U_t = 34\,kV$), the streamer velocity for $n_0 \approx 10^6\,cm^{-3}$ from formula (2.6) is close to $1.7 \times 10^6\,m/s$ and the electron density in the newly born channel portion is $n_c \approx 9 \times 10^{13}\,cm^{-3}$. Within 20% accuracy, these values coincide with the results of integration of unreduced equations in the one-dimensional model illustrated in figure 2.3 [10]. They show not more than a 2- or 3-fold disagreement with numerical simulations of streamers in a two-dimensional model developed by different workers. This type of computation is extremely complicated and not particularly advanced. So the simple formulas (2.6) are useful since they can provide rough working estimations. They are also applicable to a gas of lower density, in which similarity laws are operative. Since $\nu_i \approx N f(E/N)$, where N is the number of molecules per $1\,cm^3$, f is a function of the type given in figure 2.4, and $\mu_e \approx N^{-1}$, we have

$$E_m \sim N, \qquad r_m \sim U_t/N, \qquad V_s \sim U_t, \qquad n_c \sim N^2. \qquad (2.7)$$

The streamer velocity is independent of N and n_c of the tip potential U_t. The latter fact opens up a tempting but yet unused possibility to test experimentally the theoretical concept of the maximum tip field E_m being constant. For this, it is sufficient to measure the electron density right behind the tip of a fast streamer, in which $U_t = U_a - E_{av}l$ rapidly decreases with the channel length l (E_{av} is the average channel field). The constant data on n_c will indicate the constant values of E_m. On the other hand, the decrease in electron density with decreasing streamer velocity could become a strong argument to support the hypothesis of constant tip radius r_m, which still has advocates.

From (2.6) and (2.3), the velocity of a fast streamer is proportional to its tip potential, because its radius is proportional to the potential at $E_m = $ const. The velocity V_s becomes lower than the electron drift velocity $V_e(E_m) \approx 4 \times 10^5\,m/s$ at $U_t \approx 5$–$8\,kV$. At lower voltages, the streamer moves slower than the drift electrons, so that the formulas become invalid. The analysis of equations presented in [10] shows that the streamer velocity drops with further decrease in U_t at $V_s < V_{em}$. The electron density at the wave front decreases, all electrons are drawn out of the tip, and space charge fills it up. However, the final plasma density behind the ionization wave does not decrease. The tip radius becomes very small, and the streamer stops at low U_t. This result agrees with experiments: no streamers with a velocity less than $(1.5–2) \times 10^5$ m/s have ever been observed in air under normal conditions.

2.2.3 Current and field in the channel behind the tip

As the streamer develops, its channel is under high potential which changes from the anode potential U_a at the starting point to a certain value U_l at the front end, close to the tip potential U_t (the difference between U_l and U_t of about $E_m \Delta x \ll U_t$ is due to a small potential drop in the tip). The channel is electrically charged, since the potential at any point x along it is higher than the unperturbed potential of space $U_0(x)$ created by electrode charges in the absence of streamer. Current must flow through the channel to supply charge to the new portions of the growing streamer. When setting ourselves the task of estimating this current and the current through the external circuit (this is the streamer current to be measured), we must first find the channel charge, for it is the time variation of this charge that produces the current. Suppose a streamer has started from an anode of small radius r_a. Let us examine the stage when the streamer length becomes $l \gg r_a$ (l is much larger than the channel radius r). We can then neglect the time variation of the anode charge, because its capacitance is small, and take the external current to be close to the current i_a entering the channel through its base at the anode. Besides, a streamer conductor can be regarded as being solitary, and the unperturbed potential U_0 far from the anode can be ignored.

Assume first that the channel is a perfect conductor. From a well-known electrostatics formula, the capacitance of a long solitary conductor is $C = 2\pi\varepsilon_0 l/\ln(l/r)$. Its charge is $Q = CU$, because a perfect conductor is under only potential U. Introduce now the concept of capacitance per unit length of the conductor, C_1, which is frequently used in electro- and radio-engineering to analyse long lines. The average capacitance per unit length

$$C_1 = \frac{C}{l} = \frac{2\pi\varepsilon_0}{\ln(l/r)} = \frac{5.56 \times 10^{-11}}{\ln(l/r)} \text{F/m} \qquad (2.8)$$

has a nearly constant value which only slightly varies with l and r. Calculations show that the local capacitance $C_1(x)$ practically coincides with the average value from (2.8) along the whole length of a long conductor, except for its portions lying close to its ends. But even at the ends, the local capacitance is less than twice the average value. This, however, does not concern capacitances of the free ends which are much larger (see below).

As an approximation justifiable by calculations, we shall use the capacitance per unit length from (2.8) and apply it to a real streamer channel. If a channel possesses a finite conductivity, then it must have a longitudinal potential gradient and $U = U(x)$, when current flows through it. The charge per unit channel length has the form

$$\tau(x) \approx C_1[U(x) - U_0(x)] = \frac{2\pi\varepsilon_0[U(x) - U_0]}{\ln(l/r)} \tag{2.9}$$

which allows for the fact that the local charge of a channel raises its potential relative to unperturbed potential, $U_0(x)$. A similar refinement should be introduced in formula (2.3), which generally looks like

$$U_t - U_0(l) = 2E_m r_m, \tag{2.10}$$

as well as in (2.7). If the channel radius r varies along its length, a characteristic value may be substituted into (2.9), because the capacitance varies with r only logarithmically.

Now turn to channel current. When a channel elongates by dl, its new portion acquires charge $\tau_l\, dl$; index l will denote parameters of the front channel end, $x = l$. This charge is supplied directly by local current i_l over time $dt = dl/V_s$. Therefore, at any stage of streamer development, we have

$$i_l = \tau_l V_s = \frac{2\pi\varepsilon_0[U_l - U_0(l)]V_s}{\ln(l/r)}, \qquad U_l \approx U_t. \tag{2.11}$$

The current at the tip is defined mainly by the tip potential and streamer velocity. At the anode, the current is

$$i_a = \frac{dQ}{dt}, \qquad Q = \int_0^l \tau(x)\, dx = \int_0^l C_1[U(x) - U_0(x)]\, dx \tag{2.12}$$

where, Q is the total channel charge. Strictly, Q should be supplemented by the tip charge $Q_t = 2\pi\varepsilon_0 r_m[U_t - U_0(l)]$, but it is relatively small in a long streamer.

Currents i_a and i_l at the opposite ends of a streamer channel do not generally coincide. Of course, their values may be very close or differ considerably, depending on particular conditions. For example, if the electrode voltage is raised during the streamer development, the potential and charge distributed along the channel increase. Some of the anode current is used to supply an additional charge to the old channel portions, so that only the remaining current reaches its front end: $i_a > i_l$. But if the electrode voltage is decreased, the 'excess' charge of the old channel goes back to the supply through the anode surface, so that the current decreases nearer to the anode (positive current is created by charges moving away from the anode): $i_a < i_l$.

A long streamer can develop at constant voltage when the electric field in the channel, $E(x, t)$, does not vary much with time. The potential at any point of the existing channel $U(x) = U_a - \int_0^x E(x)\, dx$ and $\tau(x)$ vary slightly with time, which means that current does not branch off on the way from the anode to the channel tip. In this case, the anode current is close to the end current defined by (2.11) including potential U_t which may be much lower than U_a. Many experiments have shown that the average channel field must exceed a certain minimum value of about $5\,\text{kV/cm}$ for air in normal conditions (see sections 2.2.6 and 2.2.7) to be able to support a long positive streamer. For instance, if $U_a = 600\,\text{kV}$ at the anode and the streamer length is $l \approx 1\,\text{m}$, nearly all voltage drops in the channel and $U_t \ll U_a$, but currents i_a and i_l still do not differ much.

To get a general idea about the orders of magnitudes, let us consider a variant which seems quite realistic and is manifested by some calculations (section 2.2.6). This is the variant with constant applied voltage and slowly varying average field in the channel, when the current along it changes little and, therefore, can be evaluated from (2.11). For example, at $l = 1\,\text{m}$, $r = 0.1\,\text{cm}$, $V_s = 1.7 \times 10^6\,\text{m/s}$, and $U_l \approx U_t \approx 34\,\text{kV}$, as in the illustration in section 2.2.2 (with $U_a > 500\,\text{kV}$), we have $\ln(l/r) = 6.9$, $C_1 = 8 \times 10^{-12}\,\text{F/m}$, $\tau_l = 2.7 \times 10^{-7}\,\text{C/m}$, and $i_a = i_l = 0.46\,\text{A}$. Streamer currents of this order of magnitude (as well as much higher or much lower currents) have been registered in many experiments. These values can also be obtained from calculations with the account of possible streamer velocities from 10^5 to $10^7\,\text{m/s}$ in air, which have been found in some experiments to be even higher [4].

In a simple model of potential and current evolution in a developing streamer channel, the latter can be represented as a line with distributed parameters: the capacitance C_1 and resistivity $R_1 = (\pi r^2 e\mu_e n_e)^{-1}$ per unit length. The electron density $n_e(x, t)$ should be calculated in terms of the plasma decay kinetics (see section 2.2.5). Estimations show that self-induction effects are not essential in streamer development [4]. Then, the process is described by the following equations for current and voltage balance:

$$\frac{\partial \tau}{\partial t} + \frac{\partial i}{\partial x} = 0, \qquad -\frac{\partial U}{\partial x} = iR_1, \qquad \tau = C_1(U - U_0). \qquad (2.13)$$

A boundary condition in the set of equations (2.13) at $x = l$ is the equality

$$i_1 = C_1[U_1 - U_0(l)]V_s \qquad (2.14)$$

equivalent to (2.11). Formula (2.12) automatically follows from (2.13) and (2.14). Another boundary condition may be the setting of anode potential, since $U(0, t) = U_a(t)$. Equations (2.13) and (2.14) will be used in the next section to evaluate the heating of a streamer channel. Illustrations of streamer development calculations will be given in sections 2.2.6 and 2.2.7 after a discussion of the plasma decay mechanism. A complete set of equations for a long line, generalized by taking self-induction into account, will be applied in section 4.4 to the treatment of a lightning return stroke.

Equality (2.11) allows evaluation of longitudinal field E_c in the channel behind the streamer tip, where the electron density is still as high as that created by an ionization wave. The current behind the tip is conduction current $i_l = \pi r_m^2 e n_c \mu_e E_c$. By equating this expression to (2.11) and using (2.6) and (2.10) with $U_t = U_l$, we find

$$E_c \approx \frac{\beta E_m}{\ln(n_m/n_0)\ln(l/r_m)}, \qquad \beta = \frac{4k}{2k-1} \approx 2.5. \qquad (2.15)$$

For a 1 m streamer, the product of logarithms in the denominator of (2.15) is close to 100. Therefore, the field in the front end of a streamer channel in normal density air is $E_c \approx 4.2\,\text{kV/cm}$ ($E_m = 170\,\text{kV/cm}$ from section 2.2.2). Within the theory accuracy, this value does not contradict the average measured channel field of $5\,\text{kV/cm}$ necessary to support the streamer. There is no ionization in such a weak field, therefore electrons are lost in attachment and electron–ion recombination processes.

Current i_l near the channel end is lower than that of the tip adjacent to the channel, because the charge per unit tip length $\tau_t = Q_t/r_m$ is larger than τ in the channel. This is a typical consequence of end effects for long conductors, well-known from electrostatics. The surface charge density at the free end of a conductor is much higher than on its lateral surface. In our simple model, in which a channel tip has been replaced by a hemisphere of radius r_m and charge Q_t written after formula (2.12), the average charge per unit length is $\tau_t \approx 2\pi\varepsilon_0[U_t - U_0(l)]$. It is $\ln(l/r_m)$ times larger than τ_l at the channel end (see (2.9)). The tip current i_t much exceeds i_l. This does not affect the total charge balance, because the charge Q of a long channel is much larger than the tip charge Q_t.

Note that current perturbation in the tip region has a local character. It cannot be detected by current registration from the anode side. The streamer here makes use of its own resources – the charge of the 'old' tip has moved on into the gap with the elongating streamer. It is the charge overflow that creates current i_t. If a current detector were placed at the site of a newly born portion of the channel, it would register current $i \approx i_t$ for a very short period of time $\Delta t \approx r_m/V_s \approx 10^{-9}$ s; then the current would decrease to i_l and evolve as the solution of equations (2.13) and (2.14) indicates.

2.2.4 Gas heating in a streamer channel

A streamer process is accompanied by current flow and, hence, by Joule heat release. As was mentioned above, the viability of a plasma channel depends primarily on temperature, so this issue is of principal importance. The initial heating of a given gas volume occurs when a streamer tip with its high current and field passes through it. As the channel develops, the gas is heated further by streamer current flowing through it. Let us evaluate both components of released energy.

The energy released in 1 cm³ per second is $jE = \sigma E^2$, where $j = \sigma E$ is the current density and σ is the plasma conductivity in a given site in a given moment of time. The energy released in $1 \, \text{cm}^3$ as a result of ionization wave passage is

$$W = \int \sigma E^2 \, dt = \int \sigma E^2 \, dx/V_s \qquad (2.16)$$

where the integrals are formally taken from $-\infty$ to $+\infty$ but actually over the ionization wave region. The principal contribution to energy release is made by a thin layer behind the wave front where the electron density and field are high. The integral of (2.16) was found rigorously to be $\varepsilon_0 E_m^2/2$, using equations for this wave region [4]. This value has the physical meaning of electrical energy density at maximum field. The contribution of the region before the wave front is $\ln (n_m/n_0)$ times, or an order of magnitude, smaller than this value. Although the field there is as high as that behind the front, the electron density is of the order of n_m and the conductivity σ is $\ln (n_m/n_0)$ times smaller (section 2.2.2). Therefore,

$$W \approx \varepsilon_0 E_m^2/2 \approx 2.6 \times 10^{-3} \, \text{J/cm}^3 \qquad (2.17)$$

where the numerical value corresponds to $E_m = 170 \, \text{kV}$.

The fact that the density of energy release in a gas is of the same order of magnitude as the energy density of the electric field is quite consistent with electricity theory. When a capacitor with capacitance C is charged through resistance R to voltage U of a constant voltage supply, half of the work $QU = CU^2$ done by the supply is stored by the capacitor as electrical energy, and the other half is dissipated due to resistance, irrespective of its value. The value of R determines only the characteristic time of the capacitor charging, RC. Something like this is valid for the case in question but, of course, without both energies being rigorously equal to each other, because this situation is much more complicated. Indeed, according to the results of section 2.2.2, the tip capacitance is $C_t = Q/U_t \approx 2\pi\varepsilon_0 r_m$, volume $V_t \approx 4\pi r_m^3/3$, and field $E_m \approx U_t/r_m$, so that the energy dissipation per unit tip volume is $W \approx C_t U_t^2/2V_t \approx \varepsilon_0 E_m^2$ (we have ignored the unessential term $U_0(l)$).

Joule heat is released directly in a current carrier gas, or an electron gas. Then electrons give off their energy to molecules in collisions. An appreciable portion of electron energy (even most of it in a certain range of E/N) is used for the excitation of slowly relaxing vibrations of nitrogen molecules. Some energy is used for ionization and electron excitation of molecules, about $w \approx 100 \, \text{eV}$ per pair of charged particles produced, i.e., $n_c w \approx 10^{-3} \, \text{J/cm}^3$ at $n_c \approx 10^{14} \, \text{cm}^{-3}$. But even without the account of these 'losses', the gas temperature rise in the wave front region appears to be negligible: $\Delta T < W/c_V \approx 3 \, \text{K}$. Here, $c_V = \frac{5}{2} k_B N = 8.6 \times 10^{-4} \, \text{J/(cm}^3/\text{K})$ is the heat capacity of cold air and k_B is the Boltzmann constant.

Let us see what subsequent gas heating can provide by the moment it is somewhere in the middle of a long streamer channel. We multiply the second equation of (2.13) by i and integrate over the whole channel length, assuming, for simplicity, that constant voltage U_a is applied to the anode. After taking a by-part integral in the left side of the equation, we substitute $\partial i/\partial x$ from (2.13) and $i(l) \equiv i_l$ from (2.14), followed by simple transformations. As a result, we have

$$U_a i_a = \int_0^l i^2 R_1 \, dx + \frac{d}{dt} \int_0^l \frac{C_1 U^2}{2} \, dx + \left[\frac{C_1 U_l}{2} - C_1 U_l U_0(l) \right] V_s \qquad (2.18)$$

which describes the power balance in the system; here, $U_0(l)$ is unperturbed potential of the external field at the streamer tip point $x = l$. The input power $U_a i_a$ into a discharge gap is used for Joule heat release in the channel (the first term on the right), for increasing the electric energy stored in its capacitance (the second term), and for the creation of new capacitance due to the channel elongation (the third term). Joule heat associated with the ionization wave is not represented here. The field burst and the tip impulse current that make up W calculated above are absent from equations (2.13) and (2.18). Having integrated equality (2.18) over the period of time from the moment of channel initiation to the moment t the channel has acquired length l, we get the equation for the energy balance in the system at the moment t:

$$U_a Q = K_{\text{dis}} + \int_0^l \frac{C_1 [U^2 - U_0^2(x)]}{2} \, dx + \left\langle \frac{C_1 U_l^2}{2} - C_1 U_l U_0(l) \right\rangle_t l \qquad (2.19)$$

where charge Q is given by (2.12). The energy input into the channel, $U_a Q$, is used to create capacity (the last term on the right), partly stored in this capacity (the integral) and partly dissipated (K_{dis}). The braces $\langle \, \rangle_t$ indicate the time averaging of the process. In case of a long channel, much of the applied voltage drops across its length, so the tip potential U_l is small most of the time, as compared with average channel potential U_{av} of about U_a. Then, the last term in (2.19) can be neglected.

If we compare the left side of (2.19) with the substituted expression for Q from (2.12) and the integral in the right side of (2.19), we can conclude that the difference between these values cannot be much smaller than their own values but rather have the same order of magnitude. Therefore, the energy K_{dis} dissipated in the channel is equal, in order of magnitude, to the gained electrical energy, which is in agreement with a similar situation discussed above.

The average energy dissipated per unit channel length is $W_{1av} \approx C_1 U_{av}^2/2$ and the average energy contributed per unit channel volume is

$$W' \approx \frac{C_1 U_{av}^2}{2\pi r_{av}^2} \qquad (2.20)$$

where r_{av} is the average channel radius. With the formation of every new portion of the channel, its radius was approximately proportional to the tip potential owing to the fact that the maximum tip field remained approximately constant. So we have $U_{av}/r_{av} \approx E_m$. Substituting this expression and (2.8) into (2.20), we find

$$W' \approx \frac{\varepsilon_0 E_m^2}{\ln(l/r_m)} < W.$$

One can see that subsequent heating of the channel gas adds little to the initial heating by an ionization wave passing through the particular channel site.

To conclude, gas heating due to streamer development is negligible if the gap voltage remains constant. Higher voltage does not change the situation because the energy dissipated in the channel grows in proportion with the channel cross section and the air volume to be heated. Specific heating remains unchanged, since it is determined by a more or less fixed volume density of electric energy.

2.2.5 Electron–molecular reactions and plasma decay in cold air

Electron loss in cold air is due to attachment to oxygen molecules and dissociative recombination. The main attachment mechanism in dry air at moderate fields is a three-body process

$$O_2 + e + O_2 \rightarrow O_2^- + O_2,$$

$$k_{at} \approx (4.7 - 0.25\gamma) \times 10^{-31}\, cm^6/s, \qquad \gamma = E/N \times 10^{16}\, V \cdot cm^2 \qquad (2.21)$$

where k_{at} is the rate constant at $T = 300\,K$. In higher fields, the dominant process is dissociative attachment $O_2 + e \rightarrow O^- + O$ with the rate constant

$$\log k_a = \begin{cases} -9.42 - 12.7/\gamma & \text{at } \gamma < 9 \\ -10.21 - 5.7/\gamma & \text{at } \gamma > 9. \end{cases} \qquad (2.22)$$

In not excessively high fields of $E < 70\,kV/cm$ at 1 atm, air is ionized at the rate constant $k_i = \nu_i/N$

$$\log k_i = -8.31 - 12.7/\gamma \quad \text{at } \gamma < 26. \qquad (2.23)$$

Since the rate of electron loss through attachment is proportional to electron density n_e and that through recombination is proportional to n_e^2, the latter is unimportant at the beginning of ionization. The equality $k_i = k_a$ valid at $\gamma \approx 12$ determines the minimum field mentioned above, which is necessary to initiate the growth of electron density in unperturbed air; $E_i \approx 30\,kV$ at $p = 1\,atm$ and room temperature.

Oxygen molecules possessing a lower ionization potential than N_2 are ionized in fields not much exceeding the ionization threshold. Electrons recombine with O_2^+ at the rate constant β, usually termed a recombination coefficient:

$$O_2^+ + e \rightarrow O + O, \qquad \beta \approx 2.7 \times 10^{-7}(300/T_e)^{1/2}\, cm^3/s \qquad (2.24)$$

where T_e is electron temperature in Kelvin degrees. However, complex ions are more effective with respect to electron–ion recombination. The most important ions in dry air are O_4^+ ions, while in atmosphere saturated by water vapour, as in thunderstorm rain, $H_3O^+(H_2O)_3$ cluster ions are more important. For these, the recombination coefficients

$$O_4^+ + e \rightarrow O_2 + O_2, \qquad \beta \approx 1.4 \times 10^{-6}(300/T_e)^{1/2} \text{ cm}^3/\text{s} \qquad (2.25)$$

$$H_3O^+(H_2O)_3 + e \rightarrow H + 4H_2O, \qquad \beta \approx 6.5 \times 10^{-6}(300/T)^{1/2} \text{ cm}^3/\text{s}$$
$$(2.26)$$

are an order of magnitude larger than for simple ions.

Complex O_4^+ ions are formed from simple ions in the conversion reaction

$$O_2^+ + O_2 + O_2 \rightarrow O_4^+ + O_2, \qquad k \approx 2.4 \times 10^{-30}(300/T_e)^{1/2} \text{ cm}^6/\text{s}. \qquad (2.27)$$

Chains of hydration reactions lead to the production of $H_3O^+(H_2O)_3$ ions. A typical chain looks like this:

$$O_4^+ + H_2O \rightarrow O_2^+(H_2O) + O_2, \qquad k = 1.5 \times 10^{-9} \text{ cm}^3/\text{s}$$

$$O_2^+(H_2O) + H_2O \rightarrow H_3O^+ + OH + O_2, \qquad k = 3.0 \times 10^{-10} \text{ cm}^3/\text{s}$$

$$H_3O^+ + H_2O + (M) \rightarrow H_3O^+(H_2O) + (M), \qquad k = 3.1 \times 10^{-9} \text{ cm}^3/\text{s}$$

$$H_3O^+(H_2O) + H_2O + (M) \rightarrow H_3O^+(H_2O)_2 + (M), \qquad k = 2.7 \times 10^{-9} \text{ cm}^3/\text{s}$$

$$H_3O^+(H_2O)_2 + H_2O + (M) \rightarrow H_3O^+(H_2O)_3 + (M), \qquad k = 2.6 \times 10^{-9} \text{ cm}^3/\text{s}$$
$$(2.28)$$

(M is any molecule, k correspond to $p = 1$ atm, $T = 300$ K); here, a hydrated ion replaces a O_4^+ ion.

Another, similar chain begins with the production of an H_2O^+ ion in ionization of water molecules by electron impact. Then comes the conversion reaction

$$H_2O^+ + H_2O \rightarrow H_3O^+ + OH, \qquad k = 1.7 \times 10^{-9} \text{ cm}^3/\text{s}$$

producing an H_3O^+ ion, followed by the reaction chain of the type (2.28).

The production of complex ions is accompanied by their decay. For an O_4^+ ion, this is the reaction

$$O_4^+ + O_2 \rightarrow O_2^+ + O_2 + O_2,$$

$$k = 3.3 \times 10^{-6}(300/T)^4 \exp(-5040/T) \text{ cm}^3/\text{s}. \qquad (2.29)$$

It is greatly accelerated by gas heating, but in cold air the reaction effect is negligible. The same is true of other complex ions, including hydrated ions.

In a cold streamer channel, simple positive ions turn to complex ions very quickly, for the time $\tau_{conv} \approx 10^{-8}$–$10^{-7}$ s. It is these ions that determine the rate of electron–ion recombination in cold air, except for a very short initial stage with $t \leqslant \tau_{conv}$.

If the ionization rate is too low and if the detachment–decay of negative ions is slow, as in a cold streamer channel, the plasma decay is described by the equation

$$\frac{dn_e}{dt} = -\nu_a n_e - \beta n_e^2 \qquad (2.30)$$

where ν_a is electron attachment frequency. Its solution at initial electron density equal to the plasma density behind the ionization wave, n_c, is

$$n_e(t) = \frac{n_c \exp(-\nu_a t)}{1 + (\beta n_c / \nu_a)[1 - \exp(-\nu_a t)]} \qquad (2.31)$$

where the time is counted from the moment the streamer tip passes through a particular point of space.

According to (2.15), we have $E \approx 4.2\,\text{kV/cm}$ and $E/N \approx 1.7 \times 10^{-16}\,\text{V/cm}$ for a streamer channel just behind the tip at $p = 1\,\text{atm}$. The electron attachment frequency from (2.21) is $\nu_a \approx 1.2 \times 10^7\,\text{s}^{-1}$ and the characteristic attachment time is $\tau_a = \nu_a^{-1} \approx 0.8 \times 10^{-7}$. Over this time, most simple O_2^+ ions in dry air turn to complex O_4^+ ions. Electrons recombine with them with the coefficient $\beta \approx 2.2 \times 10^{-7}\,\text{cm}^3/\text{s}$ corresponding to electron temperature $T_e \approx 1\,\text{eV} = 1.16 \times 10^4\,\text{K}$ at the above value of E/N. The initial electron density $n_c \approx 10^{14}\,\text{cm}^{-3}$ is so high that the parameter $\beta n_c / \nu_a \approx 2$ determining the relative contributions of recombination and attachment is larger than unity. This means that at an early decay stage with $t < \tau_a \approx 10^{-7}\,\text{s}$, electrons are lost primarily due to recombination, with attachment playing a lesser role. Later, at $t > 2\tau_a \approx 2 \times 10^{-7}\,\text{s}$, the electron density decreases exponentially, as is inherent in attachment, but as if starting from a lower initial value $n_1 = n_c/(1 + \beta n_c \tau_a) \approx 0.3 n_c$; $n_e \approx n_1 \exp(-\nu_a t)$.

The plasma conductivity decreases by two orders of magnitude, as compared with the initial value, over $t \approx 3 \times 10^{-7}\,\text{s}$. At the streamer velocity $V_s \approx 10^6\,\text{m/s}$, this occurs at a distance of 30 cm behind the tip. A microsecond later, the conductivity drops by six orders of magnitude. The streamer plasma in cold humid air decays still faster because of a several times higher rate of recombination with hydrated ions and due to the appearance of an additional attachment source involving water molecules. These estimations indicate a low streamer viability. It is only very fast streamers supported by megavolt voltages that are capable of elongating to $l \approx 1\,\text{m}$ in cold air without losing much of their galvanic connection with the original electrode. This is supported by experiments with a single streamer and a powerful streamer corona [4].

Note that a streamer plasma has a longer lifetime in inert gases, where attachment is absent and recombination is much slower. This makes it possible to heat the plasma channel by flowing current for a longer time after the streamer bridges the gap (the estimations of section 2.2.4 do not extend to these conditions). Such a process sometimes leads to a streamer (leader-free) gap breakdown [17]. Still, the formulation of the streamer breakdown problem is justified for hot air and is related to lightning (see section 4.8 about dart leader).

2.2.6   Final streamer length

When a streamer starts from the smaller electrode (anode) of radius r_a, to which high voltage $U_a \gg E_i r_a$ is applied, it propagates in a rapidly decreasing external field. It is first accelerated but then slows down after it leaves the region of length r_a where it senses a direct anode influence. If the voltage is too low, the streamer may stop in the gap, without reaching the opposite electrode (say, a grounded plane placed at a distance d). The higher is U_a, the longer is the distance the streamer can cover; at a sufficiently high voltage, it bridges the gap. In order to estimate the sizes of the streamer zone and leader cover in a long spark or lightning – a task important for their theory – we need a criterion that would allow estimation of maximum streamer length under different propagation conditions. No direct measurements of this kind have been made for single long streamers in air, because there is always a burst of numerous streamers. This, however, is quite another matter (see below). So we shall use indirect experimental results and invoke physical considerations, theory, and calculations.

It has been established experimentally that streamers comprising a streamer burst are able to cross an interelectrode gap of length d only if the relation $E_{av} = U_a/d$ exceeds a certain critical value E_{cr} which varies with the kind of gas and its state. Under normal conditions in air, this critical value is $E_{cr} \approx 4.5$–$5\,\text{kV}$ in a wide range of $d \approx 0.1$–$10\,\text{m}$. The data spread does not exceed the measurement error. Bazelyan and Goryunov [18] recommend the value $E_{cr} = 4.65\,\text{kV/cm}$ for positive streamer, averaged over various measurements. Therefore, the voltage necessary for a streamer to bridge a gap of length d is $U_{amin} = E_{cr}d$ or more. For example, a gap of 1 m length requires about $500\,\text{kV}$ ($E_{cr} \approx 10\,\text{kV/cm}$ for negative streamer in air).

At the moment of crossing a gap, all voltage U_a is applied to the streamer, so E_{av} is also the average field in the streamer. If a gap is long enough, E_{av} can be identified with the average channel field. Indeed, in critical conditions with $E_{av} = E_{cr}$, a streamer crosses a gap at its limit parameters. It approaches the opposite electrode at its lowest velocity corresponding to the minimum excess of the tip potential $U_t \approx U_l$ over the external potential, $\Delta U_l = U_l - U_0(d) \approx 5$–$8\,\text{kV}$, below which the streamer practically stops. In the case of a grounded electrode, $U_0(d) = 0$. If a gap is so long (say, 1 m) that

$\Delta U_l = U_t \ll U_a$, nearly all applied voltage drops in the channel. Therefore, E_{cr} can be treated as the lowest field limit, at which a streamer is still capable of propagating.†

This interpretation remains valid when a streamer does not bridge a gap but stops somewhere on the way. Indeed, according to (2.6) and (2.10), the streamer velocity, and hence its ability to move on, is determined only by the tip potential excess over the external potential, $\Delta U_t = U_t - U_0(l)$, and is independent of the latter. No matter where a long streamer stops, we shall have $\Delta U_t \ll U_a$, though the external potential value at this point, $U_0(l_{max})$, may be high. Generally the average field limit in the channel and the streamer length at the moment it stops, l_{max}, are interrelated as

$$E_{cr} \approx \frac{U_a - U_0(l_{max})}{l_{max}}, \qquad l_{max} \approx \frac{U_a - U_0(l_{max})}{E_{cr}}. \qquad (2.32)$$

In order to be able to use these relations in practice, we must know not only the easily registered gap voltage U_a but also potential $U_0(l_{max})$ inaccessible to measurement. In most cases, it is hard to estimate even by calculation. For a particular streamer, the external field is determined, in addition to the anode charge, by the whole combination of charges that have emerged in the gap and its vicinity. Especially important is the charge of all other streamers that were formed together with the one under study. Consequently, the field $U_0(x)$, in which the streamer is moving, represents a self-consistent field. An outburst of hundreds of streamer branches is characteristic of air; they fill up a space comparable with l_{max}. It is this maximum length, rather than the small anode radius, that will determine the external field fall along the gap length. Estimations of a self-consistent field $U_0(x)$ involve considerable difficulties and errors (we shall come back to this when evaluating the size of the streamer zone of a leader). So in reality, critical field E_{cr} can be evaluated only from experimental data that relate to a situation with streamers bridging a discharge gap. Then, the potential $U_0(l_{max})$ is known reliably because it coincides with the potential of the electrode, usually the grounded one: $U_0(l_{max}) = U_0(d) = 0$. But if it is known that $U_0(l_{max}) \ll U_a$, as in the case of a long streamer moving in a sharply non-uniform field, the criterion of (2.32) for a definite streamer length will be extremely simplified: $l_{max} \approx U_a/E_{cr}$.

The existence of critical field E_{cr} has a rather clear physical meaning. The reason for the appearance of a minimum average field in a channel is its finite

---

† The average quantities E_{av} and E_{cr} describe, to some extent, the actual field strengths in the channel even when the external field is extremely non-uniform, changing by several orders of magnitude along the gap far from the conductor. For example, if we close the gap with such a thin wire that short-circuiting current does not change the electrode voltage, a short time later, after the current along the wire is equalized, the actual gap field will become constant along its length and exactly equal to E_{av}.

resistance. A channel must conduct current necessary to support the motion of the streamer tip. This is the current which supplies charge to a new portion of the channel produced at its tip front. The nature of the streamer process is such that the current i_l just behind the tip is proportional to its velocity (see formulas (2.11) and (2.14)). Local field E_c necessary to support this current is defined by (2.15) derived from the channel resistivity per unit length for a still dense plasma. The value of E_c in (2.15) slightly depends on varying streamer parameters, such as length, tip radius, and velocity, and is largely determined by the gap gas composition, which predetermines maximum field E_m at the tip and the slope of the $\nu_i(E)$ curve (the latter was taken into account in (2.15) by the k parameter). The calculated value of $E_c = 4.2\,\text{kV/cm}$ for air appeared to be surprisingly close to the measured value $E_{cr} = 4.65\,\text{kV/cm}$. One should not give too much importance to this coincidence of the values, but the agreement in the order of magnitude is definitely not accidental.

Because of the plasma decay and conductivity decrease, the current support in other channel portions may require a stronger field than E_c. For this reason, decay processes appreciably affect the value of E_{cr}, as is indicated by experiments. An important mechanism of electron loss in cold air in a relatively low field E_{cr} is the attachment in three-body collisions (section 2.2.5). Here, the attachment frequency is $\nu_a \approx N^2$, so the conventional similarity principle $E \sim N$ for field E_{cr} is violated: the reduced field E_{cr}/N does not remain constant and the value of E_{cr} decreases more rapidly than density N [19]. When a streamer propagates through heated air, the critical field becomes lower not only due to a lower density but as a result of a direct temperature action. This was found from measurements at various p and T, up to 900 K [19, 20]. The reason is clear: on gas heating, the action of attachment and recombination becomes weaker (section 2.2.5). In electropositive gases, in which there is no attachment, the value of E_{cr} is lower than in cold air, other things being equal. For instance, in nonpurified nitrogen with an oxygen admixture up to 2%, the field is $E_{cr} \approx 1.5\,\text{kV/cm}$ at $p = 1$ atm. In inert gases where the attachment is absent and the recombination has a lower rate than in molecular gases, E_{cr} is much lower, about $0.5\,\text{kV/cm}$ [21, 22].

The channel field does not vary much in time along its length because of the compensation due to countereffects. On the one hand, the conductivity in an old channel portion is lower than in a new one because of the plasma decay. On the other, that old portion was produced by a faster ionization wave at a higher tip potential corresponding to a larger channel cross section. As a result, the resistivity per unit length $R_1 = (\pi r^2 e \mu_e n_e)^{-1}$ does not vary much along the channel. Of course, it grows in time because of electron loss, but at the same time, the streamer velocity decreases together with the channel current. For this reason, the time variation of the channel field $E(x, t) = R_1(x, t)i(x, t)$ is much slower than that of any of the cofactors.

We shall illustrate this by giving a particular analytical solution to the set of equations (2.13), (2.14), (2.6), (2.10), and (2.31), which is not far from the actual result (see below). Assume the channel field and capacitance per unit length to be constant, with the current along the channel being the same: $E(x, t) = \text{const}$, $C_1(x, t) = \text{const}$, and $i(x, t) = i(t)$. Neglect potential $U_0(x)$, inessential to a long streamer in a sharply non-uniform field, and suppose that the plasma at point x decays exponentially starting from the moment t_x of its production (as is inherent in attachment without recombination). We shall have

$$U(x) = U_a - Ex, \qquad U_l = U_a - El, \qquad V_s = AU_l,$$

$$i = C_1 V_s U_l = C_1 A U_l^2$$

where the nearly constant coefficient A is, according to (2.6) and (2.10), equal to

$$A = \frac{\nu_{im}}{2(2k - 1)E_m \ln (n_m/n_0)} \approx \text{const.} \tag{2.33}$$

The integration of $dl = V_s \, dt$ yields $l \approx l_{max}[1 - \exp(-AEt)]$, $l_{max} \approx U_a/E$, $t_x = t(l)$ at $x = l$. But the requirement $R_1(x, t) = R_1(t)$ involved in the initial assumptions can be met only for one value of the channel field: $E = \nu_a/2A = (\Delta U_t)_{min}/2V_{em}\tau_a \approx 1.2\,\text{kV/cm}$. Here, $V_{em} \approx 4 \times 10^5\,\text{m/s}$ is electron drift velocity at maximum tip field E_m, $(\Delta U_t)_{min} \approx 8\,\text{kV}$ and $\tau_a = \nu_a^{-1} \approx 0.85 \times 10^{-7}\,\text{s}$ is the attachment time. The relation for E can be interpreted as follows. The potential difference $(\Delta U_t)_{min}$ necessary to provide the minimum streamer velocity $V_{s\,min} \approx V_{em}$ must be gained in field E along the plasma decay length $V_{em}\tau_a$. A similar treatment of the streamer process will be offered in the next section when discussing the streamer motion in a uniform field. The order of magnitude of the 'critical' field E is correct. Therefore, the assumptions underlying the particular solution are not meaningless, so the solution illustrates the main idea.

The existence of a critical field has been confirmed quantitatively by numerical models of the streamer process. Let us discuss the calculations obtained from a simple, evident model. We mean the above set of equations (2.13) and (2.14) supplemented by expressions (2.6), (2.10), and (2.33), which define the streamer velocity and local channel radius, together with (2.31) for the plasma decay, in which the time is counted off from the moment t_x of its production at point x. The streamer development in air from a spherical anode of radius $r_a = 5\,\text{cm}$ at $U_a = 500\,\text{kV}$ is demonstrated in figure 2.5.[†] The calculations were made with $\tau_a = 0.85 \times 10^{-7}\,\text{s}$ and recombination coefficient $\beta = 2 \times 10^{-7}\,\text{cm}^3/\text{s}$. The general tendency in the behaviour of principal parameters is quite consistent with the qualitative picture above.

---

† The numerical simulation was made in cooperation with M N Shneider. This type of equation, but with a constant channel radius, was solved in [23].

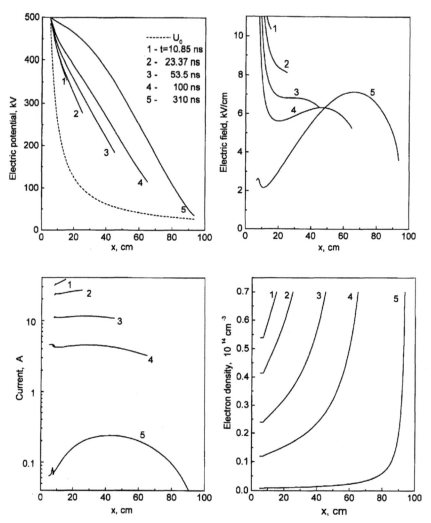

Figure 2.5. Streamer propagation in air from a spherical anode of 5 cm radius at 500 kV. The distributions of potential U, current I, field E, and electron density n_e along the channel at various moments of time until the streamer stops.

Note that the nonmonotonic character of some current distributions when the potential at a given point x grows with time is associated with a slight time decrease in capacitance (2.8) of the elongating channel. The streamer acquires its maximum velocity 10^7 m/s very soon, over 10^{-8} s; then it steadily decelerates and stops at $l_{max} = 0.94$ m.

It was found from (2.32), with the account of $U_0(l_{max})$ and $(\Delta U_t)_{min}$, that the actual average field in the channel at the moment of zero velocity was $E_{cr} = 4.9\,\text{kV/cm}$. From a simplified criterion, it was found to be $E_{cr} = U_a/l_{max} = 5.3\,\text{kV/cm}$. The agreement with the experimental value of $4.65\,\text{kV/cm}$ is quite satisfactory. Calculation with $U_a = 250\,\text{kV}$ yielded $l_{max} = 0.39\,\text{m}$ and at $U_a = 750\,\text{kV}$ it was $1.42\,\text{m}$. Equation (2.32) is satisfied at the same $E_{cr} = 4.9\,\text{kV/cm}$, i.e., the constant value of the average channel field is confirmed by calculations of the final stage of a streamer process for various streamer lengths.† When one takes into account only recombination, a streamer elongates much more, to $1.25\,\text{m}$; with no account of electron losses, it elongates to $3\,\text{m}$ ($E_{cr} = 1.7\,\text{kV/cm}$), as the qualitative picture suggests.

2.2.7 Streamer in a uniform field and in the 'absence' of electrodes

So far, we have dealt with a streamer which starts from a high-voltage electrode, to which it remains galvanically connected, and is supplied by current from a voltage source through the electrode. Such are typical experimental designs and, partly, conditions in the streamer zone of a positive leader, in which the leader channel and its tip possessing a high positive potential act as the electrode. However, a situation may arise when a streamer is initiated in the body of a gas gap, where the external field is sufficiently high. This kind of streamer develops without a galvanic connection with a high-voltage source. Such streamers seem to be present in negative leaders. Note that lightning propagating from a cloud down to the earth most often carries a negative charge, while that going up from an object on the earth is positive. In some situations, a streamer may take its origin from the electrode vicinity and begin its travel being connected to it, but later it may break off because of the plasma decay in an old channel portion. If the external field is still strong at some gap space length, the streamer will move on, having 'forgotten' about its former connection with the electrode. This behaviour is characteristic of the streamer zone of a leader.

Consider a simple case when there is a uniform electric field E_0 at some distance from the electrode and a fairly long conductor of length l and vector E_0 along the x-axis. This may be a metallic rod in laboratory conditions, or a plane or rocket going up to charged clouds, or a dense plasma entity created in this way or other. The conductor is polarized by the external field to form a charged dipole. The vectors of the dipole and external fields are summed. The total field E_{sum} in the body of a perfect conductor drops to zero, since an ideal conductor is always equipotential. In the symmetry condition, all its points take the potential of the external unperturbed field at the middle point of the conductor. Sometimes, the field in the conductor body is said to be pushed out into the external

---

† The calculation using a simplified formula without the account of $U_0(l_{max})$ at $l_{max} = 0.39\,\text{m}$ gives an error: 5.7 instead of $4.9\,\text{kV/cm}$.

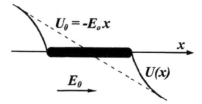

Figure 2.6. The potential distribution along a conductive rod in a uniform electric field. Broken line, the potential in the absence of a conductor.

space. The dipole charge enhances the field ($E_{sum} > E_0$) at the ends of the polarized body (figure 2.6).

The problem of field redistribution by polarization charge can be solved rigorously by numerical methods for any geometry, but simplified evaluations are also possible. In the close vicinity of a charged dipole 'tip' of radius $r \ll l$, the longitudinal field varies nearly in the same way as the field of a sphere of identical radius. Therefore, the external field perturbation by polarization charges is attenuated at a distance r from each of the two conductor ends. Let us take the conductor middle point to be the coordinate origin. The end potentials of a polarized conductor differ from that of an unperturbed one, $U_0 = -E_0x$ at the same points by $\Delta U \approx E_0 l/2$. The absolute strengths of the total field at the conductor ends rise to $E_m \approx \Delta U/r \approx E_0 l/2r$, and the field increases with increasing l/r. This estimate fits fairly well the numerical evaluation in [24].

At $l \gg r$, ionization processes and streamers may arise at the ends of a polarized conductor even at a relatively low external field E_0 (figure 2.7(*a*)). Ionization waves run in both directions, leaving plasma channels behind, in much the same way as with a streamer starting from a high-voltage electrode. If their velocity V_s appreciably exceeds the electron drift velocity, the conditions of streamer travel from the positive and negative ends will not differ much. The total charge of developing streamers is zero at any moment of time. This could not be otherwise, because none of them is connected with the electrode and, through it, with the high-voltage source. Charges do not escape the gap but are only redistributed by the streamer current. A streamer

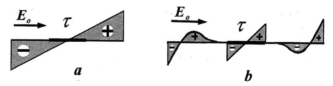

Figure 2.7. Excitation of streamers of both signs from the ends of a conductor in a uniform field. The charge distributions per unit channel length are shown schematically: (*a*) with active plasma, (*b*) with plasma decay in the older channel portions.

travelling along the vector E_0 is charged positively, while a counterpropagating streamer acquires a negative charge. Naturally, the current and the charge separation occur due to electron drift. As in a streamer starting from an electrode, the work done for plasma production and charge separation is done at the expense of the power source creating external field E_0. If it is a conventional high-voltage source, current flowing through the streamer during its propagation is due to the variation in the charge induced on the electrode surfaces when the value and distribution of polarization charges in the gap bulk change. This current takes away some of the source power which is eventually used for the streamer development.

As streamers develop, the total length of a polarized conductor increases, increasing the potential difference ΔU pushed out of the plasma channels. On the other hand, the plasma in the old, central channel decays, so that the charge overflow from one half of the conductor to the other becomes more difficult. Finally, a streamer cannot elongate any more, because the gain in length at the tip is lost due to the plasma decay at the 'tail'. What one observes now is a pair of detached plasma sections of limited length going away in both directions. At the front ends, they have a limited potential difference ΔU with respect to the external potential. The process is stabilized. It may probably go on until there is the external field.

We should like to emphasize that this issue is of principal importance. A streamer needs the external field for charge redistribution in the created plasma, i.e., behind the tip but not in front. The streamer creates its own field that contributes to the gas ionization in the tip region, while the channel field provided by an external source is necessary to support the current to the tip, without which the streamer could not move on.

When the streamer plasma in the old channel near the starting point has decayed completely, two galvanically disconnected streamers continue to move in opposite directions. Now the polarization effects of each of the conductive sections are added to the earlier polarization effects of the whole channel. As a result, there are four charged regions with alternating polarity (figure 2.7(*b*)). Nevertheless, there are still only two ionization waves moving only forward, away from the channel centre, trying to elongate the streamer. No return ionization waves arise in its central portion which has lost conductivity because of a smooth charge distribution towards the ends. The charge is 'smeared' along more or less extended 'semiconducting' regions and cannot create a sufficiently strong field to initiate ionization. High-voltage engineers are familiar with this phenomenon. They can sometimes decrease an electric field, alternating in time, at the site of its local rise between a sharp metallic edge and an insulator, by coating the dielectric at their boundary with semiconducting material.

The parameters of fast 'electrodeless' streamers are described by the same formulas (2.6), (2.10), and (2.14). The role of $U_t - U_0$ here is played by the quantity $\Delta U \approx E_0 l/2$, where l is the length of a channel section with preserved conductivity. Here, ΔU also represents the excess of the streamer tip potential over the external one. In particular, as the length l increases and the field at the conductor ends becomes as high as $E_m \approx 150$–$170\,\text{kV/cm}$ for normal density air (section 2.2.2), the growth of E_m ceases. As l and ΔU increase further, the streamer tip radius r_m, not E_m, increases because $\Delta U \approx E_m r_m$. This is accounted for by the self-regulation mechanism discussed in section 2.2.2.

For the understanding of the streamer process in the streamer zone of a leader, where the field is nearly uniform and the leader channel acts as a high voltage 'electrode', it is essential that the 'electrode' and 'electrodeless' situations should be strictly equivalent, provided that the positive and the negative streamers are identical and the external field is uniform. Let us mentally cut, at the centre, a plasma conductor developing in both directions from this centre and discard, say, the negatively charged half. Let us now replace it by a plane anode under zero potential and assume a negative potential to be applied to a remote plane cathode. A cathode-directed streamer produced at the anode by a local inhomogeneity, whose field initially supported ionization, will be identical to a positive streamer in the electrodeless case. Indeed, in both cases, the conductor potential U coincides with the external potential, $U(0) = U_0(0)$. The charge pumped from the negative half into the positive one will now be supplied by the source current from the anode. Here, the principle of mirror reflection in a perfectly conducting plane, well-known from electrodynamics, reveals itself in every detail. According to this principle, the distributions of charge, current and field in half-space do not change if the plane is replaced by the mirror reflection of half-space charges.

These considerations were used in the calculations and representation of results on streamer development in a uniform field $U_0 = -E_0 x$ from the point $x = 0$ towards lower potential (figures 2.8 and 2.9).† The solution was derived from the same set of equations (2.13), (2.14), (2.6), (2.10), (2.31) and the same plasma decay characteristics as in section 2.2.6. The calculations show that a streamer does not develop if the external field is lower than a certain minimum value. The values of $E_{0\,\text{min}}$ do not differ much from the critical channel field E_{cr} which determines the streamer length in a non-uniform field calculated with (2.32): $E_{0\,\text{min}} \approx 7.7\,\text{kV/cm}$. One may probably use for estimations the experimental value $E_{cr} \approx 5\,\text{kV/cm}$ as a realistic $E_{0\,\text{min}}$ (section 2.4.1). If the uniform external field slightly exceeds the minimum, the excess tip potential is small, the streamer velocity is low, and the channel field is close to the unperturbed external field $E_{0\,\text{min}}$. This situation is illustrated in figure 2.8.

If E_0 is appreciably higher than $E_{0\,\text{min}}$, however, the tip potential is much higher than the external potential, and the streamer develops a high velocity

† Numerical simulation was made in cooperation with M N Shneider.

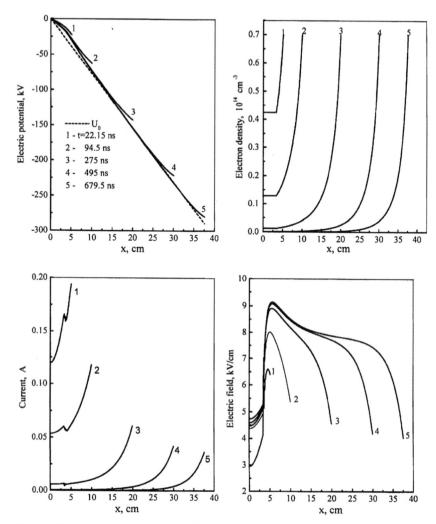

Figure 2.8. An air streamer in a uniform field $E_0 = 7.7\,\text{kV/cm}$, slightly exceeding the critical minimum, with calculated distributions of potential U, current I, field E and electron density n_e. Dashed line, applied field potential counted from the streamer origin. The oppositely charged streamer running in the opposite direction is not shown.

(figure 2.9). The current in well-conducting portions at the tip is high but decreases towards the channel centre. Owing to the high current, much positive charge is pumped into the tip region even at an early stage; in the tail, however, where the conductivity has decreased, the current is low. As a result, positive charge pumped out of it is not reconstructed; moreover,

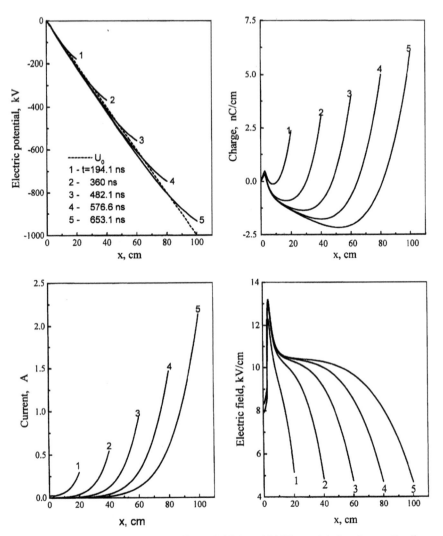

Figure 2.9. An air streamer in a uniform field $E_0 = 10\,\text{kV/cm}$ with the charge distribution τ instead of the n_e curves similar to those in Figure 2.8.

the tail becomes negatively charged. The calculated charge distributions in figure 2.9 were found to be exactly as those represented schematically in the right-hand side of figure 2.7(*b*) in terms of double polarization of the whole channel and each of its conducting sections individually. At $E_0 \approx E_{0\,\text{min}}$ and low current, the polarization effect of the conducting section is very weak (figure 2.8).

2.3 The principles of a leader process

This section is a key one for the understanding of a long spark and the first lightning component. We shall try to answer the question why a simple structureless plasma channel has no chance to acquire a considerable length in cold air of atmospheric pressure. The reader will see what is necessary for a spark to become long and have a long lifetime and how Nature realizes this possibility.

2.3.1 The necessity of gas heating

Section 2.2 dealt with the development of a simple plasma channel – a streamer – which has no additional structural details. The theoretical considerations concerning the streamer process are, in general, supported by experiment, indicating that the streamer gas is cold and the channel field is too low for ionization to occur. In these conditions, the plasma produced in the tip by an ionization wave decays later. Electrons are lost due to recombination (it exists in any gas) and attachment inherent in air as an electronegative gas. Losing its conductivity and, hence, the possibility to use current from an external source, a streamer eventually stops its development, unless it encounters a strong field on its way (section 2.2.7). Sometimes, the streamer lifetime can be made longer by a steady voltage rise, but this possibility is, naturally, limited. Even at a megavolt voltage of a laboratory generator, an air streamer can become only several metres long. Voltages of a few dozens of megavolts inducing lightning discharges are, at best, capable of increasing the streamer length to several tens of metres but not to the kilometre scale characteristic of lightning. At high altitudes, however, the air density is low and a streamer may cover a longer distance. This probably accounts for vertical red sprites above powerful storm clouds dozens of kilometres above the earth's surface [25], which were found to travel downwards.

The only way of preventing or, at least, slowing down air plasma decay in a low electric field is by increasing the gas temperature in the channel to several thousands of Kelvin degrees and, eventually, to 5000–6000 K or more. In a hot gas, electron loss through attachment is compensated by accelerated detachment reactions, and recombination slows down. The mechanism of associative ionization comes into action, and electron impact ionization is enhanced because the gas density decreases on heating. These processes make it possible for a plasma channel to support itself, or, at least, to approach this condition, in a relatively low field. A hot spark looks like a hot arc or a glow discharge column after contraction [26]. We shall not discuss here the details of these processes (for this, see section 2.5). It suffices to take for granted the statement that gas heating does maintain plasma conductivity, making the spark viable.

It follows from section 2.2 that an increase of potential U at the streamer front does not contribute to gas heating. Total energy release grows as U^2, but the streamer cross section πr_{m}^2 also increases as U^2. So the released energy density proportional to $(U/r_{\mathrm{m}})^2$, which defines the heating, remains low. To increase the channel temperature considerably, it is necessary to accumulate a much higher energy in a much narrower plasma column. For this, the functional relation providing the low U/r_{m} ratio must be violated. This is impossible in a primary ionization wave but becomes possible in a differently organized channel development. Let us try to approach this problem by considering the final result and estimate voltages and plasma channel radii, at which the gas temperature would become sufficiently high. This can be done in terms of the general energy considerations discussed in section 2.2.4.

Suppose there is a charged space with characteristic size R in the front region of a developing plasma channel. Its capacitance is $C \approx \pi\varepsilon_0 R$ with $C_1 \approx \pi\varepsilon_0$ per unit length along the channel axis. If the tip potential is U, the energy dissipated per unit length of a new portion of the system including the channel and the space being charged is $C_1 U^2/2$, provided the spark develops steadily. Since the capacitance per unit length of a new system portion is independent of its radius R or of any other geometrical dimension, we are free to assume any nature, size, and volume of this charged space. (The specific capacitance of the central portion of a long system does vary with total length and radius but only logarithmically, as is clear from formula (2.8).) The dissipated energy includes all expenditures for the creation of a new channel portion and space charge. Attribution of this energy to the various expenditures is a special problem which requires details of the process to be specified. But we can estimate the upper limit of air mass that can be heated to the necessary temperature, say, to $T = 5000\,\mathrm{K}$. For this, let us assume that all energy has been used heating an air column of initial radius r_0. This will be an estimate of the upper radius limit. This temperature will lead to considerable thermal gas expansion, because a hot channel, as will be shown later, develops much more slowly than a cold streamer channel. Current must have enough time to heat the gas, because it is eventually the released Joule heat of current that does the heating. If the heating rate is not high enough, pressure in the gas space is equalized, so that the gas of a thin channel becomes less dense. The air heat capacity does not remain constant within a wide temperature range, so energy calculation should be made in terms of specific enthalpy $h(T,p)$. Therefore, the expression to a maximum radius $r_{0\,\mathrm{max}}$ of a cold air column that can be heated to temperature T is

$$\pi r_{0\,\mathrm{max}}^2 \rho_0 h(T) \approx \pi\varepsilon_0 U^2/2. \qquad (2.34)$$

Here, $h(T)$ is specific enthalpy for air at $p = 1\,\mathrm{atm}$ and ρ_0 is its density at $p = 1\,\mathrm{atm}$ and $T_0 = 300\,\mathrm{K}$.

With tip potential $U_{\mathrm{t}} = 1\,\mathrm{MV}$ and $T = 5000\,\mathrm{K}$, when $h(5) = 12\,\mathrm{kJ/g}$, an air column that can be heated must have an initial radius less than $r_{0\,\mathrm{max}} = 0.054\,\mathrm{cm}$. The maximum radius due to thermal expansion will be less than $r_{max} = r_{0\,max}[\rho_0/\rho(5)]^{1/2} \approx 0.26\,\text{cm}$, where $\rho(5)$ is the air density at $T = 5000\,\text{K}$ and pressure 1 atm. A channel of this thickness has been observed in laboratory spark leaders. At $U_t \approx 100\,\text{MV}$, characteristic of very powerful lightning, the radius estimated from formula (2.34) must be two orders of magnitude larger. A lightning leader, however, has a temperature higher than 5000 K and $h \sim T^2$ approximately $(h(10) = 48\,\text{kJ/g})$, so that the radius does not grow as much as U_t and remains as small as several centimetres. It may seem surprising, but a leader channel is thinner than a streamer channel at the same tip potential (their radii are established due to different reasons: a leader radius follows the heating conditions, while a streamer radius is such that the lateral field is too low for intensive ionization).

2.3.2 The necessity of a streamer accompaniment

The existence of a long spark and lightning are due to two main mechanisms, even in the presence of a very high voltage source. One is the mechanism of current contraction in a thin channel which can practically be heated. The other is the attenuation of a very strong radial field that arises at the lateral surface of a very thin conducting channel under a very high potential relative to the earth. We shall begin with the second mechanism, because it opens the way for the first one. In reality, the tremendous value of $U/r \approx 10$–$100\,\text{MV/cm}$ is not the field scale near the channel tip of radius r. Nor does the value of $U/[r\ln(l/r)] \approx 1$–$10\,\text{MV/cm}$, which is somewhat less, determine field E_r at the lateral surface of a channel of length l behind the tip, as could be suggested from formula (2.9) and the Gaussian theorem, $E_r = \tau/(2\pi\varepsilon_0 r)$. This would be valid only for such a simple structureless channel as a streamer, but its lateral field cannot maintain a high strength for a long time. Lateral ionization expansion would immediately increase the channel radius. On the other hand, a channel cannot be heated to the necessary high temperature unless its radius is small. This is the reason why a single simple channel cannot be heated.

A long-living spark requiring megavolt voltages will inevitably have a complex structure. The reader has, no doubt, guessed that we mean the streamer zone in front of a leader tip and its production – the leader cover representing a thick charged envelope around the channel (figure 2.2). The space charge of a streamer zone and leader cover, having the same sign as that of the channel potential, greatly reduces the field at the channel surface. Roughly, owing to the field redistribution by space charge, the huge potential U now drops across a much longer length R of the streamer zone and the charge cover radius, rather than across a length nearly as short as the channel radius r. In this case, the field scale is a moderate magnitude U/R but not U/r, because even a laboratory spark has R of about a metre long.

Indeed, the radius of a streamer zone and, hence, of a leader cover is defined by the maximum distance streamers may cover when they travel away from the leader tip. We already know (sections 2.2.6 and 2.2.7) that the average field necessary for streamer development in atmospheric air must be at least $E_{cr} \approx 5 \, kV/cm$. Since streamers stop at the end of the streamer zone, the voltage drop along the zone length R is about $\Delta U_s \approx E_{cr}R$ (cf. formula (2.32)). About as high voltage drops outside the streamer zone, because the field there, i.e., in a zero-charge space, drops from about E_{cr} to zero, as for a solitary sphere of radius R. Hence, we have $U \approx 2\Delta U_s$ and $R = U/2E_{cr}$. At $U \approx 1 \, MV$, the streamer zone radius is $R \approx 1 \, m$, in agreement with laboratory measurements.

It follows from both calculations and measurements that the current, field, electron density, and conductivity of a heated leader channel are generally comparable with respective parameters of a fast streamer. If they are somewhat larger, the difference is not orders of magnitude. So the heating time to achieve a much higher gas temperature must be much longer. This explains why a leader propagates much more slowly than a fast ionization wave.

The capacitance per unit length of a leader system (the channel plus a charged cover) will be described by the same formula (2.8) if l is substituted by leader length L and the conducting channel radius r by cover radius R, the actual radius of a charged volume. This follows directly from electrostatics. Similarly, the current i_L at the leader channel front is related to the tip potential U and leader velocity V_L by the same expression (2.11)†

$$i_L = \frac{2\pi\varepsilon_0 U V_L}{\ln (L/R)}. \tag{2.35}$$

For a laboratory leader of length $L \approx 10 \, m$ and $R \approx 1 \, m$, the logarithmic values are several times smaller, while the linear capacitance is larger than in a streamer with $r \approx 10^{-1} \, cm$.

The linear capacitance of a conventionally semispherical streamer zone is $C_1 \approx 2\pi\varepsilon_0$, like the capacitance of a streamer tip. The tip current flowing into the streamer zone

$$i_t = 2\pi\varepsilon_0 U V_L \tag{2.36}$$

is by a factor of $\ln (L/R)$ higher than i_L, again like in a streamer. But since the leader logarithm is closer to unity, currents i_L and i_t do not differ as much as for a streamer. If the current along a leader channel does not vary much, as in a fairly short leader at constant voltage, the tip current will not differ much from experimental current i in the external circuit. A typical laboratory leader has $i \approx i_L \approx i_t \approx 1 \, A$, $U \approx 1 \, MV$, and from (2.36) $V_L \approx 2\times10^4 \, m/s$ which is close to numerous measurements, in which $V_L \approx (1–2.2) \times 10^4 \, m/s$ [27, 28]. Formula (2.36) or (2.35) permits the estimation of any of the three parameters

---

† Here and below, the external field potential $U_0(x)$ is omitted for brevity. It is indeed small in laboratory leaders normally observed in a sharply non-uniform field.

– i, U, or V_L – from the other two. This is especially useful in studying lightning leaders when actual data are very scarce.

Thus, a key condition for a long-term spark development is the formation of a thick space–charge cover around it, having the same sign as the channel potential. The charge reduces the field on the channel surface, depriving the channel of its ability to expand due to ionization. It is only a channel with a small cross section that can preserve the ability to be heated. A charge cover also contributes somewhat to the linear leader capacitance, because it is now determined by the much larger cover radius R rather than by the small channel radius r. An increase in linear capacitance is accompanied by an increase in the energy input into the channel.

If Nature were a living being and decided to make a spark or lightning travel as large a distance as possible, it would do this by organizing the streamer zone and charge cover. In actual reality, everything happens automatically: the huge voltages that create long sparks produce numerous streamers at the front end (figure 2.10).

This reminds us of a high voltage electrode creating, under suitable conditions, a multiplicity of streamer corona elements. This kind of corona can be registered in laboratory experiments.

Currents of all streamers starting from a leader tip are summed up, heating the spark channel. This total current charges the region in front of the tip, neutralizing the charge of the old tip, and when a new tip is formed, the spark elongates by a length of about the tip length, as in a

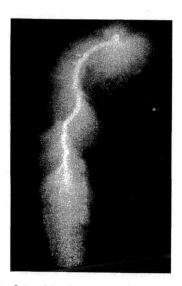

Figure 2.10. Photograph of a positive leader in a rod-plane gap of 9 m length at 2 MV; the electronic shutter was closed at the moment of contact of the streamer zone and the plane.

single streamer. Part of the streamer zone appears to be behind the tip, transforming to a new cover for the newly born leader portion. But this does not decrease the streamer zone length, because meanwhile the zone has moved forwards together with the tip. Note only that if there are many streamers they are very close to one another, and they travel in a self-consistent field close to the critical field (sections 2.2.6 and 2.2.7). Such streamers are slow and have a low current [4], so that the leader current is, indeed, a sum of numerous low streamer currents.

2.3.3   Channel contraction mechanism

The mechanism of current contraction in the front region of a leader channel is not quite clear, especially quantitatively. One may assume the existence of ionization–thermal instability. This effect looks like the one leading to glow discharge contraction [26], but it has its own specificity [4]. The instability is associated with the dependence of electron impact ionization frequency on field and molecular number density: $\nu_i(E, N) = Nf(E/N)$, where $f(E/N)$ is a rapidly rising function at small E/N (figure 2.4). This is the ionization component of the instability. Its thermal component is due to the fact that the gas pressure p rapidly equalizes in small volumes at a moderate heating rate. With $p \sim NT = \text{const}$, a more heated site proves less dense, and the reduced field E/N, determining the ionization frequency, increases there.

   As was mentioned above, numerous streamers start from the front end of a developing leader. The frequency of streamer emission has been shown experimentally to exceed $10^9\,\text{s}^{-1}$ at a typical laboratory spark current of 1 A [29]. Younger streamers have not lost their conductivity yet. The streamers at the leader tip are so close to each other that they form a continuous conducting channel of radius r_{sum}. Current i_t flows along this and the initial leader channel. It is external current relative to the tip, because it is created by the whole combination of charges exposed and displaced by the streamer zone bulk. This current is practically independent of the tip conductivity. In terms of electric circuit theory, the streamer zone acts as a current source (an electric power generator with an inner resistance $\Omega \to \infty$) relative to the leader tip. Its actual value is very large: $\Omega \approx \Delta U_s/i_t \approx U/2i_t$, where ΔU_s is the voltage drop across the streamer zone. At $U \approx 1\,\text{MV}$ and $i \approx 1\,\text{A}$, the value is $\Omega \approx 0.5\,\text{M}\Omega$. No matter what happens to the leader tip or its short front portion, the current there does not change. What changes is the electric field, because it depends on the conductivity and radius of the region loaded by current (in a glow discharge, the field is fixed and the current can vary during the instability development).

   Suppose the current density, whose average cross section value is $j = i_t(\pi r_{sum}^2)^{-1}$, has increased, for some reason or other, in a thin current column of radius $r_0 \ll r_{sum}$. Then the released energy density jE and gas temperature T will also increase. The gas density N will become smaller and E/N larger due to thermal expansion. As the ionization frequency is a steep function of reduced field, it will grow much faster than E/N. So, the electron density n_e and conductivity $\sigma \sim n_e/N$ will rise. As a result of this long chain of cause–effect relationships, the current density $j = \sigma E$ in the fluctuation region will become still larger, etc. The process may begin with any link in the chain. In any case, the current density in a particular fluctuation region will be rising without limit until all current i_t accumulates there. At the initial stage of instability development, the perturbed current density does not exceed much an average value. But as current concentrates within a small cross section πr_0^2, the gas heating rate there rises sharply. The instability now develops very quickly, acquiring an explosion-like character. The acceleration effect is manifested better in a thinner column with high density current.

A perturbation region, however, cannot be infinitely thin, and this sets a limit to the rate of instability development. The matter is that non-uniformities of electron density n_e are dispersed by diffusion, which is ambipolar at very high density values. The characteristic time for perturbation dispersion is $\tau_{amb} = r_0^2/4D_a$, where $D_a = \mu_+ T_e$ is an ambipolar diffusion coefficient (μ_+ is ion mobility and T_e is electron temperature in volts). In addition to charge diffusion, non-uniformity dispersion is due to heat conduction with a characteristic time $\tau_{th} = r_0^2/4\kappa$, where κ is thermal diffusivity. The former mechanism appears to be more effective in initial air plasma, since $D_a \approx 4\,\text{cm}^2/\text{s}$ ($\mu_+ \approx 2\,\text{cm}^2/\text{s} \cdot \text{V}$, $T_e \approx 2\,\text{eV}$) is an order of magnitude larger than $\kappa \approx 0.3\,\text{cm}^2/\text{s}$. If a non-uniformity takes less time for dispersion than for development, i.e., if τ_{amb} is smaller than the instability lifetime τ_{ins}, the latter is suppressed at its origin.

The scale for τ_{ins} is the characteristic time of, say, gas temperature doubling in a perturbed plasma column, as compared with initial temperature T_0. This time is $\rho_0 c_p T_0/jE$, where jE is the power of Joule heat release and c_p is specific heat at constant pressure with the account of thermal expansion. But this is not all. The higher the instability development rate is, the greater is the steepness of the ionization frequency dependence on reduced field $E/N \sim ET$, i.e., on gas temperature. For instance, if a 10% increase in T raises the ionization rate by 20%, the instability will, generally, double its rate, as compared with a 10% increase in the ionization rate. This circumstance brings the factor $\hat{\nu}_i \equiv \mathrm{d}\ln\nu_i/\mathrm{d}\ln(E/N)$ into the theoretical formula for τ_{ins} [26], which characterizes the $\nu_i(E/N)$ function steepness. This yields the following expression to be used for estimations:

$$\tau_{ins} \approx \frac{\rho_0 c_p T_0}{\hat{\nu}_i jE} = \frac{\rho_0 c_p T_0 \sigma}{\hat{\nu}_i j^2}. \tag{2.37}$$

For calculations, we shall take laboratory leader current $i \approx 1\,\text{A}$ and conductivity $\sigma \approx 10^{-2}\,(\Omega\,\text{cm})^{-1}$ corresponding to the electron density $n_e \approx 10^{14}\,\text{cm}^{-3}$ of air ionization by a streamer zone; $\hat{\nu}_i \approx 2.5$. Suppose the current density in a perturbation region is $j \approx 40\,\mathrm{A/cm^2}$, i.e., somewhat higher than the average value of $30\,\mathrm{A/cm^2}$ along a channel with the initial radius taken to be $r_{\mathrm{sum}} = 0.1\,\mathrm{cm}$. We shall obtain $\tau_{\mathrm{ins}} \approx 10^{-6}\,\mathrm{s}$. From the condition $\tau_{\mathrm{amb}} \geqslant \tau_{\mathrm{ins}}$, under which the instability has a chance to develop further, we find that the initial radius of a column with accumulated leader current must exceed $r_{0\,\mathrm{min}} \approx 3 \times 10^{-3}\,\mathrm{cm}$. Taking into account the upper limit $r_{0\,\mathrm{max}} \approx 5 \times 10^{-2}\,\mathrm{cm}$ derived from energy considerations, we conclude that a probable leader radius prior to thermal expansion is about $r_0 \sim 10^{-2}\,\mathrm{cm}$. For details, the reader is referred to [4], but reservation should be made concerning the result accuracy, which cannot be too high in the present state of the art.

2.3.4 Leader velocity

Streamers generated at the leader channel front cover a distance of several metres and stop. As was mentioned in section 2.3.2, such streamers are weak and their propagation is slow; their velocity is close to its low limit of $V_s \approx 10^5\,\mathrm{m/s}$, which means that their lifetime is $R/V_s \approx 10^{-5}\,\mathrm{s}$. This time is so long that the streamer plasma decays considerably. Only young streamers, whose lifetime is about the electron attachment time $\tau_a \approx 10^{-7}\,\mathrm{s}$ (section 2.2.5), can preserve good conductivity. A young streamer length is $l_t \approx V_s\tau_a \approx 1\,\mathrm{cm}$. A dense fan of such plasma conductors starts from the channel front. It is this young streamer fan that seems to be registered in photographs as a bright spot with a radius of $r_t \sim 1\,\mathrm{cm}$ in order of magnitude (figure 2.11) and is generally considered as a leader tip.

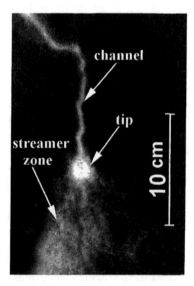

Figure 2.11. An instantaneous photograph (0.1 µs exposure) of the tip region of a leader.

This suggestion is supported by the fact that the radius of a leader travelling through air pre-heated to 900 K is $r_t \approx 10$ cm [20]. Indeed, the plasma decay slows down and the values of τ_a and l_t become higher.

Thus, a necessary condition for leader propagation is the tip region contraction to a very small radius. This results from instability development, taking a time of about τ_{ins}. Over this time, all short young streamers supplying the leader with current transform to the leader channel. Therefore, over the time of the process providing a steady propagation of the leader tip, the latter must cover a distance of about its size, i.e., a young streamer length. Only in this case can a new front region be formed to replace the old one. Hence, the leader velocity can be evaluated from the respective parameters as

$$V_L \sim r_t/\tau_{ins} \sim l_t/\tau_{ins} \sim \tau_a V_s/\tau_{ins}. \qquad (2.38)$$

In the absence of attachment or if its rate is low, the role of τ_a is performed by the time of another plasma decay process – recombination. The evaluation with (2.36) gives a correct order of magnitude for the velocity of laboratory leaders: at $\tau_{ins} \sim 10^{-6}$ s and $l_t \sim 1$ cm, we obtain $V_L \sim 10^4$ m/s. We should like to note that these qualitative and, probably, questionable considerations have not yet been substantiated by a more rigorous treatment.

Some of the above problems of the leader process will be discussed in more detail in the subsequent sections of this chapter and further. Here, our aim was only to give a general idea of the propagation of a long spark and, presumably, of the first lightning component. A reader interested exclusively in lightning hazards may find this information sufficient.

2.4   The streamer zone and cover

We have shown above that a streamer zone plays the key role in a leader process. It is here that a space charge cover is formed which stabilizes the leader channel, preventing its ionization expansion which would otherwise exclude plasma heating. A streamer zone is the site of current generation for heating the leader, providing its long life. In this section, we shall deal, in some detail, with processes occurring in the streamer zone and leader cover, defining the priorities in the causative relationships among leader parameters. We shall show how the process of streamer generation from a leader tip becomes automatic.

2.4.1   Charge and field in a streamer zone

The tip of a long leader possesses a very high potential: $U_t \sim 1$ MV for laboratory sparks and $\sim$10 MV or, probably, more for lightning. Streamers are continuously produced in a leader tip, which means that the field at its surface E_t exceeds the ionization threshold $E_i \approx 30$ kV/cm (under normal conditions). This excess cannot be very large, otherwise a streamer flux would become too intensive. The excessive charge of the same sign as U_t introduced into the space would create a much stronger reverse field which would reduce E_t to a level close to E_i. Therefore, the field E_t does not exceed much E_i and has the same order of magnitude. An automatic field stabilization is inherent in any continuous threshold process of charge generation by an electrode, for example, in a steady-state corona. Measurements have shown that the field near a corona-forming electrode is stabilized with high accuracy and does not respond to voltage rise across the gap; what changes is the corona intensity, i.e., its current. A leader tip, too, is the site of corona formation, with an intensity high enough to support a quasi-stationary state in the tip and streamer zone, corresponding to potential U_t. The field at the corona electrode E_{in} is shown by stationary corona experiments to be by a factor of 1.5 higher than E_i, if the electrode radius is about the leader tip radius $r \sim 1$ cm.

At $E_t = E_{in} \approx 50$ kV/cm and $r_t \approx 1$ cm, the leader tip charge $q_t = 4\pi\varepsilon_0 r_t^2 E_t \approx 5 \times 10^{-8}$ C is capable of creating only a small portion of $E_t r_t = 50$ kV of an actually megavolt potential U_t. The main potential source is, therefore, the space charge of the streamer zone and cover surrounding the tip. But the value of U_t is primarily determined by characteristics external relative to the tip. This is the electrode (anode) potential minus the voltage drop across the leader channel. Consequently, the charge Q_s and the size R of a streamer zone, as well as respective cover parameters, are established such that they correspond to the proper potential U_t. The mechanism by which a leader 'chooses' the values of Q_s and R are directly related to streamer properties. There are many streamers present in the zone at every moment of time. They are emitted by the tip at a high frequency (see below), have different lengths at any given moment and are at different stages of evolution, with their charges filling up the zone space. Every single streamer moves in a self-consistent field created by the whole combination of streamers. The contribution of the leader tip itself (or of its channel) to the total field has just been shown to be small. One exception is the region around the tip with a size of its radius.

There are experimental and theoretical grounds to believe that the field strength in the streamer zone, except for the tip vicinity, is more or less constant and close to the minimum at which streamers can grow. This is indicated by measurements of streamer velocity, which does not vary along the streamer zone. (Attempts to measure a single streamer in the tip region, where the streamer density is high, have so far failed.) Experiments show that until the streamer zone of a laboratory leader touches the opposite electrode, streamers move slowly, at a nearly limit velocity of about 10^5 m/s. This is possible only in a uniform field close to $E_{0\,min}$ (section 2.2.7). Streamers can travel for such a distance R, at which the field $E_{0\,min}$ still exists, but they stop on entering the region with $E < E_{0\,min}$.

Suppose, for simplicity, that the streamer zone is a hemisphere with the centre in the leader tip. The hemisphere changes to a cylindrical cover of the same radius R with the same order of the space–charge density. A thin conducting leader channel goes along the cylinder axis as far as the hemisphere centre. When evaluating the zone parameters, one should take into account the cover charge at the leader end, which also affects the zone field. We can do this simply by connecting the hemisphere, simulating a streamer zone, to another hemisphere by mentally cutting it out of the cover space. Let us assume that there is a uniform radial field $E = E_{0\,min}$ in the sphere. As was mentioned in section 2.2.7, the theoretical limit of $E_{0\,min}$ is close to the experimental critical average field in the streamer channel, below which a streamer cannot propagate. For air, therefore, we have $E \approx E_{cr} \approx 5\,kV/cm$. A uniform field in sphere geometry corresponds to the space charge density $\rho = 2\varepsilon_0 E/r$. If the leader tip is far from the earth and grounded electrodes, its potential is

$$U_t = \frac{1}{4\pi\varepsilon_0} \int_0^R \frac{4\pi r^2 \rho \, dr}{r} = 2ER. \tag{2.39}$$

The sphere charge Q and its surface potential U_R are

$$Q = \int_0^R 4\pi r^2 \rho \, dr = 4\pi\varepsilon_0 R^2 E = 2\pi\varepsilon_0 R U_t \tag{2.40}$$

$$U_R = U_t - ER = U_t/2.$$

For example, for $U_t = 1.5\,MV$, we have $R = 1.5\,m$, the charge of a hemispherical streamer zone equal to $Q_s = Q/2 = 6.2 \times 10^{-5}\,C$. The leader tip charge $q_t = Q(E_t/E)(r_t/R)^2 \sim 10^{-3}Q$ is indeed negligible, as compared with the zone charge. Its physical role, however, is very important: the high field it creates near the tip, $E_t > E_i \gg E_{cr}$, is capable of generating streamers.

As the streamer zone approaches the grounded plane, its length increases because its boundary potential U_R decreases under the action of charge of opposite polarity induced in the earth. Now it is most of the voltage U_t, rather than its half, which drops across the streamer space. At the moment of streamer contact with the 'earth', potential $U_R = 0$ and the zone length $L_s = U_t/E$ is doubled relative to the value of R from formula (2.39). This is clearly seen in streak pictures of a laboratory spark (figure 2.12). It is at the moment of streamer contact with a grounded plane that the critical field $E_{cr} \approx 5\,kV/cm$ was registered experimentally. The measurements make sense only for short (compared with the interelectrode distance) leaders, when the voltage drop across the channel could be neglected. By equating the potentials of the anode U_a and of the tip, we can write: $E_{cr} = U_t/L_s \approx U_a/L_s$.

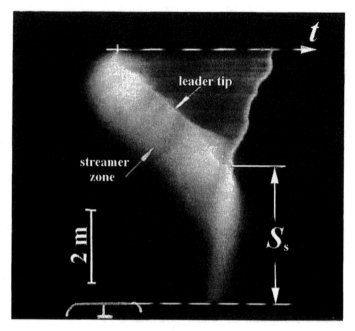

Figure 2.12. Streak photograph of the positive leader channel bottom in an air rod–plane gap of 12 m in length. The streamer zone is seen to elongate when approaching the plane cathode.

2.4.2 Streamer frequency and number

The number of streamers present in a streamer zone at every moment of time is $N_s = Q_s/q_s$, where q_s is the average charge of a streamer. Both charges were measured experimentally [29, 30], the first from the integral of conduction current through the anode for short leaders with as yet small cover charge and the second from the integral of current through a cathode measurement cell with such a small radius that only one streamer could touch its surface (with good luck). After a successful contact, the charge of a conductive streamer section flew into the cathode and through an integrating circuit. The charge averaged over many registrations was found to be $q_s = 5 \times 10^{-10}$ C. For the illustration mentioned in the previous section of this chapter, we find that the number of streamers in a streamer zone of length 1.5 m and charge $Q_s = 6.2 \times 10^{-5}$ C is $N_s \approx 1.2 \times 10^5$. Similar data for q_s can be derived from the calculations presented in figure 2.8 or from a simple theoretical treatment, we shall just perform.

A streamer produced by a leader tip crosses the streamer zone over the time $t_s = R/V_{s\,min} \approx 10^{-5}$ s, which is by two orders of magnitude larger than the attachment time $\tau_a \approx 10^{-7}$ s. Therefore, all electrons are lost from older streamer portions comparable in length with R. Conductivity is preserved only along the length $l_s \sim V_{s\,min}\tau_a \sim 1$ cm behind the streamer tip. During the motion of the streamer tip, the charge of the older streamer portions flows into the new ones located closer to the tip before conductivity turns to zero. Under steady-state conditions, when the tip goes far away from the start, the charged portion of length l_s moves together with the tip, supporting by its charge the minimum excess of tip potential over external potential, $(\Delta U_l)_{min} \approx 5$ kV, necessary for the streamer propagation (section 2.2.7). The conductor of length l_s and radius $r_m \approx 0.1$ cm carries the charge

$$q_s \approx \frac{2\pi\varepsilon_0(\Delta U_l)_{min}l_s}{\ln(l_s/r_m)} \approx 10^{-9} \text{ C}. \qquad (2.41)$$

As for the charge accumulated in the streamer tip

$$q_{st} \approx 2\pi\varepsilon_0(\Delta U_l)_{min}r_m \approx \pi\varepsilon_0(\Delta U_l)^2_{min}/E_m \approx 5 \times 10^{-11} \text{ C},$$

which was not taken into account in the calculation or (2.39), it is by one order less than that distributed along the channel. At $N_s \approx 10^5$, the average interstreamer space is $R/N_s^{1/3}$, i.e., about several centimetres. With this large separation of streamer tips, the streamers can really be considered solitary and propagating in an average self-consistent field.

When a streamer reaches the end of the streamer zone, it stops because it enters a field lower than $E_{0\,min}$. Since the streamer zone approaches this field fairly slowly, at leader velocity V_L an order of magnitude lower than streamer velocity $V_{s\,min}$ (for laboratory streamers), the streamer loses its conductivity entirely. The ions of its space charge are gradually repelled (and diffuse), reducing the field near the charge trace, so that the streamer trace becomes lifeless 'forever'. The streamer zone still passes by for a time $t_L = R/V_L \sim 10^{-4}$ s, after which the immobile charged trace, which is now behind the leader tip, becomes a cover component. Viable streamers fly across the streamer zone over time $t_s = R/V_{s\,min} \sim 10^{-5}$ s, an order of magnitude shorter. Therefore, if the frequency of streamer production is ν_s, the number of viable streamers in a streamer zone is $N_1 \sim \nu_s t_s$, while the number of non-viable traces, practically coinciding with the total number of charged streamer portions, is $N \sim \nu_s t_L$. Hence, the streamer generation frequency is $\nu_s \sim N/t_L \sim NV_L/R \sim 10^9$ s^{-1}.

2.4.3 Leader tip current

The streamer production frequency is directly related to the leader tip current: $i_t = q_s\nu_s$. With $q_s = 10^{-9}$ C and $\nu_s = 10^9$ s^{-1}, we get $i_t \approx 1$ A, a value typical for laboratory leaders in the initial stage while the streamer zone has not yet reached the cathode. This current value has been registered in many experiments [27–29], and the relation $\nu_s = i_L/q_s$ has been confirmed by direct measurements. The streamers were counted by piece from current impulses in small cathode measurement cells after the streamer zone boundary had touched its surface [29]. The counts were integrated over the area. Measurements made at different currents showed that the relation $i_L/\nu_s = q_s \approx 5 \times 10^{-10}$ C remained constant. It is consistent with measurements and the estimations of average streamer charge presented here.

The formula for leader tip current can be given in a conventional form of the type $i = \pi r^2 e n_e V_e = \tau_e V_e$, when current is expressed as the number of charge carriers (electrons) per unit current column length τ_e and electron velocity V_e. This formula can also be changed to the phenomenological expression (2.36) describing the result of the current process without indicating the nature of carriers. Although the current carriers in this case are electrons, we can also speak of 'macroscopic' carrier-moving charged streamer sections. With what we mentioned at the end of section 2.3.2 and formula (2.36), we have

$$i_t = q_s \nu_s \approx \frac{q_s N V_L}{R} \approx \frac{Q_s V_L}{R} = \tau_s V_L = \tau_1 V_{s\,min} \qquad (2.42)$$

where $\tau_s = Q_s/R \approx \pi \varepsilon_0 U_t$ is the linear charge in a streamer zone and $\tau_1 = q_s N_1/R$ is the linear charge of 'macroscopic' carriers. The three expressions for current are equivalent to one another but reflect different aspects of the current process. The second expression in the chain of equalities (2.42) indicates the current origin while the latter shows the actual process of charge transport; the penultimate expression is phenomenological and describes the result of travel of the streamer zone as a whole.

The mechanism of leader current production just described is valid until a streamer zone touches the electrode of opposite polarity (the earth or a grounded object in the case of lightning). Then the situation changes radically. In the final jump, the charges of all streamers 'hitting' the electrode leave the gap through its surface. Nonviable streamers are no longer produced, and they possess an ever-decreasing portion of total streamer zone charge. The last expression in (2.42) is valid in this case, too, $i_t = \tau_1 V_{s\,min}$, but the value of τ_1 is no longer equal to the portion $V_L/V_{s\,min}$ of the total streamer zone charge Q_s. In the limit, when the leader tip reaches the electrode, all zone charge will be provided by moving streamers, so the current will be $i_t \approx \tau_s V_s$. The linear charge of the streamer zone in the final jump remains the same in the order of magnitude as in the initial stage. Of course, the streamer zone has now a different shape – it is elongated, looking more like a cylinder than a hemisphere. But still, its longitudinal L_s and transverse R_s dimensions are comparable, and the value of $\ln(L_s/R_s)$ which appears in the denominator of the expression for linear capacitance in cylindrical geometry (2.8) and (2.35) is about unity. Therefore, leader currents after the transition to the final jump and before it are related as V_s/V_L, the ratio being above 10.

Moreover, the streamer velocity rapidly increases as the streamer zone is reduced. The potential difference between the leader tip and the grounded electrode remains the same, U_t, whereas the length L_s becomes shorter. The average field in the streamer zone $E = U_t/L_s$ rises, together with streamer velocity (section 2.2.7) and leader current $i_t \sim V_s$. The final jump current was shown by laboratory leader experiments to rise by a factor of tens or hundreds, from about 1 A to 10^2–10^3 A. This fast current rise lasting for several microseconds is a prelude to a still higher current of the return stroke. The latter begins when the leader channel reaches the electrode. The current rise of the final jump is stimulated by the fact that fast streamers cross a shorter streamer zone much faster than before, so that the zone plasma is unable to decay as much, preserving the streamer conductivity. At the end of the final jump, the leader channel appears to be linked to the opposite electrode by numerous streamer filaments with current (for details, see [4]).

2.4.4 Ionization processes in the cover

A leader cover contains a large number of non-viable charged traces of earlier streamers. They were produced and developed when a streamer zone was passing through this site. The axial field in the cover is very low, much lower than in the streamer zone. No ionization can occur in it, so it is of no interest to us. What is important is the radial field created by the leader surface charge and all cover charges, as in a streamer zone, the only difference being in geometry. In contrast to a channel cover, however, a leader tip with a streamer zone is formed as a self-consistent system from the very beginning. The tip pumps into the zone as much charge as necessary to maintain the tip field at the level E_{in} providing the production of the necessary number of streamers. A leader channel 'inherits' a ready-made cover. The charge amount and distribution in the former quasi-spherical zone is unsuitable for the cylindrical geometry and channel potentials $U(x)$ different from U_t. But the inherited charge of dead streamer traces is invariable, which means that there must be a mechanism to make the channel–cover system self-consistent and controllable, since the potential distribution $U(x)$ and linear leader capacitance vary in time.

We mentioned in section 2.3.2 that the 'intrinsic' charge of a conducting channel of length L and radius $r_L \approx 0.1$ cm would create at its surface a huge radial field $E_{r_0} = U/[r_L \ln(L/r_L)] \approx 1$ MV/cm at channel potential $U \approx 1$ MV. This critical situation is unfeasible owing to the presence of a cover. The cover charge induces in the conductor an opposite charge which is to be subtracted from the intrinsic surface charge. As a result, the field E_r created by the resultant charge has a moderate value. It is hard to imagine, however, that the inherited charge will be as large as is necessary for confining the resultant surface field in the narrow range $-E_{in} < E_r < E_{in}$ with $E_{in} \approx 50$ kV/cm $\ll E_{r_0} \approx 1$ MV/cm. No doubt, the cover charge will turn out to be either too large or too small. In the first case, the channel will be charged negatively (if the leader is positive), the field around it will become negative, its module exceeding E_{in}. A negative corona will be excited and introduce into the cover as much charge as is necessary to reduce $|E_r|$ to E_{in}. This situation becomes feasible when the gap voltage is constant or decreases, since then the channel 'enters' the cover with a too-large charge. Indeed, suppose the charge of a streamer hemisphere of radius R is distributed as $\rho = 2\varepsilon_0 E/r$ (section 2.4.1), creating potential $U_t = 2ER$ (2.39) in its centre. The cover inherits a charge of the same radial density distribution $\rho = 2\varepsilon_0 E/r = 2\varepsilon_0 U_t/Rr$ and amount $\tau' = 2\pi\varepsilon_0 U_t$ per unit length. At point x far enough from the channel ends, $L \gg R$, this charge will create potential

$$U'(x) = \frac{1}{\gamma\pi\varepsilon_0} \int_0^R \int_0^L \frac{2\pi r\rho \, dr \, dz}{[r^2 + (z-x)^2]^{1/2}} \approx \frac{\tau'}{2\pi\varepsilon_0} \ln\frac{Le}{R} = U_t \ln\frac{Le}{R} \quad (2.43)$$

where e is the natural logarithmic base. The potential U' is by the logarithmic factor larger than the actual value of U_t. The excess cover charge which has created excessive potential is $U' - U_t$ and must be compensated by introducing a charge of opposite sign.

The other situation, when the inherited charge is too small, is usually feasible if the gap voltage rises appreciably during the leader evolution. The channel potential $U(x)$ at a given point increases, and the cover must be charged up.

It is this mechanism of direct or reverse corona display by a 'wire', such as a leader channel, which leads to a self-consistent channel–cover system. The system is controlled and corrected automatically but not very quickly. It is sensitive to the slightest variation in potential distribution along the channel due to the field E_{in} being too small compared with E_{r_0} of the 'intrinsic' channel charge. A slight effect on the cover is sufficient to change the field value, and even its direction, at the channel surface. The cover of a developing leader with a reverse corona acquires a double-layer structure: outside is the charge inherited from the streamer zone and inside is the new charge of opposite sign, introduced by the corona. For example, at $U_t = 1.5\,\text{MV}$, $R = 1.6\,\text{m}$, and $L = 10\,\text{m}$, the linear cover capacitance from (2.43) is $C_1 = \tau'/U' = 2 \times 10^{-11}\,\text{F/m}$. It is defined by the same formula (2.8) but with the effective radius R_{eff} varying with the radial charge distribution; for $\rho \approx 1/r$, $R_{eff} = R/\text{e}$ and for $\rho(r) = \text{const}$, $R_{eff} = R/\text{e}^{1/2}$, etc. With $U = U_t$, the corrected steady state charge in the cover is $\tau_L = C_1 U = 3 \times 10^{-5}\,\text{C/m}$. Since the values of C_1 and τ_L gradually decrease with the leader length, the surface field must support a negative corona; hence, $E_r \approx -50\,\text{kV/cm}$. The actual channel charge (intrinsic charge minus induced charge) is found to be $\tau_{Lc} = 2\pi\varepsilon_0 E_r = -8.2 \times 10^{-7}\,\text{C/m} \ll \tau_L$. Therefore, it is easy to control the value of τ_{Lc} and even to reverse its sign.

We have deliberately considered the mechanism of cover–leader self-regulation in so much detail, because a reverse corona neutralizes the cover charge in a laboratory spark and lightning during the return stroke, when the channel potential becomes equal to the earth's zero potential (see section 4.4).

2.5   A long leader channel

All ionization processes responsible for the leader development are localized in the streamer zone, leader tip and a short channel section behind it. In the latter, gas heating is completed and a quasi-stationary state characteristic of a long spark is established. In this sense, the rest of the channel plays a minor role, simply connecting the operating part of the leader to a high-voltage source. High potential and current vital to the ionization and energy supply are transmitted through the channel. But how much voltage reaches the leader tip depends on the channel conductivity which, in turn, is determined by the channel state. For this reason, what is going on in a developing channel is as important to the leader process as the mechanisms described above.

2.5.1   Field and the plasma state

There are no direct experimental data on the state of a lightning leader channel. Therefore, of special value is the information derived from laboratory spark experiments, since it can serve as a starting point in lightning treatments. Here we present some values derived from experimental data [27] with a minimum number of assumptions. Streak photographs were taken continuously of a leader propagating from a rod anode to a grounded plane. Pulses of voltage U_0 with the microsecond risetime were applied to gaps of various length d. By measuring the streamer zone length L_s in the photographs at the moment the zone touched the grounded electrode and assuming the average zone field to be $E_{cr} = 4.65\,\text{kV/cm}$ (section 2.4), one can find the leader tip potential $U_t = E_{cr}L_s$ and evaluate the average field in the leader channel as $E_L = (U_0 - U_t)/L$, where $L = d - L_s$ is the channel length (table 2.1).

The accuracy of E_L evaluation, however, is not high, because one calculates a small difference between large values. Besides, measurements of the streamer zone length at the moment of contact with the cathode contain errors as large as those of the channel length. When determining the latter from streak pictures, one can hardly take into account all channel

Table 2.1. Leader parameter derived from experimental data.

d, m	U_0, MV	L_s, m	L, m	U_t, MV	E_L, V/cm
5	1.3	2.3	2.7	1.1	750
10	1.9	3.2	6.8	1.5	590
15	2.2	3.6	11.4	1.7	440

bendings, which may increase the length by 20–30%. Finally, the accuracy of E_{cr} is not as high as is necessary for such a delicate operation. Experimental researchers know that the streamer zone field varies with air pressure, humidity, and temperature but they do not know the respective corrections. Nevertheless, the data in table 2.1 demonstrate a decrease in the average field with increasing channel length. This is also evident from experiments with superlong sparks. A voltage of 3–5 MV is sufficient to create a spark 100 m long or longer. The tip potential necessary for the development of a streamer zone of several metres in length is 1–2 MV, as is clear from table 2.1, therefore the average field in such a long channel will be as low as 200–250 V/cm.

These values are more applicable to older, remote channel sections which have acquired a quasi-stationary state, but the fields close to the tip are much higher. This follows from many experiments indicating a regular increase in average field with decreasing leader length. More explicitly this was shown by supershort spark experiments, when the channel length was only a few dozens of centimetres [31]. The field in a supershort leader and, therefore, the field at the respective distance from the tip of a long spark may be 2–4 kV/cm. At the site of ionization–thermal instability, where current is accumulated within a thin column, the field was found (section 2.3) to be 20 kV/cm [4]. But far from the tip, it is nearly two orders of magnitude lower.

At a typical experimental leader current of $i \approx 1$ A, its velocity is $V_L \approx 1.5–2$ cm/µs, and the lifetime of a leader section at a distance of 3 m from the tip is at least 150 µs. This time is long enough for the relaxation processes in the channel to be nearly completed and for a nearly steady-state to be established. The thermal expansion of the channel, very fast at the beginning, is also completed by that time. Measurements made in a 10 m gap between a cone anode and a grounded plane [28] (voltage 1.6–1.8 MV, average current about 1 A, and average leader velocity 2 cm/µs) showed that the channel expansion rate was 100 m/s at first but 100 µs later the rate dropped to 2 m/s. Measurements made by the shadow technique showed the average thermal expansion radius to be $r_L \approx 0.1$ cm. According to spectroscopic measurements, the temperature of a channel which has reached the gap middle is 5000–6000 K. Some other experimental data on laboratory leaders can be found in [4, 27, 28, 31, 32].

The air ionization mechanism changes radically at temperatures $T \approx 3000–6000$ K and relatively low reduced fields (for example, at $E = 450$ V/cm, $T = 5000$ K, and $p = 1$ atm, we have $E/N = 3 \times 10^{-16}$ V/cm^2).† In cold air,

---

† We should like to warn against the commonly used postulate that the field in a leader channel has a constant value of $E/N \approx 8 \times 10^{-16}$ V cm^2. Some authors use it for the calculation of breakdown voltages in air gaps, including long ones. At $T = 5000$ K, we have $E = 1.15$ kV/cm, which disagrees with experimental data for more or less long leaders (table 2.1) and contradicts the physics. The underlying implicit suggestion is that the only consequence of air heating is a change in its density. We shall show that this is not the case.

it is ionization of O_2 molecules by electrons gaining energy in a strong field, but at the above value of E/N, the ionization rate of unexcited oxygen and nitrogen molecules and atoms by electron impact is negligible. Electrons are mostly produced in the associative ionization reaction

$$N + O + 2.8eV \rightarrow NO^+ + e. \qquad (2.44)$$

Due to a low ionization potential of NO (9.3 eV), the reaction requires a small activation energy and occurs at a large rate constant. Recent data [33] give

$$k_{i\,as} = 2.59 \times 10^{-17} T^{1.43} \exp\left(-31\,140/T\right) \text{ cm}^3 \text{ s}^{-1}, \qquad T\,[\text{K}] \qquad (2.45)$$

Direct NO ionization by electron impact may compete with associative ionization (2.44) but at an electron temperature higher than $T_e \approx 10^4$ K.

Both estimations and kinetic calculations [34] show that thermodynamically equilibrium concentrations of N, O, NO and electrons at $T \approx 4000$–6000 K (table 2.2) are established for 20–50 μs.

During this time, a leader elongates only by 20–100 cm, i.e., the process of establishing a thermodynamic equilibrium in the channel seems to occur concurrently with the transitional process of channel formation and heating to a quasi-stationary state. Although the electron density does not practically differ from the density due to streamer generation, $n_e \approx 10^{14}$ cm^{-3}, the ionization degree at $T = 5000$, $n_e/N \approx 3.3 \times 10^{-5}$, is an order higher than that in the streamers, $n_e/N \approx 4 \times 10^{-6}$. Therefore, intensive ionization occurs during the evolution of ionization–thermal instability and subsequent heating to the final temperature.

Thus, a long laboratory leader channel can be subdivided into two unequal parts. First, there is a relatively short (about 1 m) transitional portion just behind the tip where the gas is gradually heated and additionally ionized. This is accompanied by a change in the plasma density and conductivity. Second, there is the rest of the channel heated to 5000–6000 K, which has reached a quasi-stationary state. The suggestion of an equilibrium electron density in this part of the channel generally leads to a correct value of its radius, $r \approx 0.13$ cm, close to the measured thermal radius [28]. It can be

Table 2.2. Equilibrium air composition at $p = 1$ atm.

T, K	4000	4500	5000	5500	6000
N, 10^{18} cm^{-3}	1.79	1.60	1.48	1.35	1.27
n_e, 10^{13} cm^{-3}	0.63	1.70	4.90	11.2	21.4
N_O, 10^{17} cm^{-3}	4.70	4.90	4.60	4.35	3.81
N_N, 10^{16} cm^{-3}	0.25	1.15	3.67	9.92	20.6
N_{NO}, 10^{16} cm^{-3}	7.62	4.54	2.73	1.67	1.03
n_e/N, 10^{-5}	0.35	1.06	3.31	8.30	16.8

derived from the relation for current $i = \pi r^2 e n_e \mu_e E$ if we take electron mobility to be $\mu_e \approx 1.5 \times 10^{22} N^{-1}\,\mathrm{cm^2(V \cdot s)^{-1}}$, $i \approx 1\,\mathrm{A}$, $E \approx 250\,\mathrm{V/cm}$ and if we use the values of n_e and N from table 2.2, corresponding to $T = 5000\,\mathrm{K}$.

In reality, with the T and E/N values characteristic of remote channel portions, the electron temperature T_e may differ considerably from the gas temperature: T_e may be as high as $10\,000\,\mathrm{K}$ at $T = 5000\,\mathrm{K}$. This slightly shifts the quasi-stationary values of n_e relative to the thermodynamic equili- brium values corresponding to T (as in table 2.2). Stationary n_e corresponds to the equality of forward and reverse reaction rates in (2.44). The forward reaction rate is independent of T_e, while the reverse reaction rate at $T_e \approx 10^4\,\mathrm{K}$ is proportional to $T_e^{-3/2}$. Hence, the stationary value of n_e will be larger than that in table 2.2 by a factor of $(T_e/T)^{3/4} \approx 2$.

As for the less heated, recent channel sections, the difference between the electron and gas temperatures is greater. The reduced field E/N and T_e must be higher in the unheated channel to provide impact ionization, since there is no other source of electron production. At temperatures $T < 2500\,\mathrm{K}$, this is O_2 ionization by electron impact. As the channel is heated further, NO ionization requiring lower T_e and E/N begins to domi- nate, and only at $T > 4000\text{–}4500\,\mathrm{K}$ requiring a still lower field does the reaction of (2.44) become important. Clearly, the channel field cannot follow the condition $E/N = \mathrm{const}$ because of the change of ionization reactions with different energy thresholds. Calculations show [34] that the value of E/N drops from 55 to $1.5\,\mathrm{Td}$ with heating from 1000 to $6000\,\mathrm{K}$ (figure 2.13).

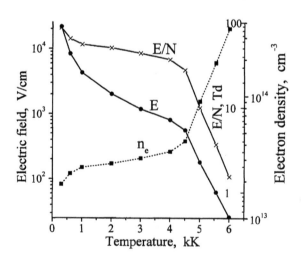

Figure 2.13. Parameters of the initial leader channel right behind the tip as a function of the gas temperature (model calculations of [34]; $1\,\mathrm{Td} = 10^{-17}\,\mathrm{V \cdot cm^2}$).

2.5.2   Energy balance and similarity to an arc

The older leader portions are similar to an arc in atmospheric air. Current 1 A and temperature 5000 K correspond approximately to the minimum arc values and the field equal to 200–250 V/cm is only by a factor of two or three stronger than that in a low current arc. So it is natural to look at the leader channel as an arc analogue.

All characteristics of a long stationary arc at atmospheric pressure when the plasma is usually quasi-stationary (maximum temperature T_m along the axis, longitudinal field E, current channel radius r_0) are defined only by one 'external' parameter, normally, by current i. Joule heat released in the current channel is carried out by heat conduction. Radiation is essential only for very intensive arcs when the channel temperature exceeds 11 000–12 000 K. The further fate of the energy depends on the arc cooling providing its steady-state. Heat can be removed via heat conduction through the cooled walls of the tube containing the arc. It can be carried away by the cooling gas flow or due to a natural convection if the arc burns in a free atmosphere. Definite relations among E, T, and r_0 with current i are obtained if the heat release mechanism is known. None of the above mechanisms (convection does not seem to have enough time to develop) are operative in a leader channel. One may suggest that heat is carried farther away from the channel via heat conduction, gradually heating an ever increasing air volume. Strictly, this is not a steady state process, so it is not a simple matter to find all relations.

However, the state of the channel itself is close to a stationary one. This is due to a small and definite temperature variation in the current channel owing to an exceptionally strong dependence of equilibrium plasma conductivity on its temperature. So one can find the relation between the leader channel temperature T_m and the power $P_1 = iE$ released per unit length. Using the available experimental values for T and i, one can find E to see that a fairly low field is sufficient to support plasma in a well developed leader channel.

The electron density in an equilibrium plasma is $n_e \approx \exp(-I_{eff}/2kT)$, where I_{eff} is an effective 'ionization potential' of the gas. The relation with an actual ionization potential of atoms is strictly valid for a homogeneous gas (the Saha equation). For the temperature range $T \approx 4000$–6000 K, we have, in accordance with table 2.2, $I_{eff} = 8.1$ eV and $I_{eff}/K = 94\,000$ K, which is close to $I_{NO} = 9.3$ eV. Since $I_{eff}/2kT \approx 10$, the conductivity $\sigma \sim n_e$ is strongly temperature dependent and decreases with radius much more than the temperature. Therefore, we can use the concept of a current channel with a more or less fixed boundary – the radius r_0 (figure 2.14). By denoting the channel boundary temperature as T_i and bearing in mind that the temperature variation in the channel is $\Delta T = T_m - T_i \ll T_m$, we can write an approximate expression for the channel energy balance:

$$P_1 = -2\pi r_0 \lambda_m \left(\frac{dT}{dr}\right)_{r_0} \approx 4\pi r_0 \lambda_m \frac{\Delta T}{r_0} = 4\pi \lambda_m \Delta T \qquad (2.46)$$

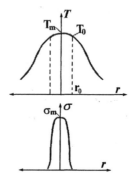

Figure 2.14. Schematic arc channel with nearly the same distributions of T and σ at the axis of the 'old' portion of a leader channel.

where λ_m is heat conductivity at $T = T_m$. The refined factor 4 appears instead of 2 if the heat conduction equation for a uniform distribution of heat sources is integrated over the range $0 < r < r_0$.

An arbitrary channel boundary should be set such as to allow an adequate current through the channel. The current should not be too low, because another channel will appear 'outside'; it should not be too high either, because a current-free periphery will arise inside the current channel. Assume, for the sake of definiteness, that the channel boundary conductivity $\sigma(T_i)$ is by a factor of e less than the axial value $\sigma(T_m)$. With the exponential dependence of $\sigma(T)$, this approximately yields $\Delta T \approx 2kT_m^2/I_{eff}$. By substituting this expression into (2.46), we find the desired relation:

$$P_1 = iE \approx 8\pi\lambda_m \frac{kT_m^2}{I_{eff}}, \qquad T_m = \left(\frac{I_{eff}}{8\pi\lambda_m k}P_1\right)^{1/2}. \qquad (2.47)$$

Expression (2.47) does not contain the radius r_0, its account requires a consideration of the channel environment [26]. The channel temperature T_m grows more slowly with power than $P_1^{1/2}$ because the air heat conductivity rises rapidly with temperature in the range of interest. At $T_m = 5000\,\text{K}$, $\lambda_m = 0.02\,\text{W/cm K}$; so we have $P_1 \approx 130\,\text{W/cm}$. For current $i \approx 1$, for which this temperature seems to be characteristic, $E \approx 130\,\text{V/cm}$. The values of E and P_1 are only by a factor of two smaller than those found from experimental evaluations of the field in older leader channel portions. It is possible that the channel instability and the necessity to heat an increasing air volume require a higher power and a higher field at the given current. This problem remains unsolved and deserves close study. When applied to arcs, formula (2.47) gives a fairly good agreement with experiments.

These considerations of the thermal balance of a leader channel permit establishing the 'current–voltage' characteristic (CVC) $E(i)$ that we shall need in the next section. Heat flow from the channel grows with temperature but not very rapidly. But the conductivity σ and current $i = \pi r^2 \sigma E$ would grow very quickly if the field remained constant. As fast would be the growth of energy release $P_1 = iE$, which would set the system out of heat balance. The balance is maintained because the field drops with rising current, while the power and temperature do not change much. In an ideal case with $T_m = $ const, which is close to the low temperature conditions for a leader and low current arc, we have $E \approx i^{-1}$ from (2.47). The arc CVC is indeed a descending curve, though its goes down somewhat more slowly because T_m and P_1 slightly rise with current [26].

A similarity between the states and CVC curves for quasi-stationary leaders and arcs was established in model experiments [4]. Sparks 7 cm long were generated in air between rod electrodes. The circuit parameters were chosen such that the channel current was stabilized at a level characteristic of a long laboratory spark at the moment of gap bridging. The stabilizing mode lasted several milliseconds. During this time, a quasi-stationary state was established under energy supply conditions close to those of the leader process. The CVC thus measured is approximated by the expression

$$E \approx 32 + 52/i \ \text{V/cm}, \qquad i \ [\text{A}]. \qquad (2.48)$$

The obtained field appears to be lower than in a leader (84 V/cm at 1 A) and closer to the arc field.

2.6 Voltage for a long spark

The problem of minimum voltage, at which a spark can develop to a certain length is of primary importance for high-voltage technology. This quantity characterizes the electric strength of an air gap, since its bridging by a leader results in a breakdown. This problem also applies to lightning, because it is interesting to know the minimum cloud potential at which a lightning discharge is possible.

Experiments show that the leader process has a threshold character. An initial leader cannot survive in normal air at gap voltages less than 300–400 kV. A leader can only be formed at low voltages in a short gap when it develops as a final jump from the very beginning. Then streamers immediately reach the opposite electrode, and the energy supply mode differs from that of the initial stage, with the streamer zone isolated from the grounded electrode. The reason for a threshold is easy to understand in terms of the discussion in section 2.2.3 and formula (2.34). A leader channel has a minimum possible radius. The radius of a cold air column, in which current can accumulate, is $r_0 > 10^{-2}$ cm. A thinner current channel is immediately enlarged by ambipolar diffusion. To heat a column of such initial radius to 5000 K, the leader tip potential from (2.34) must be at least 200 kV. If we consider the inevitable energy expenditure for ionization and gas excitation in the streamer zone, this value will increase by, at least, a factor of 1.5 [4]. Therefore, the tip potential in the initial leader stage will be several hundreds of kilovolts even under favourable conditions.

The voltage U_0 applied to the gap drops across the leader channel and is partly transported to the tip. The general formula is

$$U_0 = EL + U_t \qquad (2.49)$$

where E is the average field in a leader channel of length L. We showed in section 2.5 that in a long channel, most of which is in a quasi-stationary state, E is a more or less definite value varying with current i. The channel field decreases with increasing current. But current growth requires that the tip potential determining the leader velocity and current $i = C_1 U_t V_L$ should be raised. At a fixed length L, the $U_0(i, L)$ function has a minimum, since it is the sum of a falling component and a component rising with i. Minimum voltage $U_{0\,\mathrm{min}}(L)$ corresponds to current $i_{\mathrm{opt}}(L)$ optimal for a leader of length L. It is hardly possible, in the present state of the art, to find the $U_{0\,\mathrm{min}}(L)$ function theoretically. We shall try to define its character using semi-empirical data.

Many experimental physicists have measured the leader velocity variation with applied voltage U_0. Much work has been done on short leaders because one can neglect the voltage drop across the channel, assuming $U_0 \approx U_t$. With the account of this approximate equality, Bazelyan and Razhansky [35] suggested an empirical formula: $V_L \approx a U_t^{1/2}$, where $a \approx 1.5 \times 10^3\,\mathrm{V}^{-1/2}\,\mathrm{cm\,s}^{-1}$. Physically, the velocity increase with voltage looks quite natural (though this variation is not strong). We also know that the tip current is defined by (2.36). This gives the relation $U_t = A i^{2/3} (V_L \sim i^{1/3})$ with $A = (C_1 a)^{-2/3} = (2\pi\varepsilon_0 a)^{-2/3}$. Let us use the analogy between a well developed leader and an arc and take the CVC $E = b/i$ typical for a low current arc. Let us put $b = 300\,\mathrm{V\,A/cm}$ for numerical calculations and ignore the difference between the tip and channel currents. We shall then get $U_0 = Lb/i + A i^{2/3}$ and after differentiation

$$i_{\mathrm{opt}} = (3Lb/2A)^{3/5}, \qquad U_{0\,\mathrm{min}} = \tfrac{5}{3} A^{3/5} (3bL/2)^{2/5} = \tfrac{5}{3} U_{t\,\mathrm{opt}} \qquad (2.50)$$

If a leader develops under optimal conditions, the applied voltage is shared by the tip and the channel in comparable proportions. The mode with a low tip potential close to the limit admissible from the energy criteria, is unprofitable for a long leader, because it corresponds to low current leading to a considerable voltage drop across the channel. The long spark parameters in table 2.3 found from (2.50) with semi-empirical constants are quite reasonable: these orders of magnitude for current, voltage, and velocity meet the requirements on the optimal experimental conditions for long leader development. Besides, the experiment requires a nonlinear, slow dependence of minimum breakdown voltage on the gap length. It is generally known that increasing the length of a multi-metre gap is not a particularly

Table 2.3. Long spark parameters.

L, m	$U_{0\,min}$, MV	$U_{t\,opt}$, MV	i, A	V_L, 10^4 m/s	E_L, V/cm
50	3.3	2.0	1.1	2.1	260
100	4.3	2.6	1.3	2.4	170
3000	17	10	17	4.7	22

effective way of raising its electrical strength. This is a key challenge to those working in high-voltage technology.

The results of extrapolation of formula (2.50) to a lightning leader ($L = 3$ km) also lie within reasonable limits.

What is the rate of gap voltage rise necessary for the optimal mode of spark development? Clearly, the gap voltage must be raised as the spark length becomes longer according to (2.50), where L is an instantaneous leader length. The existence of an optimal mode of spark development has been confirmed experimentally [36–38]. It has been shown that for a breakdown to occur at minimum voltage, the pulse risetime t_f must increase with the gap length d. The authors of [39] recommend the following empirical formula for the evaluation of an optimal risetime:

$$t_{f\,opt} \approx 50d \; [\mu s], \qquad d \; [m] \qquad (2.51)$$

Generally, optimal voltage impulses have a fairly slow risetime. Their values vary between 100 and 250 μs in modern power transmission lines with the insulator string length of 2–5 m. We shall return to this issue in chapter 3, when considering the diversity of time parameters of lightning current impulses. The minimum electric strength of an air gap with a sharply non-uniform field can be found from the formulas [4]

$$U_{50\%\,min} = \frac{3400}{1 + 8/d} \; [kV], \qquad d < 15 \text{ m}$$

$$U_{50\%\,min} = 1440 + 55d \; [kV], \qquad 15 \leqslant d \leqslant 30 \text{ m}$$

$$(2.52)$$

2.7 A negative leader

Most lightnings carry a negative charge to the earth because they are 'anode-directed' discharges. It is always more difficult to break down a medium-length gap between a negative electrode and a grounded plane. A negative leader requires a higher voltage. The difference between leaders of different polarities is due to the streamer zone structure, while their channels and voltage drop across them are quite similar. Indeed, a gap of about 100 m long, in which an appreciable part of voltage drops across the channel, is bridged by leaders of both signs at about the same voltages [2, 3].

The streamer zone formation in a negative leader requires a higher tip potential for the same reason as a single anode-directed streamer needs a higher voltage for its development. Fast streamers with the velocity V_s much higher than the electron drift velocity V_e do not exhibit much difference associated with polarity. But streamers in a leader streamer zone are slow: $V_s \approx V_e$. It is of great importance whether the components of electron velocity relative to the streamer tip are summed, $V_s + V_e$, as in a cathode-directed streamer, or subtracted, $V_s - V_e$, as in an anode-directed one. In the former case, electrons produced in front of the tip move towards it, and the ionization occurs in a strong field near the tip. In the latter, electrons tend to 'run ahead' of the moving tip and spend most of their time in a lower field, so that the ionization occurs under less unfavourable conditions.

The fact that negative streamers generally require a higher field and voltage has been supported by many experiments. They show that the average critical field, which defines the maximum streamer length in formula (2.32), is twice as high for an anode-directed streamer as for a cathode-directed one: $E_{cr} \approx 10\,\text{kV/cm}$ against $E_{cr} \approx 5\,\text{kV/cm}$. We shall illustrate this with figure 2.15 for a gap of length $d = 3\,\text{m}$ between a sphere of radius $r_0 = 50\,\text{cm}$ and a grounded plane. The streamers stopped, having covered the distance l_{max} at negative sphere potential $U_c = 1.5\,\text{MV}$ (the unperturbed potential at the stop with the account of the sphere charge reflection in the plane is $U_0(l_{max}) \approx 0.25 U_c$). Under these conditions, cathode-directed streamers practically cross the whole gap.

The propagation mechanism and streamer zone structure of a negative leader are much more complicated than those of a positive leader and are still

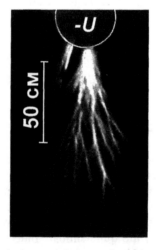

Figure 2.15. Anode-directed streamers from a spherical cathode of 50 cm radius at a negative voltage impulse of 1.8 MV and a 50 μs front duration.

poorly understood. In the 1930s, when Schonland started his famous studies of lightning [41], a negative leader was found to have a discrete character of elongation, so it was termed stepwise. Streak photographs exhibit a series of flashes, indicating that the leader propagates in a stepwise manner. Later, a similar process was found in a long negative leader produced in laboratory conditions [42, 43]. With every step, a negative leader elongates by dozens of centimetres, or by several metres in superlong gaps [3]; steps of a hundred metres have been registered in negative lightning discharges. Every step of a laboratory leader is accompanied by a detectable current overshoot which quickly vanishes during the time between two steps.

Without going into theoretical explanations of this mechanism, based on an unverified hypothesis, let us see what information can be derived from streak photographs of the process, made during laboratory experiments [44]. These are naturally more informative than streak photographs of a stepwise lightning leader. It is seen from figures 2.16 and 2.17 that in the intervals between the steps, the tip of a negative leader slowly and continuously moves on together with its streamer zone made up of anode-directed streamers. The main events occur near the external boundary of the negative streamer zone. It seems that a plasma body elongated along the field arises there and is polarized by the field (compare with the discussion in section 2.2.7). The positive plasma dipole end directed towards the main leader tip serves as a starting point for cathode-directed streamers. They move towards the tip, thus elongating the conducting portion of the channel and enhancing the negative field at its end directed to the anode. Almost at the same time, the plasma body generates an anode-directed streamer. This nearly mystic picture of streamer production in the gap space is clearly seen in a streak photograph in figure 2.18. Nothing like this has ever been observed with a positive leader.

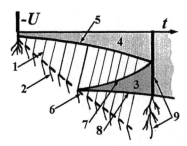

Figure 2.16. A schematic streak picture of a negative stepped laboratory leader: (1,2) secondary cathode- and anode-directed streamers from the gap interior; (3) secondary volume leader channel; (4) main negative leader channel; (5) its tip; (6) plasma body; (7), (8) tip of secondary positive and negative leader (9) leader flash concluding step development.

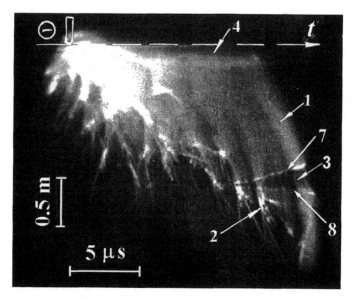

Figure 2.17. A streak photograph of the initial stage in a negative laboratory leader. Marking numbers correspond to figure 2.16.

The polarized plasma section becomes the starting point not only of streamers but also of secondary leaders which follow them. They are known as volume leaders. A positive cathode-directed volume leader grows intensively. Normally, its streamer zone almost immediately reaches the main negative leader, so it looks as if the secondary positive leader develops

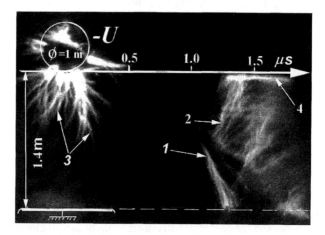

Figure 2.18. The origin of anode-directed (1) and cathode-directed (2) streamers from the gap interior; (3) initial flash of a negative corona (static photograph) which trigger a streak photograph regime; (4) arisen negative leader.

in the final jump mode, i.e., very quickly. The negative volume leader moves towards the anode somewhat more slowly. When the tips of the main negative and of the positive volume leaders come into contact, they form a common conducting channel, giving rise to the process of partial charge neutralization and redistribution. As a result, the former volume leader acquires a potential close that of the main negative leader tip. This process looks like a miniature return stroke of lightning, accompanied by a rapidly rising and just as rapidly falling current impulse in the channel and external circuit. The intensity of the channel emission increases for a short time. It is hard to say what exactly stimulates this increase – the short temperature rise or the ionization in the channel cover, which changes the cover charge, thereby getting ready for a potential redistribution along the channel (see section 2.4.4). The negative portion of the plasma dipole turns to a new negative tip of the main leader. This is the mechanism of step formation and stepwise elongation of the main channel. Then the story is repeated.

The picture just described gives no ground to draw the conclusion about a stepwise character of negative leader development. The motion of a negative leader is continuous, but secondary positive volume leaders, also continuous, produce a stepwise effect. Discrete is the final result of their 'secret activity', but only if the observer is equipped with imperfect optical instruments. In other words, what is generally known as a step is an instant result of a long continuous leader process. As for gap bridging by a main negative leader, one should bear in mind that most of the channel is created by auxiliary agents – by a succession of positive volume leaders.

This picture has been reconstructed from streak photographs. But we still do not know how polarized plasma dipoles are formed far ahead of the main leader tip. Their appearance is hardly a result of our imagination. Steps can be produced deliberately by making a volume leader start from a desired site in the gap. For this, it suffices to place there a metallic rod several centimetres long (figure 2.19). A series of rods placed in different sites of a gap will create a regular sequence of volume leaders. The work [45] describes an experiment with a negative leader 200 m long. Its perfectly straight trajectory was predetermined by seed rods suspended by insulation threads at a distance of 2–3 m from each other. A volume leader started from a rod when it was approached by the negative streamer zone boundary of the main leader. Clearly, the rods are polarized by the streamer zone field to serve as seed dipoles instead of natural (hypothetical) plasma dipoles.

There are many hypotheses concerning the stepwise leader mechanism, but they are so imperfect, lacking strength, and, sometimes, even absurd that we shall not discuss them here. We are not ready today to suggest an alternative model either. Additional special-purpose experiments could certainly stimulate the theory of this complicated and challenging phenomenon. It would be desirable to take shot-by-shot pictures of a negative leader tip region with a short exposure. A sequence of such pictures would

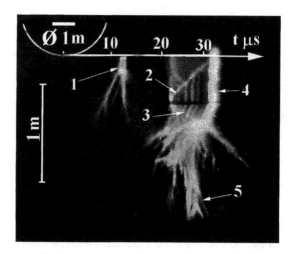

Figure 2.19. An artificially induced step: (1) initial flash of a negative corona from a spherical cathode; (2, 3) cathode- and anode-directed leaders from a metallic rod 2.5 cm long, placed in the gap interior; (4) leader flash concluding the step development; (5) new streamer corona flash from the tip of the elongating channel.

form a film more accessible to unambiguous interpretation than continuous streak photographs with confusing overlaps of many details.

References

[1] Lupeiko A V, Miroshnizenko V P *et al* 1984 *Proc. II All-Union Conf. Phys. of Electrical Breakdown of Gases* (Tartu: TGU) p 254 (in Russian)
[2] Baikov A P, Bogdanov O V, Gayvoronsky A S *et al* 1998 *Elektrichestvo* **10** 60
[3] Gayvoronsky A S and Ovsyannikov A G 1992 *Proc. 9th Intern. Conf. on Atmosph. Electricity* 3 (St Peterburg: A.I. Voeikov Main Geophys. Observ.) p 792
[4] Bazelyan E M and Raizer Yu P 1997 *Spark Discharge* (Boca Raton, New York: CRC Press) p 294
[5] Loeb L B 1965 *Science (Washington D.C.)* **148** 1417
[6] D'aykonov M I and Kachorovsky V Yu 1988 *Zh. Eksp. Teor. Fiz.* **94** 32
[7] D'aykonov M I and Kachorovsky V Yu 1989 *Zh. Eksp. Teor. Fiz.* **95** 1850
[8] Shveigert V A 1990 *Teplofiz. Vys. Temperatur* **28** 1056
[9] Bazelyan E M and Raizer Yu P 1997 *Teplofiz. Vys. Temperatur* **35** 181 (Engl. transl.: 1997 *High Temperature* **35**)
[10] Raizer Yu P and Simakov A N 1996 *Piz. Plazmy* **22** 668 (Engl. transl.: 1996 *Plasma Phys. Rep.* **22** 603)
[11] Dutton J A 1975 *J. Phys. Chem. Ref. Data* **4** 577
[12] Cravith A M and Loeb L B 1935 *Physics (N.Y.)* **6** 125
[13] Raizer Yu P and Simakov A N 1998 *Piz. Plazmy* **24** 700 (Engl. transl.: 1996 *Plasma Phys. Rep.* **24** 700)

[14] Vitello P A, Penetrante B M and Bardsley J N 1994 *Phys. Rev. E* **49** 5574

[15] Babaeva N Yu and Naidis G V 1996 *J. Phys. D: Appl. Phys.* **29** 2423

[16] Kulikovsy A A 1997 *J. Phys. D: Appl. Phys.* **30** 441

[17] Aleksandrov N L, Bazelyan E M, Dyatko N A and Kochetov I V 1998 *Fiz. Plazmy* **24** 587 (Engl. transl. 1998 *Plasma Phys. Rep.* **24** 541)

[18] Bazelyan E N and Goryunov A Yu 1986 *Elektrichestvo* **11** 27

[19] Aleksandrov N L and Bazelyan E M 1998 *J. Phys. D: Appl. Phys.* **29** 2873

[20] Aleksandrov D S, Bazelyan E M and Bekzhanov B I 1984 *Izv. Akad. Nauk SSSR. Energetika i transport* **2** 120

[21] Bazelyan E M, Goryunov A Yu and Goncharov V A 1985 *Izv. Akad. Nauk SSSR. Energetika i transport* **2** 154

[22] Aleksandrov N L and Bazelyan E M 1999 *J. Phys. D: Appl. Phys.* **32** 2636

[23] Gayvoronsky A S and Razhansky I M 1986 *Zh. Tekh. Fiz.* **56** 1110

[24] Kolechizky E C 1983 *Electric Field Calculation for High-Voltage Equipment* (Moscow: Energoatomizdat) p 167 (in Russian)

[25] Raizer Yu P, Milikh G M, Shneider M N and Novakovsky S.V. 1998 *J. Phys. D: Appl. Phys.* **31** 3255

[26] Raizer Yu P 1991 *Gas Discharge Physics* (Berlin: Springer) p 449

[27] Gorin B N and Schkilev A V 1974 *Elektrichestvo* **2** 29

[28] 'Positive Discharges in Air gaps at Las Renardieres – 1975' 1977 *Electra* **53** 31

[29] Bazelyan E M 1982 *Izv. Akad. Nauk SSSR. Energetika i transport* **3** 82

[30] Bazelyan E M 1966 *Zh. Tekh. Fiz.* **36** 365

[31] Bazelyan E M, Levitov V I and Ponizovsky A Z 1979 *Proc. III Inter. Symp. on High Voltage Engin.* (Milan) Rep. **51.09** p 1

[32] Meek J M and Craggs J D (eds) 1978 *Electrical Breakdown of Gases* (New York: Wiley)

[33] Makarov V N 1996 *Zh. Prikl. Mekh. Tekhn. Fiz.* **37** 69

[34] Aleksandrov N L, Bazelyan E M, Dyatko N A and Kochetov I.V. 1997 *J. Phys. D: Appl. Phys.* **30** 1616

[35] Bazelyan E M and Razhansky I M 1988 *Air Spark Discharge* (Novosibirsk: Nauka) p 164 (in Russian)

[36] Stekolnikov I S, Brago E N and Bazelyan E M 1960 *Dokl. Akad. Nauk SSSR* **133** 550

[37] Stekolnikov I S, Brago E N and Bazelyan E M 1962 *Conf. Gas Discharges and the Electricity Supply Industry* (Leatherhead, England) p 139

[38] Bazelyan E M, Brago E N and Stekolnikov I S 1962 *Zh. Tekh. Fiz.* **32** 993

[39] Barnes H and Winters D 1981 *IEEE Trans.* **Pas-90** 1579

[40] Gallet G and Leroy J 1973 IEEE Conf. **Paper C73-408-2**

[41] Schonland B 1956 *The Lightning Discharge. Handbuch der Physik* **22** (Berlin: Springer) p 576

[42] Stekolnikov I S and Shkilev A B 1962 *Dokl. Akad. Nauk SSSR* **145** 782

[43] Stekolnikov I S and Shkilev A B 1963 *Dokl. Akad. Nauk SSSR* **145** 1085; 1962 *Intern. Conf.* (Montreux) p 466

[44] Gorin B N and Shkilev A V 1976 *Elektrichestvo* **6** 31

[45] Anisimov E I, Bogdanov O P, Gayvoronsky A S *et al* 1988 *Elektrichestvo* **11** 55

Chapter 3

Available lightning data

Scientific observations of lightning were started over a century ago. Much factual information has accumulated about this natural phenomenon since that time. Most of it, however, has been obtained by remote observational techniques which can reveal only external manifestations of lightning. This is not the researchers' fault. Even a long laboratory spark keeps the experimenter at a respectful distance: there have been single and mostly unsuccessful attempts to study the leader interior and the ionization region in front of its tip. No attempts of this kind have yet been made with lightning. Nevertheless, the accumulated material is being analysed and systematized, so that our knowledge about atmospheric electricity is gradually expanding.

A number of carefully written books has made the results of field studies of lightning accessible to specialists. Among them, of great interest is the recent book by Uman [1] and the co-authored work edited by Golde [2]. The reader will find there nearly all available data on lightning, so there is no need to discuss them in this book. We have set ourselves a different task – to select the few data available on the lightning discharge mechanism and to try to build its theory. In addition, we shall make a detailed analysis of lightning characteristics important from the practical point of view. The nature of hazardous effects of atmospheric electricity on industrial objects will be considered in much detail and lightning protection principles will be offered.

This task cannot be solved completely, because many lightning parameters have never been measured or, more often, even estimated in order of magnitude. One hope is a method similar to the identical text analysis used in cryptography to read a text written in a dead language. If there is at least part of the text written in an accessible, better, related, language, the task is not considered hopeless. With patience and ingenuity, the researcher has a chance if he compares these texts carefully. In this respect, we expect much from long spark studies. Clearly, a spark and lightning are phenomena of different scales, but it is also clear that both have a common nature. For this reason, we shall often compare the parameters of lightning with those of a long spark. We should like to emphasize that this will be a comparison rather than a direct extrapolation, because there is no complete analogy between the two phenomena.

3.1 Atmospheric field during a lightning discharge

There is no strict answer to this physically ambiguous question. It is necessary to specify what part of the space between the cloud and the earth is meant. One thing is clear – the electric field at the lightning start must be high enough to increase the electron density by impact ionization. This value is $E_i \approx 30\,kV/cm$ for normal density air and about $20\,kV/cm$ at an altitude of 3 km (the average altitude for lightning generation in Europe). Such a strong field has never been measured in a storm cloud. The maximum values were recorded by rocket probing of clouds ($10\,kV/cm$, Winn *et al*, 1974 [7]) and during the flight of a specially equipped aeroplane laboratory ($12\,kV/cm$). The value obtained by Gunn [4] in 1948 during his flight on a plane around a storm cloud was about $3.5\,kV/cm$. The values between 1.4 and $8\,kV/cm$ were obtained from some similar measurements [3–9]. It is hard to judge about the accuracy of these measurements, especially those made in strong fields, because parts of the field detector or the carrier-plane parts close to it can produce a corona discharge. In any case, the corona space charge will not allow the strength in the region being measured to go beyond a threshold value (for details, see [20]). There are reasons to believe, however, that a corona on hydrometeorites (water droplets, snow flakes, ice crystals) keeps the field at a level below E_i in the whole of the cloud. If this is indeed so, a field can be enhanced above E_i only in a small volume for a short time, say, as a result of eddy concentration of charged hydrometeors. This enhancement will be reduced to zero by a corona for less than a second. The experimenter has no chance to guess where the field may be locally enhanced to be able to introduce a probe detector there.

Theoretically, it is also important to know the average gap field capable of supporting a lightning leader. The field decreases in the charge-free space from the cloud towards the earth. At the earth, the storm field was found to be 10–$200\,V/cm$. Such a low field did not prevent the lightning development. Lightnings were deliberately produced in numerous experiments described by Uman [1, 10–16]. A rocket was launched from the earth, pulling behind it a thin grounded wire. A lightning leader was excited at 200–$300\,m$ above the earth's surface. The near-surface field during a successful launching was usually 60–$100\,V/cm$.

Strictly, measurements made at two points, at the earth and in the cloud, are insufficient for an accurate evaluation of an average electric field. The

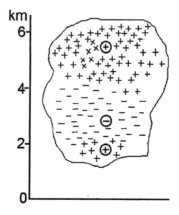

Figure 3.1. The 'dipole' model of the charge distribution in a storm cloud.

space between the cloud and the earth should be scanned, and this must be done for some fractions of a second, just before a lightning discharge, in the vicinity of its anticipated trajectory. Unfortunately, such attempts have not been quite successful. More successful were measurements made at points on the earth's surface separated at a distance of hundreds and thousands of metres [17–19]. These have been used to reconstruct the charge distribution within a storm cloud, invoking the results of direct cloud probing. The reconstruction procedure and its possible errors are discussed in [20]. Generally, with simultaneous field measurements made at *n* points, one can write a closed set of equations for the same number of parameters of charged regions. Its solution provides the parameters, for example, the average space charge densities in pre-delineated regions.

Most often, the number of points is too small, so the results obtained only permit the construction of simplified models with point charges. Very common is the dipole model with a negative charge at 3–5 km above the earth with the same value of the positive charge raised at a double altitude. Sometimes, a small positive point charge is added to them, which is placed at a distance by 1–2 km closer to the earth than the negative charge. All point charges are assumed to be located along the same vertical line (figure 3.1).

The concept of a cloud filled by charged layers of different signs is based on probe measurements of charge polarity in hydrometeors. In this respect, this model raises no doubt. But as for the field distribution, the measurement error is too large, especially for the space in the cloud between two point charges. Luckily, the descending lightning trajectory lies mostly outside of the cloud, in the air free from charged particles. For this part of the trajectory, the average field evaluation in terms of a simple model makes sense.

We shall illustrate the procedure of deriving information from such measurements. Suppose we have at our disposal the values of field E_1 at the earth's surface just under an anticipated charged centre of a storm cloud, as well as the values for field E_2 at the lower cloud boundary (also under the charged centre) measured during a plane flight around the cloud. The altitude of the lower boundary, h, is also known. Assume the centre of the main lower charge q to be above the lower cloud boundary at an unknown distance r. If we ignore the effects of the remote upper charge and of the additional charge lying under the main one, we can write

$$E_1 = \frac{2q}{4\pi\varepsilon_0(h+r)^2}, \qquad E_2 = \frac{q}{4\pi\varepsilon_0 r^2} + \frac{q}{4\pi\varepsilon_0(2h+r)^2}. \qquad (3.1)$$

The factor 2 in E_1 and the second term in E_2 are due to the action of charge induced in the earth's conducting plane (mirror reflection). Since in reality $E_1 \ll E_2$, it is natural to suggest that $r \ll h$. Then, with the second term in E_2 ignored, we find

$$q = 2\pi\varepsilon_0(h+r)^2 E_1, \qquad r \approx \alpha h, \qquad \alpha = (E_1/2E_2)^{1/2}. \qquad (3.2)$$

Substituting the values above, i.e., $E_1 = 100\,\text{V/cm}$, $E_2 = 3000\,\text{V/cm}$, and $h = 3\,\text{km}$, we shall find $q = 6.3\,\text{C}$, $\alpha = 0.13$, and $r \approx 390\,\text{m}$. The average field in the region between the lower cloud boundary and the earth, equal to the lower boundary potential $\varphi_2 \approx q/4\pi\varepsilon_0 r$ divided by the cloud distance from the earth, h, $E_{\text{av}} \approx (E_1 E_2/2)^{1/2} \approx 390\,\text{V/cm}$. Allowance for the second term in E_2 (the effect of mirror charge reflection by the earth) can hardly be justified, because our model did not take into account the effect of the upper charge of opposite sign. This charge is closer to the cloud edge than that reflected by the earth and has, therefore, a greater effect on E_2. Its consideration, however, simple though it may seem, would require field measurement at another point of space and another equation for finding a new unknown – the altitude of the upper charge centre of the dipole.

A larger scale correction would, probably, be necessary to account for the effect, on the near-earth field, of the space charge induced by coronas from pointed grounded objects (tree branches, high grass, various buildings, etc.) [21]. Estimations made just at the earth's surface show that this charge reduces the actual field of a storm cloud by half. So one cannot say that lightning moves in an unusually low electric field. These are just the values at which superlong sparks are excited in laboratory conditions (see chapter 2). Therefore, there is no need to invent a special propagation mechanism for lightning, different from that of a long laboratory spark, if we deal with average electric fields capable of breaking down the cloud–earth gap.

3.2   The leader of the first lightning component

The leader of the first component of lightning develops in unperturbed air, so only this leader behaviour should be compared directly with laboratory data on long spark leaders. The comparison can be carried out along two lines. We can first compare the leader structure in lightning and a spark and, second, their quantitative parameters, primarily velocities.

3.2.1   Positive leaders

Streak pictures of a positive leader are easy to interpret. So we shall begin with positive lightnings, though their occurrence is not frequent. Many books and papers refer to the successful streak photographs of a descending positive leader taken by Berger and Fogelsanger in 1966 [22]. Its schematic diagram is reproduced in figure 3.2. The leader became accessible to photography at 1900 m above the earth's surface. It moved down in a continuous mode, without an appreciable intensity variation. The average leader velocity over the registration time was 1.9×10^6 m/s, increasing somewhat as the leader tip approached the earth. Such a leader is much faster than a long laboratory spark, whose velocity is 50–100 times lower at minimum breakdown voltage before the streamer zone contacts a grounded electrode.

In the streak picture, the leader tip looks much brighter than its channel, but no signs of a streamer zone can be identified. We cannot say how the original negative looks, but in the published photograph the resolution threshold is hardly less than 50 m. It is a very large value for a streamer

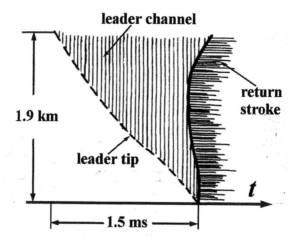

Figure 3.2. Schematic streak picture of a positive descending lightning leader registered on the San Salvatore Mount in Switzerland [22].

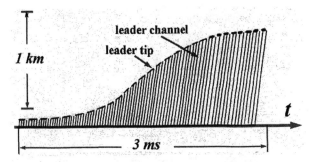

Figure 3.3. Schematic streak picture [22] of a positive ascending leader from a 70-m tower on the San Salvatore Mount.

zone. With the data of section 2.4.1, we can calculate the leader tip potential U_t, at which the streamer zone will exceed at least twice the length of about 100 m, a threshold value for the measuring equipment. For the average streamer zone field $E_s \approx 5\,\mathrm{kV/cm}$, the potential derived from (2.39) is $U_t \approx 100\,\mathrm{MV}$. Such a high value cannot be typical of lightning with average parameters. There is another circumstance preventing streamer zone registrations – different radiation wavelengths of a hot leader channel and a cold streamer zone. Violet and ultraviolet radiation from streamers is dissipated by water vapour and rain droplets in the air much more than long wavelength radiation characteristic of a mature channel. At a distance of about a kilometre between the lightning and the registration site (closer distances are practically unfeasible), a streamer zone may become quite invisible to the observer's equipment. Note that this is totally true of optical registrations of an ascending positive leader.

A schematic streak picture of an ascending positive leader, based on 18 successful registrations [22], is shown in figure 3.3. All lightnings started from a 70-m tower on the San Salvatore Mount near the Lake Lugano. The leader does not exhibit specific features that would distinguish it from a long laboratory spark. On the whole, it developed in a continuous mode with irregular short-term enhancements of the channel intensity. Normally, they did not accelerate the leader development. Something like this has been observed in a long laboratory spark. The streak picture in figure 3.4 demonstrates this with reference to a positive leader in a sphere–plane gap 9 m long. But this phenomenon has nothing to do with a stepwise elongation of a negative leader channel.

The velocity of an ascending positive leader near the starting point is close to that of a laboratory spark, about $2 \times 10^4\,\mathrm{m/s}$. From some data [22], it was in the range of $(4-8) \times 10^4\,\mathrm{m/s}$ for a channel length of 40–100 m; but when the leader tip was at a height of 500–1150 m, it increased by nearly an order of magnitude, to 10^5–$10^6\,\mathrm{m/s}$.

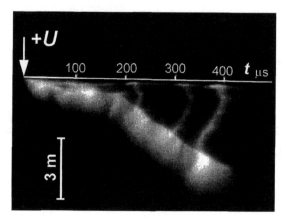

Figure 3.4. A streak photograph of the initial stage of a positive leader in a 9-m rod–plane gap, displaying short flashes of the channel.

3.2.2 Negative leaders

Negative lightnings occur more frequently than positive, and their registration is more common. The main distinguishing feature of the negative leader of the first lightning component is its stepwise character. The leader tip leaves a discontinuous trace in streak pictures which look like a movie film (figure 3.5). One can sometimes find such pictures in a sports magazine illustrating the successive steps in a sportsman's performance. The bright flash of the tip and the channel right behind it are followed by a dead zone with practically zero intensity. This is followed by another flash showing that the tip has moved on for several dozens of metres. Such negative leader behaviour was observed by Schonland and his group as far back as the 1930s [24, 25]. According to their registrations, the average pause between the steps was close to 60 µs, with a spread from 30 to 100 µs, and the step

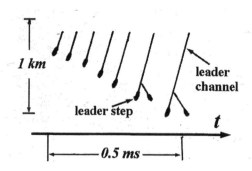

Figure 3.5. Schematic streak picture of a descending negative leader.

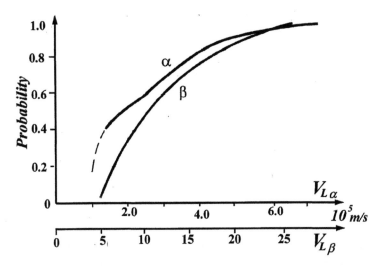

Figure 3.6. Typical integral velocity distributions for descending leaders of the α- and β-types (from [24, 25]).

length varied between 10 and 200 m with the average value being 30 m. The duration of a step is likely to be within several microseconds. The available streak photographs are not good enough to identify the details of a step. In any case, it is hard to decide whether it is similar to a step of the long negative spark described in section 2.7.

A stepwise negative leader approaches the earth at an average velocity of 10^5–10^6 m/s. Two descending leader types can be identified in terms of their velocity: slow α-leaders and fast β-leaders. The former travel at their step-averaged velocity; it varies with the discharge in the range of $(1$–$8) \times 10^5$ m/s with the average value of 3×10^5 m/s. The respective β-leader parameters are 3–4 times higher. This can be seen in figure 3.6 showing the integral velocity distributions described in [24, 25]. Usually, β-leaders are more branched and their steps are longer. They abruptly slow down when they approach to the earth, after which they behave as α-leaders.

An ascending negative leader also has characteristic steps. Most of the 13 registered leaders ascending from a 70-m tower on the San Salvatore Mount [22] were identified as α-leaders. They have relatively short steps (5–18 m) and a velocity of $(1.1$–$4.5) \times 10^5$ m/s. Two of the discharges were referred to β-leaders because their velocity was $(0.8$–$2.2) \times 10^6$ m/s and the step length up to 130 m. On the whole, ascending and descending stepwise leaders do not show significant differences.

The registrations of ascending discharges from the San Salvatore Mount provide direct evidence for the existence of a streamer zone in a lightning leader. Registrations made at a sufficiently close distance, which became

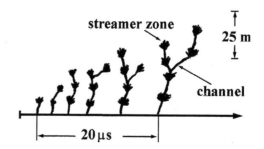

Figure 3.7. Schematic diagram of the initial development of a leader ascending from a 70-m tower on the San Salvatore Mount, as viewed from close distance streak photographs.

possible due to the tower top being the only starting point, show streamer flashes arising at the moment a new step begins. Streamers were initiated not only from the tip of the main channel but also from its branches (figure 3.7).

3.3   The leaders of subsequent lightning components

Leaders of lightning components following the first one are known as dart leaders because of the absence of branches. The streak photograph in figure 3.8 shows the trace of only one bright tip looking like a sketch of an arrow or dart. A dart leader follows the channel of the previous lightning component with a velocity up to 4×10^7 m/s. Averaging over many registrations gives the value $(1–2) \times 10^7$ m/s, with the minimum values being an order of magnitude less than the maximum one [23, 25]. The dart leader velocity does not vary much on the way from the cloud to the earth.

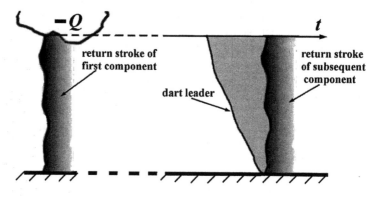

Figure 3.8. A schematic streak picture of a dart leader.

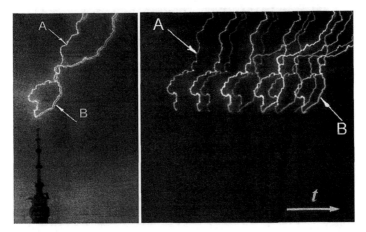

Figure 3.9. A streak photograph of a well-branched lightning striking the Ostankino Television Tower; the components along the branches A and B are formed at different moments of time.

Similar results have recently been obtained from 23 streak photographs of fairly good quality showing dart leaders taken in Florida by a camera with a time resolution 0.5 µs [26]. The average velocity of a dart leader varied from 5×10^6 to 2.5×10^7 m/s in some registrations; it was $(1.6-1.8) \times 10^7$ m/s for three typical pictures presented in the publication.

It is clear that a dart leader somehow makes use of the previous channel with a different temperature, gas density and composition. There are several indications to this. First, there is a tendency for the dart leader velocity to decrease with increasing duration of the interleader pause. This is because the gas in the trace channel is gradually cooled to return to the original condition. If a pause lasts longer, the subsequent component may take its own way. Figure 3.9 shows a lightning discharge which struck the Ostankino Television Tower in Moscow. Some of its initial components followed a common channel but then the discharge trajectory changed. Naturally, there is nothing like a dart leader in unperturbed air – the leader of each next component develops in a step-wise manner.

Second, it has been found in triggered lightning investigations [15, 27] that a dart leader requires a current-free pause for its development. The long-term current, which supports the channel conductivity in the period between two subsequent components, must entirely cease to allow the channel to partly lose its conductivity. Only then will the channel be ready to serve as a duct for a dart leader. But if a new charged cloud cell is involved and raises the potential of the channel with current, an M-component is produced instead of a dart leader. Its distinctive feature is a higher intensity of the existing channel lacking a well defined tip (figure 3.10). The absence of

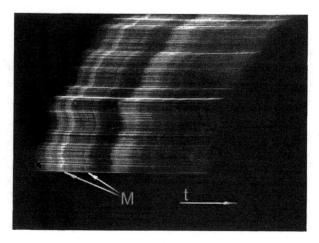

Figure 3.10. A streak photograph of a well-branched lightning with M-components.

a clear luminosity front is a serious obstacle to the measurement of its velocity. Investigators often point to a nearly simultaneous increase of the light intensity in the whole channel. This suggests either an almost sub-light velocity of the leader or an exceptionally smeared boundary of its front. In contrast to a dart leader, an M-component is never followed by a distinct return stroke with a high (10–100 kA) rapidly rising current impulse.

Both the external view and photometric data obtained from streak photographs reveal clearly a dart leader tip. Some authors [26] made an attempt to measure the time variation of the leader light intensity. Although the measurements were performed near the time resolution limit, they indicate that the light pulse front at the registration point rises for 0.5–1 μs and then is stabilized for 2–6 μs. Therefore, with the dart leader velocity of 1.5×10^7 m/s, the extension of the front rise is 7.5–15 m with the full pulse front length of 35–105 m. It can be mentioned, for comparison, that the M-component has a pulse front, if any, of a kilometre length.

It is important for the theory of dart leaders that they always move from a cloud down to the earth. This means that the voltage source that excites them is 'connected' to the trace channel of the previous component right in the cloud. The direction of the previous leader does not matter much because their channels are equally suitable for the development of a dart leader.

3.4   Lightning leader current

We can only guess about the values of descending leader currents or estimate them from indirect data. We shall make such estimations in section 3.5, using leader charge data, also obtained indirectly. Ascending leader currents are

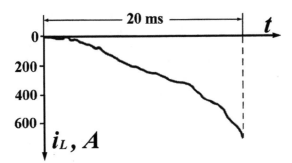

Figure 3.11. A schematic oscillogram of the leader current in an ascending lightning.

not difficult to measure, and there have been many measurements of this kind. Normally, a current detector is mounted on top of a tower dominating the locality [28–30]. The current impulse of an ascending leader, registered on an oscillogram, lasts for about 0.1 s, corresponding to the time of ascending leader development. The current nearly always rises in time (figure 3.11). The current supplies an elongating leader with charges. Physically, these charges are induced by the electric field of a cloud. When a leader approaches a cloud, going through an increasingly higher field, the linear density of induced charge τ increases. Besides, the leader goes up with an increasing velocity V_L, reducing the time for the charge supply. A combination of these factors raises the current $i = \tau V_L$. At the moment an ascending leader starts its travel, its current is lower than 10 A, whereas at the end of the travel, it may rise to 200–600 A, with an average value of about 100 A. Sometimes, just before the leader begins its continuous elongation, impulses with an amplitude of several amperes may arise against the background of a milli-ampere corona current.

Current oscillograms of an ascending leader triggered from a thin wire elevated by a small rocket to 100–300 m [13, 31] give a similar picture. They show the same slowly rising impulse with an amplitude of 100–200 A and duration 50–100 ms. It has no overshoots at the front, even if the leader goes up in a stepwise mode.

There are no reasons to suggest any principal difference between average currents of ascending and descending leaders. In both cases, the leader is supplied by charges induced by the electric field of a storm cloud, and the leader lifetimes are comparable because they move at approximately the same velocity.

Qualitatively, the current variation of the first component leader is similar to that of a laboratory spark. When the gap voltage is raised slowly, one can observe initial leader flashes at the high-voltage electrode, followed by distinct current impulses [32]. As for long spark steps, they practically do not change the current at the leader base. It has been shown

[20] that this is to be expected if the charge perturbation region is separated from the registration point by an extended channel section with high resistivity and distributed capacitance (section 4.4). The perturbation wave travelling along the channel towards the detector is attenuated. Of course, the current of a laboratory spark rarely exceeds a few amperes, but such a difference is predictable. it follows from the expression for the current $i = \tau V_L$ cited above. A lightning leader has an order higher velocity V_L and, at least, an order larger linear charge τ (due to the voltage being 10–20 times higher). All in all, this increases the current to within the anticipated two orders of magnitude.

Now one can judge about the current of a dart leader. There are no direct registrations of this current. One exception was an attempt at its measurement in a triggered lightning just before its contact with the earth. This is principally possible since the point of contact is known exactly – this is the point of wire fixation to the earth. The wire evaporates completely, having passed the current of the first lightning component. Using the still hot trace channel, a dart leader follows the path of the wire. A current detector can be placed at the wire grounding site.

It is much more difficult to interpret the recorded oscillograms, because it is unclear at what moment of time the development of a dart leader stops and the return stroke with the high current begins. Nevertheless, the published current measurements vary from 0.1 to 6 kA with the average value of 1.7 kA [33]. The lower limit of the range is more typical for the first component (this may be the next component, too, but after a long current-free pause, when the previous trace channel has nearly totally decayed). The value of several kiloamperes seems reasonable, since the velocity of a dart leader is 30–50 times higher than that of the first component.

3.5   Field variation at the leader stage

The subdivision of experimental data between this and the previous section is somewhat arbitrary. Electric field measurements provide information about leader charge, while charge and current are related by leader velocity. On the whole, this is a general problem. If the observations were arranged properly and the data analysis was made carefully, relatively simple field measurements can add much to our knowledge of electrical parameters of lightning. The knowledge of the field itself is rarely of importance, probably, except in some applied problems of lightning protection of low voltage circuits. For this reason, it is not the measurements but, rather, methodological approaches to their treatment which are significant. So we shall begin with these approaches and the general principles underlying a treatment of most lightning stages.

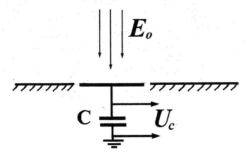

Figure 3.12. Measurement of fast variations in the electric field during the lightning development.

Suppose the electric field at an observation point near the earth's surface is $E_0(0)$ at the moment of the lightning start. When the discharge is completed, the field takes a new value $E_0(t_m)$. The field has changed by the value $\Delta E_0 = E_0(t_m) - E_0(0)$ over the time t_m. When the measurement time is relatively short, it is more convenient to register the field change rather than to measure its values. This is usually done with electrostatic antennas, i.e., metallic conductors (normally, flat) grounded through a reservoir capacitor C (figure 3.12). If the capacitor and the measurement circuits connected to it have an infinitely high leakage resistance R_l, the capacitor voltage, at any moment of time, is

$$U_C(t) = \frac{q_C(t)}{C} = \frac{\varepsilon_0 E_0(t) S_a}{C} \tag{3.3}$$

where S_a is the area of a flat antenna and q_C is the charge induced on it. If R_l is finite (which is always the case due to the input resistance of the circuit taking voltage readings from the capacitor), the use of (3.3) requires the condition $R_l C \gg t_m$, which can be easily met for the lightning duration $\sim 10^{-2}$ s but becomes problematic for a time interval of several minutes between two flashes.

An accurate measurement of the field change $\Delta E_0(t_m)$ requires the necessary time constant of the measurement circuit $R_l C$ and the account of effects of external field variation in the atmosphere, E_0, by making allowance for local effects. (The antenna may be raised above the earth, say, mounted on a building roof, so that the field there will be higher than on the earth. On the other hand, a nearby high construction may reduce the field, acting as an electrostatic screen.) The field value obtained is not particularly informative. In order to get information about the lightning discharge, we have to make certain assumptions concerning the distribution of charges which have changed the field.

Let us begin with a simple illustration. Suppose a lightning leader passing from a spherical volume has changed the charge of only one sign in a storm cloud cell. If there are other changes in the sphere charge during the leader travel, they are assumed to have been completely neutralized later, at the return stroke stage. If this assumption is correct, the measured value of $\Delta E_0(t_m)$ can give an idea about the quantity of charge transported by the leader from the cloud to the earth:

$$\Delta Q_M = \frac{2\pi\varepsilon_0(H^2 + R^2)^{3/2}\Delta E_0(t_m)}{H}. \tag{3.4}$$

Here, H is the altitude of the charged storm centre and R is its radial displacement relative to the registration point. Both parameters should be measured by an independent method or simultaneous field registrations at two more points should be made at given distances from the first one. This will provide additional equations for the unknown values of H and R.

Such an unambiguous treatment results from the simple model we have chosen, which contains no geometrical parameter except for the distance to the charge. However, a slightly more complicated, dipole model deprives the measurement treatment of this advantage. Still, electric field measurements have always been attractive to lightning researchers owing to their simplicity. Interest in such measurements increased with the application of lightning triggering by small rockets raising a grounded wire to 150–300 m above the earth's surface (triggered lightning). The first component of such lightning is genuinely artificial, but then the first trace channel is used by practically natural dart leaders travelling to the earth. Their point of contact with the earth is predetermined, so field detectors can be placed at any distance from the leader. This registration system is quite sensitive and capable of responding to the linear charge density not far from the leader tip when it approaches the earth.

To illustrate our analysis, we shall use the field measurements described in [34, 35]. The authors of this work kindly made them available to us after their discussion at the IXth International Conference on Atmospheric Electricity, held in St. Petersburg in 1992. The files contained detailed records of electric fields, taken during the flight of dart leaders, and of their return stroke currents. Detectors were placed at the distance of $R = 500$ m and 30 m from the contact point. Regretfully, the recordings at these distances were not simultaneous but made in different years. Their comparison is still possible because the fields were recorded at the same time as the return stroke currents. By sorting out identical current oscillograms, one can select lightning discharges with about the same leader tip potentials. This provides close values of leader velocity and linear charge density in the charge cover not too far from the leader tip. Some representative oscillograms of $\Delta E(t)/\Delta E_{max}$ normalized by their amplitudes are shown in figure 3.13. They correspond to discharges with really close currents in the return strokes ($I_M = 6$ kA at point $R = 500$ m and 7 kA at point 30 m). The amplitude values of field variation ΔE_{max} over the time of the dart leader

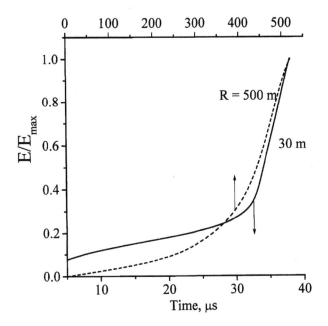

Figure 3.13. Oscillograms taken in Florida, USA [35], from the vertical field component during the development of the dart leader in the subsequent component of a triggered lightning. The detectors were positioned at 30 and 500 m from the point of strike; the pulses are related to their maximum amplitudes.

travel were 6.9 V/cm and 120 V/cm, respectively. Note that the measurements in [35] result in $\Delta E_{max}/I_M \approx$ const at every point. There is no geometrical similarity between the pulses $\Delta E(t)/\Delta E_{max}$ at the different points. On the contrary, there is a sharp difference in the rates of strength rise, as a dart leader was approaching the earth. The field increase in the range $(0.5-1.0)\Delta E_{max}$ took $\Delta t_{1/2} = 76\,\mu s$ for point $R = 500\,m$ and only $5\,\mu s$ for point $R = 30\,m$.

These data will be treated in terms of a simple model, in which a dart leader is represented as a uniformly charged axis with linear charge density τ_L. Naturally, the real cover radius R_c can be ignored in the field calculation at a distance R. We shall show below that field calculations can only take into account the charge distribution along a relatively short length behind the tip, comparable with R. This will justify the assumption of τ_L being constant, because it actually refers to a short length of about R near the tip. Therefore, the field change due to the leader charge at point R at the earth, with the allowance for its mirror reflection by the earth, is described as

$$\Delta E = \frac{\tau_L}{2\pi\varepsilon_0}\int_h^H \frac{z\,dz}{(z^2 + R^2)^{3/2}} = \frac{\tau_L}{2\pi\varepsilon_0}\left[\frac{1}{(R^2 + h^2)^{1/2}} - \frac{1}{(R^2 + H^2)^{1/2}}\right] \quad (3.5)$$

where h is the height of the leader tip from the earth at the moment of registration and H is the height of the leader base. The field change is maximum
when the tip contacts the earth, and for $R \ll H$ this gives

$$\Delta E_{max} \approx \frac{\tau_L}{2\pi\varepsilon_0}\left(\frac{1}{R}-\frac{1}{H}\right) \approx \frac{\tau_L}{2\pi\varepsilon_0 R}. \qquad (3.6)$$

It indeed follows from (3.6) that only the charge distribution along a short
length comparable with R is important for field evaluation. (For example,
at $H > 5R$, the error of the model with $\tau_L = $ const will be less than 20%
for any charge distribution, unless τ_L grows rapidly from the tip toward
the base; but there is no reason for this, because the channel field E_c is
weak and the cloud potential does not vary much.)

Formula (3.6) allows charge density evaluation with a good accuracy,
since the leader is strictly vertical at the earth – it reproduces the path of
the rocket taking up the wire which has evaporated. The value calculated
from the measurements at point $R = 30$ m appears to be unexpectedly
small: $\tau_L \approx 2 \times 10^{-5}$ C/m. Nearly as much charge is transported by long
laboratory sparks (section 2.4). The potential of a lightning leader tip,
U_t, does not seem to be much larger than that of a laboratory spark.
According to (2.8) and (2.35), the linear leader capacitance is
$C_1 \approx 2\pi\varepsilon_0/\ln(H/R_L) \approx (2-10) \times 10^{-12}$ F/m, even with indefinite leader
radius R_L. From this, we have $U_t \approx \tau_L/C_1 \approx 2-10$ MV.

The velocity of a dart leader proves to be very high. For its evaluation, we
shall use the measured value of $\Delta t_{1/2}$, which is 5 µs for $R = 30$ m. Formula (3.5)
gives $\Delta E = \Delta E_{max}/2$ at $h = \sqrt{3}R$. Hence, the average velocity along a path of
length $h \approx 50$ m at the earth's surface is $V_L \approx \sqrt{3}R/\Delta t_{1/2} \approx 10^7$ m/s, quite
consistent with direct measurements. It should be emphasized that this velocity
refers to the perfectly vertical path at the earth's surface, so it is the true velocity.

Similar evaluations can be made with the measurements at the far point
$R = 500$ m but with a lower reliability, since the parameter averaging is to be
made over a leader length of about 10^3 m with an unknown path. Neverthe-
less, the values of $\tau_L \approx 2.3 \times 10^{-5}$ C/m and $V_L \approx 1.15 \times 10^7$ m/s are found to
be close to those above. It will be shown in the next section that an indefinite
trajectory may produce an error much larger than the obtained difference in
the values of τ_L and V_L. So the dart leader of triggered lightning with the
definite path at the earth is a lucky exception.

Another illustration of $\Delta E(t)$, cited in [35], characterizes a more powerful dart leader. The current amplitude in the return stroke was as high as
40 kA. The maximum field change was found to be $\Delta E_{max} = 810$ V/cm,
i.e., a little more than a value proportional to current, while the characteristic
time of the process, $\Delta t_{1/2}$, decreased to 1.8 µs. Calculations similar to those
described above give $\tau_L \approx 1.35 \times 10^{-4}$ C/m, $U_t \approx 20-30$ MV, and
$V_L \approx 2.9 \times 10^7$ m/s, thereby supporting the hypothesis of a direct, though
not very strong, dependence of the leader velocity on the tip potential. For

the calculated values of linear charge and velocity at the earth, the leader current is found to be $i_L = \tau_L V_L \approx 3.9\,\text{kA}$, only an order of magnitude lower than the current amplitude in the return stroke.

3.6 Perspectives of remote measurements

What we described in the previous section is a very favourable situation, in which the point of leader contact with the earth is fixed and its final path is strictly vertical, at least, at a length of 150–300 m above the earth. One should not expect such favourable conditions for natural lightnings, especially for their first components. Still, one should take quietly and with some scepticism the idea of indirect remote measurements of lightning parameters. The experimeter resorts to them because, otherwise, his life would turn out too short to bring his experiment to a conclusion. Reconstruction of an electromagnetic field source from strength measurements made at definite points is an incorrect solution to a fairly common problem of electrodynamics in various areas of science and technology. Lightning is not an exception to the rule. We shall consider critically the treatments of results obtained from solutions to such problems and discuss inverse electrostatic problems, as applied to the lightning leader.

Generally, the density of space charge $\rho(x, y, z)$ between some boundary surfaces can be found if the electric field in the whole confined volume is known. Experimentally, this means simultaneous field measurements at an infinitely large number of points, which is practically unfeasible. A well organized service for field lightning observation has, at best, several synchronized field detectors. A theoretical treatment of the field records always suggests an *a priori* construction of a simplified field source model. The inverse problem can be solved if the number of unknown parameters in this model does not exceed the number of registration points. What follows is quite obvious. One writes down a set of equations with the measurements on the right and the expression for field at a given point (derived from the model with yet unknown charge parameters) on the left. The solution defines the parameters as rigorously as the measurements permit. One should always remember, however, what has been found from the equations, since these are parameters of a speculative model rather than a real phenomenon. How much they coincide is not a matter of accuracy of measurements or calculations but that of the model adequacy to the phenomenon under study. Most often, it is here that possible errors originate.

3.6.1 Effect of the leader shape

Without claiming a general analysis, we shall consider a special but frequently used model of near-earth field variation at a large distance from a

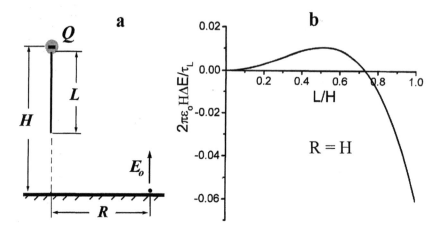

Figure 3.14. Electric field variation at the earth, evaluated for a large distance from the vertical channel of a descending leader.

descending lightning leader. In this model, the leader is represented by a thin vertical uniformly charged thread with a linear charge density τ_L, and the storm cell, from which the leader started, is taken to be so small that it is replaced by a point charge Q at height H. The value of Q may be unknown, because the analysis uses the time variation of the field rather than its absolute value [1].

The field at point R near the earth (figure 3.14(a)) varies in time for two reasons. The absolute field decreases because of the charge reduction in the storm cell by the charge $\Delta Q = \tau_L L$ carried away by the leader on its way to the earth.† The second component is due to the charge accumulation on the leader of length L; as the leader moves on, this charge goes down, enhancing the field at the earth. As a result, for the change of the vertical field component at moment t with $L = V_L t$, where V_L is average leader velocity, we have

$$\Delta E(L) = -\frac{\Delta Q H}{2\pi\varepsilon_0 (H^2 + R^2)^{3/2}} + \frac{\tau_L}{2\pi\varepsilon_0} \int_0^L \frac{(H - x)\,\mathrm{d}x}{[(H - x)^2 + R^2]^{3/2}}. \tag{3.7}$$

This expression takes into account the doubled field associated with the earth-induced charge. The evaluation of the integral gives the known expression

$$\Delta E(L) = \frac{\tau_L}{2\pi\varepsilon_0} \left\{ \frac{1}{[(H - L)^2 + R^2]^{1/2}} - \frac{1}{(H^2 + R^2)^{1/2}} - \frac{LH}{(H^2 + R^2)^{3/2}} \right\}. \tag{3.8}$$

† This component may be ignored in field measurements at a small distance from the leader, as described in section 3.5.

When analysing this expression, an experimenter will find it difficult to avoid a temptation. In the range of $R/H < 1.4$, the function $\Delta E(L)$ has an extremum (figure 3.14(b)) which can be easily recorded by an oscillogram. What remains to be done is to find the height H at which the lightning started (it may be taken to be equal to the average height of the storm front), to measure the distance between the observation point and the leader (e.g., from the thunder peal delay time, since the sound velocity is known and the point of sound wave excitation is near the earth's surface), and to find the moment of time when the leader tip descended to the height $H - L_m$ by calculation, from (3.8), the leader length L_m corresponding to the extremal point. At least one more moment of time is registered exactly by the oscillogram – the moment of the leader contact with the earth, giving rise to the return stroke. The oscillogram indicates this moment by a field strength overshoot. The time interval Δt in figure 3.14(b) defines the leader average velocity along the path length $H - L_m$ at the earth: $V_L \approx (H - L_m)/\Delta t$. Note that one of the boundaries of the measured length might also be found from the moment of sign reversal of the field being registered. It follows from the analysis of (3.8) that the curve $\Delta E(L)$ intercepts the abscissa if the registration point lies at a distance $R \approx (0.8-1.4)H$ from the vertical path axis. Technically, the reference point is easier to find than the extremum.

It is known that the appetite comes with eating. If one substitutes the geometrical parameters used and the measured values of $\Delta E(L)$ into (3.8), one can find the average linear charge density τ_L. Together with the velocity, this provides the average current in the leader for the final period of time Δt: $i_L \approx \tau_L V_L$. The calculation of charge density was replaced in [36] by graphical differentiation of the oscillogram $E(t)$ at the point corresponding to the moment of leader contact with the earth; this, however, gave a low accuracy. Therefore, the electric field registration only at one point on the earth's surface seemed to be sufficient to evaluate one of the least accessible parameters – the leader current in a descending lightning discharge.

Let us now try to assess this situation without considering the measurement errors. Obviously, the main error is associated with finding the starting point of a lightning spark and the distance to it. A common 10% error in measurements gives much larger errors in evaluations of leader velocity and current. This always happens when one deals with the difference of two comparable parameters. We shall focus on errors of the model itself. The problem of leader branching effects will be ignored. After all, one can always consider a dart leader which has no branches. The representation of a real leader as a vertical axis is quite another matter. Any photograph shows numerous bendings of a lightning trajectory, so a straight vertical leader is nothing more than a speculative mathematical concept. To assess its implications, let us make another step and consider a tilted straight

3.6.2 Effect of linear charge distribution

The model of a leader with a uniform charge distribution and the concept of a storm cell as an electrode with a capacitor battery supplying conduction current to the leader are extremely far from reality. A storm cloud does not look like a giant capacitor plate, to which a lightning leader is connected galvanically during its motion to the 'plate' of opposite sign, i.e., to the earth. In actual reality, the cloud charge is concentrated on hydrometers which do not contact one another and their assemblage does not possess the properties of a metallic electrode. A better analogy would be that of an electrode-free spark, rather than of a spark starting from a high voltage electrode of a laboratory generator.

To illustrate this, consider a small metallic rod suspended along the field vector in an inter-electrode gap, where the field is supported by a high-voltage generator. The rod has no contact with the generator poles. Two sparks of opposite sign are excited simultaneously at the rod ends, i.e., in the region of enhanced local field (figure 3.16). The charges appearing on the sparks must be regarded as polarization charges. This is the way a

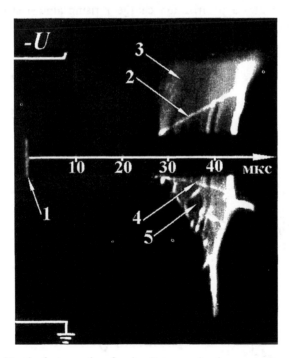

Figure 3.16. Streak photographs of a simultaneous development of a positive and a negative leader from the ends of a metallic rod of 50 cm in length in a uniform electric field; (1) rod; (2, 3) channel and streamer zone of a positive leader; (4, 5) the same for a negative leader.

metallic conductor is polarized when it is introduced into an electric field. Similarly, a charged storm cloud possessing no conductivity is only a field source in the space extending to the earth. A plasma conductor arising in this way or other is polarized in the field and grows, being supplied by polarization current. This system is definitely not a *perpetuum mobile*. Its energy source is the electric energy of the cloud field. As the leader develops, this energy decreases, in accordance with the conservation law. If the external field and the conductor are homogeneous, the linear density of polarization charge is equal to zero exactly at the conductor centre and its absolute value rises towards the ends of different polarities. As long as the conductor has no contact with the high-voltage generator terminals, its total charge, naturally, remains equal to zero. The latter is also valid when the field and conductor are inhomogeneous. Using numerical methods, one can find the polarization charge distribution for any electric field. The distributions presented in figure 3.17 have been found by the equivalent charge method [37]. This method is simple and convenient for long conductors, like those used to simulate lightning leaders.

Numerical computations show that a uniform field in a perfectly conducting rod creates a polarization charge τ rising almost strictly linearly from the rod centre towards its ends. The ends are an exception, because

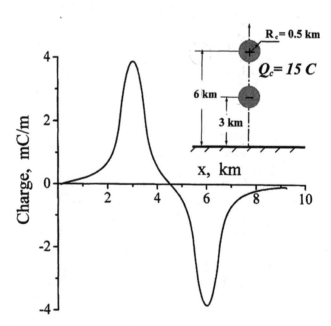

Figure 3.17. Polarization charge distribution along a straight conductor (a leader system) in the cloud dipole field, with allowance for a dipole reflection in the conducting earth.

the charge density here (along a length of the rod radius) rises rapidly. The charge distribution at the rod base can be approximated as $\tau(x) = ax$, where the coordinate origin x is at the rod centre. The reader will soon see that the contribution of the end charges $\pm q$ is small at larger distances, if the rod length $2d$ is much greater than its radius r.

For simplicity, let us assume that the charge is concentrated on the outer rod surface, as is the case when it has a conductivity, and that the potential will be calculated along the longitudinal axis. The rod centre will be taken as the zero point of the external field E_0 potential. The potential φ at the point x is a sum of potentials created by the external field, $-E_0 x$, the end charges, φ_q, and the charges distributed along the rod, φ_τ:

$$\varphi_\tau(x) = \frac{a}{4\pi\varepsilon_0} \int_{-d}^{d} \frac{y\,dy}{[(y-x)^2 + r^2]^{1/2}}$$

$$= \frac{a}{4\pi\varepsilon_0} \left\{ [(d-x)^2 + r^2]^{1/2} \right.$$

$$\left. - [(d+x)^2 + r^2]^{1/2} + x\ln\frac{(d-x)+[(d-x)^2+r^2]^{1/2}}{-(d+x)+[(d+x)^2+r^2]^{1/2}} \right\}$$

$$\approx \frac{ax}{4\pi\varepsilon_0} \left[\ln\frac{4(d^2-x^2)}{r^2} - 2 \right]. \tag{3.10}$$

Here, the last approximate expression refers to the rod sites lying far from its ends, $|d \pm x| \gg r$. Here, the term φ_q can be neglected, and we shall approximately have $\varphi_\tau - E_0 x \approx 0$. With the actual charge distribution $\tau(x)$ and the end charges providing φ_q, the rigorous equality $\varphi(x) = 0$ should be valid along the whole rod length. By relating the approximate equality to the centres of the semi-axes $x = \pm d/2$, we find

$$a \approx \frac{2\pi\varepsilon_0 E_0}{\ln\left(\sqrt{3}d/r\right) - 1} = \frac{2\pi\varepsilon_0 E_0}{\ln\sqrt{3}d/er}. \tag{3.11}$$

The potential at the rod ends must be calculated with the account of their higher charges. Assuming this charge to be concentrated along the end circumference, the potential at the centre of the end plane (at the points $x = \pm d$ on the axis) can be described as

$$\varphi(\pm d) = \varphi_q + \varphi_\tau(\pm d) - (\pm E_0 d) \approx 0, \qquad \varphi_q = \pm\frac{q}{4\pi\varepsilon_0 r}. \tag{3.12}$$

The potential φ_τ must now be calculated from the unsimplified formula (3.10). It follows from (3.12) that the end charge is approximately equal to $q \approx 2\pi\varepsilon_0 r E_0 d$ and by a factor of

$$K = \frac{ad^2}{2q} \approx \frac{d}{2r\ln\left(\sqrt{3}d/er\right)} \tag{3.13}$$

smaller than the charge distributed over each half (it is an order of magnitude smaller for $d/r \approx 100$). Therefore, the account of localized tip charge may be necessary only for the calculation of electric field in the region close to the tip, at a distance less than $10r$ from it. In a remote region, where measurements are usually made, such a subtlety is unnecessary – it is sufficient to consider only the charge distribution along the leader channel. Clearly, this is not a uniform distribution, taken for granted by some researchers.

It is time to look at the shape of a field strength pulse at the earth, determined by the charge of a linearly polarized vertical axis with charge $\tau(x) = \pm ax$ per unit length. It is defined by the algebraic sum of terms from the positively and negatively charged semi-axes and is equal to

$$\Delta E(L) = \int_{-L}^{L} \frac{a}{2\pi\varepsilon_0} \frac{x(H-x)\,\mathrm{d}x}{[(H-x)^2 + R^2]^{3/2}}$$

where L are the lengths of leader sections which have moved away from the starting point to the earth and upwards. Integration with (3.11) gives

$$\Delta E(L) = \frac{E_0}{\ln\left(\sqrt{3}L/er\right)} \left[\frac{L}{[(H-L)^2 + R^2]^{1/2}} - \frac{L}{[(H+L)^2 + R^2]^{1/2}} \right.$$

$$\left. - \ln \frac{[H + (H^2 + R^2)^{1/2}]^2}{\{H - L + [(H-L)^2 + R^2]^{1/2}\}\{H + L + [(H+L)^2 + R^2]^{1/2}\}} \right]$$

$$(3.14)$$

Here, H is the height of the leader start, r is its radius, and R is the distance between the leader axis and the observation point. It can be shown that the function $\Delta E(L)$ of (3.14) rises smoothly with L and has no extrema.

The linear charge distribution assumed in the above illustration is, of course, another speculation (section 4.3). Moreover, a leader goes up and down non-uniformly, and the field in the earth–cloud gap is far from being uniform: its strength decreases towards the earth. This limits the linear charge growth from the start downward. The finite channel conductivity exhibits similar behaviour, reducing the tip potential. So it is impossible to find the actual charge distribution exactly without knowing these parameters. Thus, a processing of field oscillograms can give nothing more than what they actually show. The field at a point is an integral effect of the whole combination of charges created or transported by a given moment of time. It is probably worth speculating about registrations but one should assess the results soberly, considering all possible variants and insuring oneself whenever possible. The best insurance is, of course, to increase the number of registration points and parameters determined by independent methods.

3.7 Lightning return stroke

All lightning hazards are associated with the return stroke, and this accounts for the great effort of investigators to learn as much as possible about this discharge stage. It has been established that the contact of a descending lightning leader with the earth or a grounding electrode produces a return wave of current and voltage. It travels up along the leader channel, partially neutralizing and redistributing the charge accumulated during the leader development (figure 3.18). The travel is accompanied by an increased light intensity of the channel, especially at the wave front. At the earth, the wave front intensity acquires its maximum over 3–4 μs [31]. As the wave goes up to the cloud, the wave intensity steepness and amplitude decrease many-fold, indicating a considerable decay. Judging by streak pictures, the region of a high light intensity at the wave front extends to 25–110 m. The whole wave travel takes 30–50 μs. This time is especially convenient for electron–optical methods of streak photography. However, available attempts to use such methods can hardly be considered successful. A serious obstacle is the exact synchronization of a streak camera and lightning contact with the earth. Although there are many synchronization methods, they have no simple technical solutions and are seldom used in lightning experiments. Continuous (e.g. sinusoidal) electron streak photography has not justified hopes. Basic results on return stroke velocities have been obtained using cameras with a mechanical image processing, which do not need synchronization (Boyce camera).

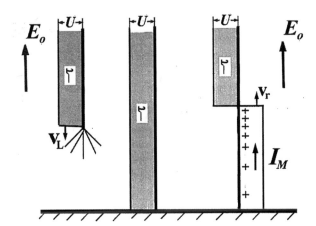

Figure 3.18. Scheme of the return stroke propagation after the contact of a descending leader with the earth (at moment $t = 0$). A leader brings potential $U < 0$; I_M is return stroke current.

3.7.1   Neutralization wave velocity

The measurements made half a century ago [25, 39] and those performed recently [40] indicate a high velocity of a return current–voltage wave. The minimum measured values are close to $(1.5-2) \times 10^7$ m/s and the maximum ones are an order of magnitude higher, reaching 0.5–0.8 of light speed c. A velocity comparable with light speed does not mean that we deal with relativistic particles or purely electromagnetic perturbations. The wave velocity is the phase velocity of the process.

There are not so many successful optical registrations of the return stroke, the number of really good ones being about 100. Most of the available data concern subsequent lightning components. This is natural because every successfully registered discharge includes the return strokes of several components. The wave velocities of subsequent components are somewhat higher than those of the first ones. According to [40], the first component has an average velocity $V_r \approx 9.6 \times 10^7$ m/s while the subsequent ones are a factor of 1.25 higher. Similar data are cited by other authors for subsequent components of lightning discharges triggered from a grounded wire elevated by a rocket.

To illustrate the statistical velocity spread in individual measurements and in those made by different researchers, figure 3.19 shows integral distribution curves for the data of [25] and [40]. The first and subsequent

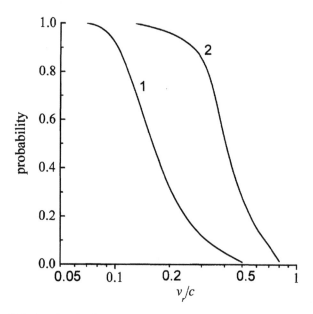

Figure 3.19. Velocity distribution of the lightning return stroke: (1) averaged over the visible channel length [25]; (2) averaged over 1.3 km above the earth [40].

components were not separated. Within a 50% probability, there is a 2-fold difference between the velocities. Earlier measurements generally give lower return stroke velocities. The point is that most measurements performed during the 1980s were two-dimensional, usually providing higher velocities, whereas the earlier data had allowed conclusions only about the vertical component of velocity. Moreover, the application of improved optics and photographic materials, as well as higher relative motion rates of the image and film, improved the time resolution of streak photographs. As a result, the velocity value obtained in the 1980s was more accurate and higher because the measurements were averaged over the initial stroke length of about 1 km at the earth's surface, where the wave moves 1.5–2 times faster, rather than over the whole stroke length.

All measurements show that the return stroke velocity gradually decreases and that the velocity V_r drops abruptly when the wave front passes through the point of leader branching. The latter fact suggest a certain relation between the stroke velocity and the current transported by the wave: at the branching point, the current is divided among the branches, so the velocity becomes lower. The knowledge of this relation could improve the calculation accuracy of overvoltages in electrical circuits during lightning discharges. Unfortunately, the available data are insufficient to allow finding this relation reliably. Simultaneous registrations of current and velocity have been made only for return strokes of subsequent components of triggered lightnings but they cannot provide a representative statistics. With reference to [12, 41], there is note in [1] about a satisfactory agreement between these registrations and Lundholm's semi-empirical formula $V_r/c = (1 + 40/I_M)^{-1/2}$, where I_M is a return stroke current amplitude expressed in kA (see section 3.7.2). The lack of factual data is sometimes compensated by a superposition of distribution statistics. It is assumed that the values of current and velocity characterized by an equal probability correspond to each other. There are no serious arguments in favour of this operation but it is used for the lack of a better method.

3.7.2   Current amplitude

The current amplitude is an important lightning parameter. Most hazards of lightning are associated, directly or indirectly, with stroke currents, whose registration has taken much time and effort. Very few of them were made by direct methods, using a shunt and a Rogovski belt [28, 29, 42–46]. Still fewer direct measurements were made by equipment with a wide dynamic range, which can register both powerful impulses with an amplitude to 200 kA and low currents of a few hundreds of amperes, which are equally important for the understanding of the lightning physics.

A large number of measurements have been made by magnetic detectors. Such a detector represents a rod several centimetres in length, made from magnetically hard steel. Preliminarily demagnetized rod detectors were placed at a fixed distance from a conductor aimed at leading lightning current to the earth. This could be a grounding lead of a lightning conductor or a metallic tower of a power transmission line. With the appearance of lightning current, the detector proves to be within the range of its magnetic action and becomes magnetized. One measures the residual steel magnetization and calculates the current by solving the inverse problem. The advantages of this method are its simplicity and low cost. Usually, magnetic detectors are installed by the thousand to obtain the necessary statistics. However, they can yield nothing else but a current impulse amplitude. Of course, by marking the ends of the detector, one can also determine the direction of current and attribute it to the lightning type (positive or negative). The accuracy of current measurements is very low for several reasons.

First, there are few objects with a simple system of current spread over metallic constructions. A single conductor would be ideal in this respect, because it excludes current branching. In reality, lightning current is distributed among many conductors, the distribution pattern being unpredictable since it varies with temporal parameters of the impulse. We shall illustrate this situation with reference to a simple system consisting of two parallel inductively connected branches with their own inductances L_1 and L_2, mutual inductance M, and resistances R_1 and R_2. Suppose a rectangular current impulse I with a short risetime is applied to the system. The current distribution between the two branches is described as

$$R_1 i_1 + L_1 \frac{di_2}{dt} + M \frac{di_2}{dt} = R_2 i_2 + L_2 \frac{di_2}{dt} + M \frac{di_1}{dt}, \qquad i_1 + i_2 = I. \quad (3.15)$$

Initial currents i_{10} and i_{20} at the stage when current $I(t)$ is stabilized to I are generated over a very short time equal to the I risetime. The branch currents, therefore, rise from zero very quickly. The reactive components of voltage drop $\sim di/dt$ produced by them are much larger than the ohmic ones $\sim i$ that can be neglected for the time being. Hence, we have $i_{10}/i_{20} = (L_2 - M)/(L_1 - M)$, and the initial current, say, in the first branch is $i_{10} = I(L_2 - M)/(L_1 + L_2 - 2M)$. When the transitional process, whose duration is defined by the time constant $\Delta t = (L_1 + L_2 - 2M)/(R_1 + R_2)$, is over, currents $i_{1\infty} = IR_2/(R_1 + R_2)$ and $i_{2\infty} = i_{1\infty} R_1/R_2$ are established in the circuits. The durations of lightning currents are usually comparable with the time constant Δt. Therefore, a magnetic detector placed in one of the branches will register a current intermediate between the initial and established values having a maximum amplitude, since the residual magnetization of the rod contains information only about the maximum magnetic field of current. For this reason, one can calibrate a magnetodetector for deriving a full current amplitude only if the impulse shape is known. This cannot be done in a real experiment, so one has to resort to a rough estimation of current distribution over metallic constructions and use it in data processing.

Second, the operating range of the rod magnetization curve is not large, and the transition from a linear to saturation region may produce additional errors in data processing. To avoid saturation, the magnetodetector is placed far from the conductor, which creates difficulties in data processing of lightnings with low current and low magnetic field. Besides, when the distance between a conductor and a detector is large, the magnetic field effects of other metallic elements with current are hard to take into account. So a 100% error does not seem too high for magnetodetectors, even when several detectors are placed at different distances from a current conductor. Their records provide sufficient material for engineering estimations or for a qualitative comparison of storm intensity in different regions, but they are insufficient for theory. Organization of direct registrations takes much time and effort. There are no more than a hundred successful registrations made over a decade. Let us see what information can be derived from them.

Current impulse amplitudes vary widely, from 2–3 to 200–250 kA. Some magnetodetector measurements give even 300–400 kA, but these amplitudes seem doubtful. According to [42, 46], the integral amplitude distributions for the first and subsequent lightning components obey the so-called lognormal law, in which it is current logarithms, rather than currents themselves, that meet the normal distribution criterion. The probability of lightning with a current larger than I_M, is defined as

$$P(I_M) = \int_{I_M}^{\infty} \frac{1}{(2\pi)^{1/2}\sigma_{lg}} \exp\left\{ \frac{-[\lg I_M - (\lg I)_{av}]^2}{2\sigma_{lg}^2} \right\} d(\lg I), \qquad I\,[\text{kA}]$$

(3.16)

where $(\lg I)_{av}$ is an average decimal logarithm of the currents measured and σ_{lg} is the mean square deviation of their logarithms. This approximation cannot be considered accurate. The relative deviation of the value of (3.16) from the real one may be several tens percent; it may be even more for practically important current ranges. Nevertheless, lognormal distributions allow measurement comparison and serve as a guide to engineering estimations. For example, about 200 current oscillograms for lightnings that struck the 70 m tower on the San Salvatore Mount in Switzerland [42] satisfactorily obey the lognormal law with $(\lg I)_{av} = 1.475$ and $\sigma_{lg} = 0.265$ for the first component currents of a negative lightning discharge. This means that the 50% current value is estimated to be 30 kA; 95% of lightnings must have currents exceeding 4 kA and 5% of lightnings 80 kA. The probability of higher currents rapidly decreases: 100 kA is expected in 2% of cases and 200 kA in less than 0.1% of cases (figure 3.20). It should be emphasized again that the distribution boundaries must be treated with caution. The curve shape in the low current range strongly depends on the sensitivity of the measuring instruments used (its left-hand limit is usually taken to be 1–3 kA in distribution plots). There are few measurements in the high current

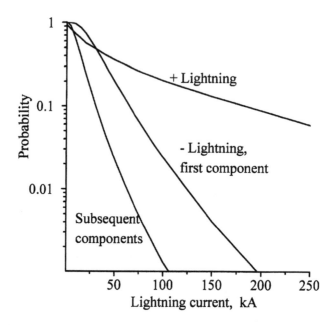

Figure 3.20. Lognormal distributions of return stroke currents: A, the first component of a negative lightning with $(\lg I)_{av} = 1.475$ and $\sigma_{1g} = 0.265$; B, subsequent components with $(\lg I)_{av} = 1.1$ and $\sigma_{1g} = 0.3$; C, positive lightnings with $(\lg I)_{av} = 1.54$ and $\sigma_{1g} = 0.7$.

range: it is considered as good luck if they provide a reliable order of magnitude. Note that negative lightning currents above 200 kA have never been registered reliably.

The approximation of data on subsequent lightning components in [42] gives a much lower integral probability for high currents. A lognormal distribution can be satisfactorily described by $(\lg I)_{av} = 1.1$ and $\sigma_{1g} = 0.3$. The calculated 50% current is 12.5 kA, 5% current is only 39 kA, and the chance for a subsequent component to exceed 100 kA is close to 0.1% (figure 3.20).

The statistics for positive lightnings, whose number is about 10% of the total registrations, is less representative. All descending positive lightnings are one-component. The integral current distribution for them has a large spread. The probabilities of both low and high currents are larger without an essential change of the 50% value. The 50% value is close to 35 kA, i.e., it is nearly the same as for the first component of negative lightning. An approximate description of the lognormal distribution of positive currents in [42] can be made with $(\lg I)_{av} = 1.54$ and $\sigma_{1g} = 0.7$ (figure 3.20). Positive high current lightnings are more frequent than negative ones. A 5% probability corresponds to 250 kA, and 100 kA can be expected with a

20% probability. Among 26 successful registrations of positive lightnings in [43], only one showed 300 kA current. It seems likely that many positive lightnings were, under the observation conditions [43], ascending ones, which may account for the large spread. Such lightnings have practically no return stroke, and the equipment seems to have registered the relatively low leader current of the final development stage. These data were used to derive the integral distribution extending to the low current region.

The great importance of lightning current statistics to applied lightning protection necessitated a unification of theoretical distribution curves. Otherwise, engineers would have been unable to compare the frequency of harmful lightning effects and protection efficiency. This work is being done within the frame of the CIGRE (Conférence Internationale des Grands Reseaux Electrique à haute tension) – an operating international conference on high-voltage networks. Data on current from all over the globe are collected and analysed. However, there is no unified approach to these data: different data are discarded for different reasons, so that the distributions obtained differ markedly. For example, a report submitted to [47] compares two log-normal laws with $(\lg I)_{av} = 1.4$ and $1.477 \, \sigma_{lg} = 0.39$ and 0.32. The latter is preferable for power transmission lines, since the measurements for objects higher than 60 m were excluded from this derivation (power transmission lines are usually lower). In his book, Uman [1] gives a table of lightning currents mostly based on the measurements of [42].

Attention to details is inevitable, since slight corrections in parameter distributions may cause manifold changes in the calculated probabilities of currents above 100 kA, especially important in lightning protection of important objects. Both theory and applications suffer from a lack of lightning current measurements. We shall list here some key issues to be discussed in more detail below.

We have mentioned the importance of an object's height. It has been known since Benjamin Franklin's experiments that high constructions attract more lightnings. It seems likely that the process of attraction depends on the potential of a descending leader. If this is so, the statistics of descending leader currents for objects of various height may prove different: there will be a kind of lightning separation. In this case, a comparison of reliable current statistics for various objects could help resolve the much debated problem of lightning–object interaction mechanism.

Of interest in this connection is the following fact. In the case of a very high construction, many first components are of the ascending type having no return stroke. But the first component is followed by subsequent descending components, whose average stroke currents are lower than in subsequent components affecting low buildings. This suggests an involvement of high ground constructions in the formation of storm clouds. It appears that an ascending leader starts from a high construction before the cloud has matured. Its charge and potential are, therefore, lower. This accounts for the fact that subsequent components discharging an immature cloud will transport lower stroke currents than a mature cloud.

Finally, an important issue is the effect of grounding resistance of objects on lightning current. This may provide information on the resistance of lightning itself. This resistance is to be introduced in equivalent circuits, when calculating overvoltages affecting various electrical circuits. This problem is still much debated: some investigators suggest the substitution of a lightning channel by a current source with an 'infinite' resistance, others ascribe to the channel the wave resistance of a common wire (about 300 Ω). It would not be hard to solve this problem if we had at our disposal reliable current statistics for objects of various height but different grounding resistances. No such statistics exist yet. To speed up the work in this area and to reduce its cost, various remote registration techniques are being employed. They register electromagnetic fields and coordinates of points where the lightning strikes (ideally, the lightning trajectory), followed by the solution of the inverse problem for the field source, i.e., lightning current (see section 3.7.4).

There is also an increasing number of direct current registrations from lightnings triggered from a wire lifted by a rocket to the height of 150–250 m. The first component of a triggered lightning (ascending leader) has no return stroke; therefore, one deals only with subsequent components. A comparison of such registrations with natural lightning currents was made in Alabama, USA [15]. The statistics were not particularly representative (45 measurements), so no principal differences were revealed. The lognormal distribution of currents corresponded to the parameters $(\lg I)_{av} = 1.08$ and $\sigma_{lg} = 0.28$, nearly the same as those obtained in Switzerland for subsequent components of natural lightnings [42]. We should like to warn the reader against a possible overestimation of this coincidence. The comparison involved measurements from geographical points separated by large distances, whereas the global variation of lightning parameters still remains unclear. A more important thing is that the lightnings studied in [42] cannot be regarded as totally natural. They struck a 70-m tower on a mountain elevated at 600 m above the earth's surface close to a lake. The conditions here are more similar to those of lightning triggering than to its natural development in a flat country. Lightning parameters are known to differ with altitude: currents registered by magnetodetectors at an altitude of 1–2 km were two times lower than in a flat country, for less than 50% probabilities [48].

3.7.3 Current impulse shape and time characteristics

Records of lightning current impulses look more like abstractionists' pictures – they are so diverse and fanciful. The conventional approximation of a impulse by two exponents $I(t) = I_0[\exp(-\alpha t) - \exp(-\beta t)]$, which is suggested in various guides to equipment testing lightning resistance, is

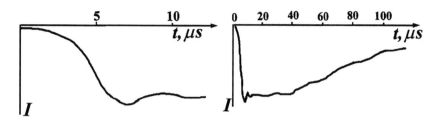

Figure 3.21. A schematic oscillogram of a current impulse in the first component of a negative lightning.

intended for currents of laboratory sources simulating lightning, rather than for natural lightning. Let us try to identify the main features of the time variation of current, essential for the understanding of the return stroke mechanism and applications.

The most reliable data have been obtained for first component currents of a negative lightning. This current is easy to register, since the impulse front takes several microseconds and an oscillographic record reproduces it in detail. A sketch of a current impulse averaged over many oscillograms is shown in figure 3.21 in two time scales. Note the concave shape of the front. An expression of the type $1 - \exp(-\beta t)$ looks least suitable for its description. The first current peak is often followed by a higher one, and evaluation of the impulse risetime t_f is associated with some reservations. For example, [42] measured the time of current rise from 2 kA, a value close to the resolution threshold, to the first maximum I_M. In this case, about 50% of negative lightnings had the risetime of the first component over 5.5 μs, 5% exceeded 18 μs, and another 5% less than 1.8 μs. The knowledge of the risetime allows calculation of the average impulse slope $A_{I\,av} = I_M/t_f$. However, the calculation of electromagnetic fields of lightning and the evaluation of possible hazards require a maximum slope $A_{I\,max} = (dI/dt)_{max}$, rather than an average one. The error in evaluations of this parameter from current oscillograms may be very large, because one has to replace the tangent to the $I(t)$ curve by a secant. Nevertheless, this operation has a sense for a fairly long impulse of the first component. The integral distribution of the values, like the current itself, is described by the lognormal law with the parameters $(\lg I)_{av} = 1.1$ and $\sigma_{1g} = 0.255$, if the slope is expressed in kA/μs. It results in 12 kA/μs for 50% current, and the slope exceeds 33 kA/μs with a 5% probability.

To describe the electromagnetic effect of lightning current, let us find the induced emf U_M in a frame of area $S = 1\,m^2$ placed at distance $D = 1\,m$ from the channel or a grounding conductor, when the first component current flows through it (the frame is in a plane normal to the current magnetic field). Even for a moderate steepness $A_{I\,max} = 33\,kA/\mu s$, we have $U_M = \mu_0 A_{I\,max} S(2\pi D)^{-1} = 6.6\,kV$, where $\mu_0 = 4\pi \times 10^{-7}\,H/m$ is vacuum magnetic permeability. The role of a frame can be performed by any metallic structure within the construction affected by lightning-wires, wall fittings, rails, metallic stripes touching each other, etc. At the site of a poor contact, induced emf will produce a spark, much more effective than that in an electric lighter. This is dangerous because the spark may come in contact with an explosive gas mixture.

Sometimes, lightning current behaves in a dual way, creating the induction emf and voltage at the resistance of the grounding electrode $U_R = IR$. It is important, therefore, to have knowledge about the relation between the current amplitude and maximum slope. Although both parameters obey the same lognormal law, no correlation has been found between them. This is bad for engineering applications, for one has to calculate the probabilities of each current with a whole set of possible slopes.

There have been attempts at a more detailed description of the current impulse front. They were initiated by the CIGRE mentioned above to handle hazardous effects of lightning on power transmission lines. A set of additional parameters has been suggested to reduce errors in current oscillogram processing and some regions of the impulse front have been described quantitatively. This is illustrated in figure 3.22 and requires no comment. Some of the results were cited in [47]. The processing technique used did not lead to considerable data refinement, since the 50% maximum slope $A_{I\,max} = 12\,\text{kA}/\mu\text{s}$ is only two times larger than the 50% average slope $A_{50\%} = I_{50\%}/t_{f50\%} = 30/5.5 = 5.5\,\text{kA}/\mu\text{s}$. But the factor of 2 is essential to the electrical strength of ultrahigh voltage insulation.

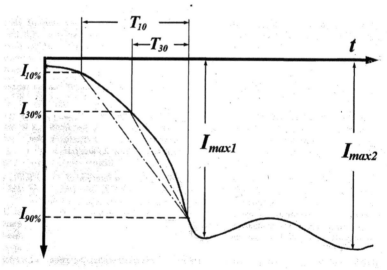

Figure 3.22. The distribution of current impulse parameters in the return stroke, based on oscillograms.

Current impulses of subsequent components have a shorter risetime. In the work cited above [42], $t_f < 1.1\,\mu s$ for a 50% probability and $0.2\,\mu s$ for a 5% probability. The latter should be treated with caution because this value is close to the resolution limit of the measuring equipment. The impulse front in subsequent components is likely to rise faster. It is mentioned in [1] with reference to other publications that in many digital registrations the current could rise to a maximum during the first detector reading (for $0.2\,\mu s$).

The maximum slope of a impulse front in subsequent components obeys, in the first approximation, the lognormal law: $(\lg A_{I\,max})_{av} = 1.6$ and $\sigma_{1g} = 0.35$. The value of $A_{I\,max}$ exceeds $40\,kA/\mu s$ with a 50% probability and is larger than $120\,kA/\mu s$ with a 5% probability. When affected by such a steep impulse, the amplitude of induced voltage would exceed $25\,kV$ in the above example of a frame.

Current of positive lightnings rises slowly. In 5% of cases, the front duration was $t_f > 200\,\mu s$. With these impulses, the electric strength of air gaps of several metres in length is close to a minimum (section 2.6, formula (2.51)). The voltage with $t_f \approx 200\,\mu s$ is much more dangerous than a 'common' lightning overvoltage impulse with a risetime of several microseconds. Minimum breakdown voltage in air gaps with a sharply non-uniform field (see formula (2.52)) is about 1.5 times lower than in a standard lightning overvoltage impulse of $1.2/50\,\mu s$ (in accordance with the conventional way of presenting time characteristics of a impulse, 1.2 is the risetime and 50 is the impulse duration at 0.5 amplitude, all in μs).

The duration of a current impulse is as important for lightning protection practice as the risetime. Impulse duration is usually characterized as a time span between its beginning and the moment its amplitude decreases by half. Since current is related to the neutralization wave travelling along the channel, the impulse duration t_p is comparable with the time of the wave travel. If its velocity is $V_r \approx 10^8\,m/s$ and the average channel length is $3\,km$, the value of t_p will be several tens of microseconds. A similar value is derived from experimental data. The impulse duration in the first component of a negative lightning is above 30, 75 and $200\,\mu s$ for the probabilities 95, 50 and 5%, respectively. For subsequent components, the impulse is much shorter: 6, 32 and $140\,\mu s$ for the same probabilities. Positive lightnings must be longer because most of the positive charge of a storm cloud is located 2–3 km higher than the negative charge. Indeed, t_p is above $230\,\mu s$ with a 50% probability. The shortest durations for positive lightnings are the same as for the first component of a negative one. 'Anomalously' long impulses stand out against this background – about 5% of positive currents decreased to half the amplitude for $2000\,\mu s$.

Today, we know nothing about the nature of superlong positive impulses. One thing is clear: they are unrelated to the wave processes in the lightning channel. One may suggest that hydrometeor charge is accumulated and descends to the earth due to an ionization process in the positively charged region of a cloud. But we can only speculate about the nature of this process producing final current of 100 kA and ask why it is manifested only in positive lightnings.

3.7.4 Electromagnetic field

Electromagnetic field of lightning is familiar to those leaving a TV or radio set on during a thunderstorm. Sound and video noises inform about a storm long before it actually begins. Lightning was the first natural radio station used by the founders of radio engineering for testing their receivers. The lightning detector designed by A S Popov in 1885 is still Russia's national pride. For many years meteorologists surveyed approaching storm fronts by registering so-called atmospherics – pulses of electro-magnetic radiation from lightning discharges occurring hundreds of kilometres away. In the late 1950s, much interest in atmospherics was due to the nuclear weapon race: suspiciously similar to radiation pulses from nuclear explosions, they interfered with the diagnostics of the latter.

It is clear from the foregoing that in a return stroke the charge accumu-lated by a leader cover varies and is redistributed rapidly along the channel, producing variation of the static component of the electric field. Charge variation occurs simultaneously with the propagation of a current wave along the channel, inducing a magnetic field. The induction emf varying in time gives rise to an induction component of the electric field. Finally, variation in the current dipole moment (a leader channel can be regarded as a dipole, with the account of its mirror reflection by the earth) gives rise to an electromagnetic wave producing a radiation component of the electric field with a concurrent magnetic radiation component. There is another mag-netic component – a magnetostatic one proportional directly to current.

It is common practice to distinguish between the near and far regions of electromagnetic radiation. In the near region, static field components may be dominant: the electric component, damped in proportion to the cubic distance r to the dipole centre, and the magnetic component, varying with distance as r^{-2}. These can be neglected for the far region, because they are much smaller than the radiation components $E, H \approx r^{-1}$. Now, after these preliminary remarks, we shall turn to experimental data showing how much the shape of a registered pulse varies with distance between a lightning discharge and a field detector.

The shapes of return stroke radiation pulses are shown schematically in figure 3.23 for the near and far regions. At large distances, where the static components of magnetic and electric fields are nearly completely damped, the pulses $E(t)$ and $H(t)$ become geometrically similar. Both are bipolar and have a high front slope, a well defined initial maximum and several smaller ones along the slowly falling pulse slope, producing the effect of damping oscillations. Note that the oscillation period is smaller than the

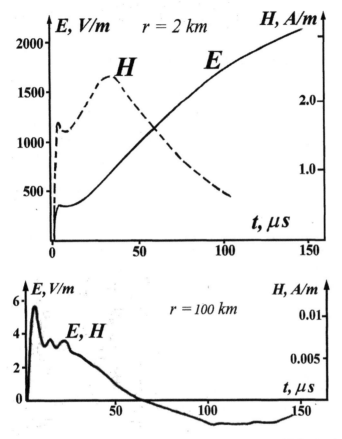

Figure 3.23. Schematic oscillograms of electromagnetic pulses of lightning in the near (top) and far (bottom) zones at the distances 2 km (top) and 100 km (bottom).

double time of the wave run along the channel. After passing the zero point, the pulse part opposite in sign rises and then decreases with nearly the same rate; its amplitude is 2–3 times smaller that the first 'half period'.

The inverse proportionality of radiation components to the distance from the radiation source was the reason why measurements are presented in the above form: they are normalized to the basic distance $r_{bas} = 100$ km as $E^*_{max} = 10^{-5}E_{max}r$ with r in metres. For the first lightning component, the average values of the initial pulse peak of the vertical component, E^*_{max}, lie within 5–10 V/cm [49–54] (compare: radio receivers detect well signals of 1 mV/m in the medium bandrange). The electric component of subsequent lightning components is 1.5–2 times smaller. The spread of measurements is as large as that of lightning currents. The standard deviation σ_E is in the range 35–70% for the first lightning component and 30–80% for subsequent ones. The horizontal component of magnetic field strength $H^*_{max} = (\mu_0/\varepsilon_0)^{-1/2}E^*_{max}$ varies respectively. Magnetic induction $B_{max} = \mu_0 H_{max}$ is about 10^{-8} T at a distance of 100 km from the lightning.

The radiation pulse of the first lightning component rises to the initial peak with an increasing rate. In oscillogram processing, the risetime is arbitrarily subdivided into two components: the initial slow one of 3–5 µs duration and the final fast one taking 1–0.1 µs. The standard deviation is also large here: 30–40% of the average value for the slow front and about 50% for the fast front. In the final stage, the signal rises for about 0.5–$1.0E^*_{max}$. With some reservations, a subdivision into a slow and fast component can be also made for radiation pulse of the return stroke of subsequent lightning components. But it would be more correct to consider that the rise to the initial peak occurs quickly there, for 0.15–0.6 µs. Note that the risetimes for the first and subsequent components are close to those of their current impulses in a return stroke.

The moment of sign reversal for radiation pulses of the first components is delayed, relative to the onset of a return stroke, by 50 µs in temperate latitudes [54] and by 90 µs in the tropics [52]. The sign reversal for subsequent components occurs by a factor of 1.3–1.5 earlier. The time for maximum field to be established after the sign reversal is of the same order of magnitude as that prior to the reversal.

The radiation components E and H are, naturally, present in the near region, too, but they are much smaller than the static component. One exception is the initial moments of time. The initial peaks in oscillograms $E(t)$ and $H(t)$ should be attributed to radiation, since the static field components did not have enough time to reveal themselves. The monotonic rise of electric field over 20–50 µs, the time long enough for the radiation component to be damped, is nearly totally due to electrostatic effect. The induced electrostatic field is quite powerful, because the charge accumulated by the stepwise leader of the first component or by the dart leader of subsequent components is neutralized during the return stroke. For example, the electrostatic field changes by several kV/m at the distance of 1 km from the channel lightning during the first 50 µs (for the subsequent component, the signal is 2–3 times lower than for the first one); a slower field rise may continue for about 100 µs. All in all, the field of the first lightning component is an order of magnitude higher than the initial radiation rise. With increasing distance r to 15–20 km, the radiation component becomes dominant over the others, and the initial radiation peak becomes an absolute maximum of the registered signal.

The magnetostatic component in the near region is not so important. Still, at a distance of 1 km, it contributes as much to the signal as the radiation component (figure 3.23). The magnetic induction here is as high as 10^{-5} T. The absolute magnetic field maximum is achieved later than the stroke current peak registered at the earth's surface. This is clear because the magnetostatic component is proportional not only to the current but to the conductor length. The length increases as a neutralization wave travels from the earth up to the cloud. For the same reason, the times for the first and subsequent components do not differ much. The duration of pulse $B(t)$ in the near region is comparable with that of current inducing a magnetic field.

3.8   Total lightning flash duration and processes in the intercomponent pauses

A. descending negative lightning flash has on average two or three components, each terminated by a more or less powerful current impulse of the return stroke. The average number of components in an ascending lightning is four. The maximum number of components in a lightning flash may be as large as 30. The pauses between the components Δt_{com} vary from several milliseconds to hundreds of milliseconds. With a 50% probability, their duration exceeds 33 ms; the integral distribution curve is described by the lognormal law with the parameters $(\lg \Delta t_{com})_{av} = 1.52$ and $\sigma_{lg} = 0.4$, at Δt_{com} [ms]. The total flash duration varies with the number of components. Negative one-component flashes are the shortest ones, since their current often ceases right after the return stroke, for less than a millisecond. An ascending one-component positive flash can pass current for a longer time, 0.5 s, in spite of the absence of a return stroke. Of course, this is a low current, less than 1 kA. The average flash duration is close to 0.1–0.2 s and the maximum is 1.5 s. These large times are discernible by the naked eye, so lightning flickering is not a physiological by-product of vision but a physical reality.

Intercomponent pauses take most of the flash time. They cannot be said to be current-free. A lightning leader is supplied by current nearly all the time, and this current is high enough to support plasma in a state close to that of a steady-state arc. Current of an intercomponent pause is referred to as continuous current, which is a fairly ambiguous term. Average continuous current varies between 100 and 200 A. Nearly as high current supplies an arc in a conventional welding set used for cutting metal sheets or for welding thick pipes. Most thermal effects of lightning are associated with its continuous current, rather than with return stroke impulses which are more powerful but shorter. The highest continuous current measured [55] was 580 A. Continuous current usually slowly decreases with time. In a one-component ascending lightning having no return stroke, the contact of the leader with the cloud is terminated by charge overflow from the cloud to the earth as a decreasing continuous current of about the same value.

Cloud discharging by continuous current can be easily registered by an electric field detector. Field varies monotonically, as long as current flows through the channel. These are appreciable changes, since current of 100 A extracts, from a cloud, charge $\Delta Q \approx 10C$ over the time 0.1 s. The field on the earth right under a cloud changes by the value $\Delta E = \Delta Q/(2\pi\varepsilon_0 H^2) \approx 200$ V/cm if the height of the charged cell centre is $H = 3$ km; at distance $r = 10$ km from the lightning axis, $\Delta E = \Delta QH/[2\pi\varepsilon_0(H^2 + r^2)^{3/2}] \approx 5$ V/cm. Similar values were registered during observations.

Continuous current flow is accompanied by slowly rising and as slowly decreasing current impulses with an amplitude up to 1 kA. These are M-components of lightning. The risetime of a typical M-component is about 0.5 ms, an average impulse duration (on the level 0.5) is twice as much, an average amplitude is 100–200 A, although M-components with current up to 750 A have also been registered [56, 57]. Pulsed current rise is always accompanied by an increase in light emission intensity of the whole channel, from the cloud down to the earth. Streak photographs (even taken slowly) do not show the propagation of a well defined emission wave front similar, say, to the tip of a dart leader. It seems as if most of the channel flares up simultaneously, although excitation, no doubt, propagates down from a cloud with a high velocity, $(2.7–4) \times 10^7$ m/s (from measurements of [58]). Two M-components were identified in [58] as ascending ones. In later measurements, the existence of ascending processes were questioned, because there were no clear physical reasons for the appearance of an inducing perturbation at the earth's surface.

Variations in current and electric field of M-components were registered in triggered lightning flashes at a short distance from the channel ($r = 30$ m) [57]. The field variation of a vertical component at the earth is shown in figure 3.24. The pulse ΔE rises to its maximum 70 μs earlier than the current impulse. The field rises and decreases at nearly the same rate. The pulse component of field perturbation is nearly completely damped while the current still has a high amplitude.

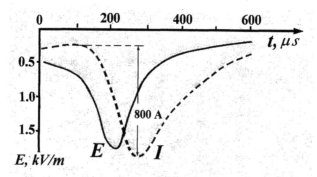

Figure 3.24. Superimposed schematic oscillograms of M-component electric field and current at the earth [57].

The number of M-components in a flash may even be larger than that of subsequent components, but they are of little interest to lightning protection practice – their charge and current are too low. Theoretically, however, these components are of great interest, because they seem to contain information on unobservable processes occurring in storm clouds. It is quite likely that these processes give rise to a dart leader with a return stroke or to a stroke-free M-component. Some authors [27] believe that an M-component is always formed against the background of continuous current, whereas a necessary prerequisite for a dart leader is a current-free pause, during which the grounded lightning channel partly loses its conductivity. This is a very important detail shedding light on processes occurring in a storm cloud after a grounded plasma channel of the first lightning component has penetrated it. The transport of the earth's zero potential to a cloud by a conducting channel, resulting in a rapid increase in the cloud electric field in the vicinity of the channel top, is a powerful stimulus for gas discharge processes there (for details, see sections 4.7 and 4.8).

3.9 Flash charge and normalized energy

During intercomponent pauses, charge is transported from a cloud to the earth by both powerful return stroke impulses and continuous current, the latter being much lower but longer-living. The contributions of these currents to the total charge effect are comparable. With a 50% probability, the stroke charge transported by the first component of a negative flash is over $4.5C$, while 5% of flashes transport over $20C$ and another 5% less than $1.1C$ [42]. The lognormal law described above is suitable for an approximate representation of the integral distribution curve with the values $(\lg Q)_{av} = 0.653$ and $\sigma_{lg} = 0.4$. The return strokes of subsequent components have, for the same probabilities, five times smaller charges due to their shorter duration and lower currents. The largest spread of charge measurements is characteristic of positive lightning, in agreement with the diversity of their shape and duration. Positive pulse charges exceed $16C$ with a 50% probability, $150C$ with a 5% probability, and are less than $2C$ with a 5% probability. These seem to be positive lightning with no return stroke. For the description of integral charge distribution for positive pulses, the lognormal parameters may be taken to be $(\lg Q)_{av} = 1.2$ and $\sigma_{lg} = 0.6$.

We have already mentioned that the charge of a lightning flash is always larger than the sum of charges transported by the return strokes of the first and subsequent components, since a substantial contribution to the total charge is made by continuous current. The total negative flash charge exceeds $7.5C$ with a 50% probability, $40C$ with a 5% probability, and is nearly the same as the first negative pulse charge in the least powerful flashes. The total positive charge is appreciably larger – with 95%, 50% and 5%

probabilities, it exceeds, respectively, 20, 80 and 350C. One cannot say that the charge transported by a flash is very large. For comparison, even a very large lightning charge of 350C flows through the arc of a conventional welding unit for 3–5 s.

Charge transport is accompanied by energy release. An average negative flash with a charge $Q = 10C$ and gap voltage 50 MV dissipates about $QU = 5 \times 10^8$ J, which is equal to the energy released by a 100 kg trinitrotoluene explosion. While most energy is released within the lightning trace, the problem of energy release and heating of metal constructions is of much interest. Normally, the resistance of metallic conductors and that of a grounding electrode are much less than the equivalent resistance of a lightning channel $R_l \approx U/I_M$ (I_M is the impulse amplitude of a return stroke); $R_l \approx 1$ kΩ if $U \approx 50$ MV and $I_M = 50$ kA. Therefore, lightning can be regarded as a current source, assuming that current I_M is independent of the object's resistance. Any conductor with lightning current flow releases the energy

$$K = R \int_0^\infty i^2 \, dt \equiv (K/R)R, \qquad K/R = \int_0^\infty i^2 \, dt$$

proportional to the conductor resistance R. For practical calculations, data on 'normalized' energies K/R characterizing lightning only are published. According to [42], 95%, 50% and 5% probabilities correspond to the measured values exceeding 2.5×10^4, 6.5×10^5 and 1.5×10^7 A^{2}s for positive flashes and 6.0×10^3, 5.5×10^4 and 5.5×10^5 A^{2}s for negative flashes, respectively. For subsequent components of negative flashes, the respective values are an order of magnitude smaller and do not contribute much to the total energy release. To get an idea about thermal potency of lightning, evaluate the heat of a steel conductor with a cross section of $S = 1$ cm^2. With resistivity $\rho = 10^{-5} \, \Omega$ cm, the energy density released by a powerful positive flash ($K/R = 1.5 \times 10^7$ A^2 s) is $(K/R)(\rho/S^2) = 150$ J/cm^3, with the conductor temperature increasing by 40°C. Owing to Joule heat, a lightning flash is capable of burning down only a very thin conductor with a cross section less than 0.1 cm^2. In many cases, however, heating just by several hundred degrees may become hazardous.

3.10 Lightning temperature and radius

Plasma temperature is usually measured by spectroscopic methods. Lightning spectroscopy is a hundred years old, and it was used even before photography and field-current measurements. Reviews of spectroscopic results can be found in Uman's books [1, 59] together with extensive references. However, direct data on lightning plasma are still very scarce. Lightning spectra, naturally, contain lines of molecular and atomic oxygen and nitrogen, as well as singly charged ions N_2, argon, cyane and some other impurities. No doubly charged ions have been detected, indicating that the temperature does not exceed 30 000 K. Measurements of time resolved N II (N^+) line intensities show that the return stroke temperature reaches 30 000 K for the first 10 µs [59, 62] and drops to 20 000 K in 20 µs. Average temperatures are estimated to be about 25 000 K. These results are obtained assuming that a plasma channel is optically transparent and that the excitation of atoms in the plasma is equilibrium (of the Boltzmann type). The estimations justify this assumption.

Electron densities found from the Stark broadening of the H_α lines are 10^{18} cm^{-3} for the first 5 µs of the stroke life. Under thermodynamic equilibrium conditions at $T = 30\,000$ K, this value of n_e corresponds to the pressure of 8 atm [63]. About 10 µs later, n_e decreases to 10^{17} cm^{-3}, corresponding to the pressure drop down to the atmospheric pressure. Then the value of n_e remains unchanged over the time of the N II line registration. This does not seem strange. Equilibrium electron density in air at $p = $ const $= 1$ atm changes only slightly in a wide temperature range 15 000–30 000 K, remaining about 10^{17} cm^{-3}. As the channel cools down, the ionization degree $x = n_e/N$ certainly decreases, but when the pressure reaches the atmospheric value, the gas density N rises simultaneously. For this reason, $n_e = xN$ does not change much. High intensity radiation is observed for about 100 µs (from 40 to 1000 µs). The first peak is often followed by another one several hundreds of microseconds later.

Spectroscopic measurements were mostly made during a return stroke, but some authors [64] managed to register the spectrum of a 2-m portion of a stepwise leader. The leader tip temperature calculated from the N II lines lies within 20 000–35 000 K. The diameter of the radiation region is less than 35 cm. More accurate evaluations are unavailable. It seems unlikely that this temperature is characteristic of the whole leader channel. Rather, the experiment registered a short temperature rise during a powerful step which was akin to a miniature return stroke (section 2.7). The step-induced perturbation involving part of the channel region is most likely to be damped rapidly along the leader length.

It is not only the plasma dynamics but the channel radius, too, which still remains enigmatic. In making evaluations of the radius, one usually relies on photographs. But in this case, it is very important to agree on the kind of radius being evaluated. This may be the radius of the channel, through which current flows during the leader and stroke stages. Clearly, such a radius will include the best conducting and, hence, the hottest core of the plasma channel. Or, one can follow another approach. When solving the problem of electric field variation during the lightning development, one has to deal with the radius of the leader cover where most of the space charge is concentrated. This is the charge radius of lightning. Therefore, each time we speak of radius, we must define exactly what we mean.

Here, we shall use the concept of channel radius as applied to the region where the lightning current is accumulated and the concept of cover radius to the region where most of the space charge is concentrated. The former can be determined, to some extent, by using optical methods, although this is a complicated task. With reference to the optical measurements [65], one usually deals with radii of several centimetres. This resolution is accessible to modern cameras at a distance of about a kilometre, but the cameras must have the highest resolution possible. Anyway, we have never heard about the application of such perfect optical equipment in lightning research.

In addition to using special-purpose optics, the experimentalist must match perfectly the sensitivity of photographic materials and exposures. A longer exposure produces a halo, increasing the actual radius. Unless special measures are taken, the error may be very large, especially for flashes with a high light intensity. For some reasons, the optical radius of a lightning channel may exceed manifold the thermal radius. Such an effect was observed in studies of spark leaders in laboratory conditions [66]. Registration of the thermal radius appears problematic even for triggered lightning, with a fixed point of contact with the earth. For natural lightning, this task is much more complicated. As for the cover radius, there is no reliable technique for its registration at all. So lightning radius measurements cannot provide unquestionable data, and the researcher is to rely on theoretical evaluations only.

3.11   What can one gain from lightning measurements?

It was not our task to review all experimental studies on lightning: this has been well done in [1, 59]. We believe that the latest experimental data will be presented in a new Uman book now in preparation. But the basic facts have been discussed here, and we can now ask ourselves whether the available data are sufficient to build lightning theory and to check it by experiment.

The situation with lightning is somewhat similar to that for a long laboratory spark, i.e., experiments give mainly external parameters of a discharge. In the laboratory, these are velocities of the major structural elements (streamers and leaders), their initiating voltages, currents, transported charges, and, possibly, some other characteristics Sometimes, we have some information on channel radii, or on the time variation of radii, or scarce data on plasma parameters. But that is all.

The arsenal of lightning researchers is much smaller. First, they have no information about the voltage in the cloud–earth gap at the lightning start, and there are no data on the initial distribution of electric field. Both literally and figuratively, the bulk of a storm cloud, where a descending leader originates, is obscure. Measurements made at the earth's surface cannot help much, because the number of registration points is too small, so it is impossible to reconstruct the initial field distribution along the whole lightning path.

The fine structure of a lightning flash is not clear either. Observations give no information about the size of the streamer zone in a lightning leader, and even the existence of such a zone is largely speculative. Nor do we know the origin and structure of volume leaders, which are responsible for the stepwise pattern of a negative leader, at least, observable in laboratory conditions. There is no information on the gas state in the track of a preceding component, when a dart leader travels along it. The only dart leader parameter that has been measured is its velocity. What has just been listed refers primarily to the return stroke. It appears that space charge neutralization – the basic process occurring in it – is related to the fast radial propagation of streamers away from the channel. This is the way the cover charge is supposed to change. But there are no experimental data on this process, nor can we hope to obtain any in the near future.

Most available findings concern lightning currents and transported charges. As in a laboratory spark, lightning currents are usually registered at the earth's surface, so we have data on leader currents for ascending discharges only. There are no direct measurements of currents for descending or dart leaders, the latter fact being especially disappointing. There are more or less detailed descriptions of currents for return strokes, but the measurements made at one point (that of contact with the earth) restrict the possibilities of both a theoretical physicist and a practical engineer. Data on the current wave damping along the leader are important for the former because then he may try to reconstruct the plasma conductivity variation. The latter needs them to be able to calculate the lightning electric field at the earth and in the troposphere, because it is hazardous to both ground objects and aircraft.

Lightning current statistics deserves special attention. Normally, they are used in calculations of the occurrence probability of lightning with hazardous parameters, e.g., a critically fast rise of the impulse front and/or amplitude. The practical requirements on the calculation reliability are extremely high. Indeed, it is impossible to provide the necessary accuracy, using lognormal parameter distributions. Any approximation of an actual distribution lognormally would be approximate, especially in the range of large values important for lightning protection. The error may be as high as 100%. One should keep this in mind when comparing calculations of hazardous lightning effects and the available experience in object protection.

This is the reality not to be ignored either by a theorist attempting to create a lightning model or by an engineer working on lightning protection. No matter how ingenious a theorist may be, he will not be able to check his model, filling the gaps by laboratory spark data or by general physical considerations. As for practical lightning protection, one usually gained

[31] *Saint Privat d'Allier Research Group* 1982 Extrait de la Revue Générale de l'Electricité, Paris, September
[32] Gorin B N and Shkilev A V 1974 *Elektrichestvo* **2** 29
[33] Idone V P, Orville R E 1985 *J. Geophys. Res.* **90** 6159
[34] Rubinstein M, Uman M A and Thomson P 1992 *Proc. 9th Intern. Conf. on Atmosph. Electricity* **1** (St Peterburg: A I Voeikov Main Geophys. Observ.) p 276
[35] Rubinstein M, Rachidi F, Uman M A *et al* 1995 *J. Geophys. Res.* **100** 8863
[36] Thomson E M 1985 *J. Geophys. Res.* **90** 8125
[37] Kolechizky E C 1983 *Electric Field Calculation for High-Voltage Equipment* (Moscow: Energoatomizdat) p 167 (in Russian)
[38] Jordan D M and Uman M A 1983 *J. Geophys. Res.* **88** 6555
[39] Schonland B and Collens H 1934 *Proc. Roy. Soc. London Ser. A* **143** 654
[40] Idone V P and Orville R E 1982 *J. Geophys. Res.* **87** 9703
[41] Idone V.P, Orville R E, Hubert P *et al* 1984 *J. Geophys. Res.* **89** 1385
[42] Berger K, Anderson R B and Kroninger H 1975 *Electra* **41** 23
[43] Berger K 1972 *Bull. Schweiz. Elekrtotech. Ver.* **63** 1403
[44] Gorin B N and Shkilev A V 1979 in *Lightning Physics and Lightning Protection* (Moscow: Krzhizhanovsky Power Engineering Inst.) p 9
[45] Gorin B N and Shkilev A V 1974 *Elektrichestvo* **2** 29
[46] Eriksson A J 1978 *Trans. South Afr. IEE* **69** (Pt 8) 238
[47] Anderson R B and Eriksson A J 1980 *Electra* **69** 65
[48] Alizade A A, Muslimov M M *et al* 1974 in *Lightning Physics and Lightning Protection* (Moscow: Krzhizhanovsky Power Engineering Inst.) p 10
[49] Master M J, Uman M A, Beasley W H and Darveniza M 1984 *IEEE Trans. PAS* **Pas-103** 2519
[50] Krider E P and Guo C 1983 *J. Geophys. Res.* **88** 8471
[51] Cooray V and Lundquist S 1982 *J. Geophys. Res.* **87** 11203
[52] Cooray V and Lundquist S 1985 *J. Geophys. Res.* **90** 6099
[53] McDonald T B, Uman M A, Tiller J A and Beasley W H 1979 *J. Geophys. Res.* **84** 1727
[54] Lin Y T, Uman M A *et al* 1979 *J. Geophys. Res.* **84** 6307
[55] Krehbiel P R, Brook M and McCrogy R A 1979 *J. Geophys. Res.* **84** 2432
[56] Thottappillil R, Goldberg J D, Rakov V A, Uman M A *et al* 1995 *J. Geophys. Res.* **100** 25711
[57] Rakov V A, Thottappillil R, Uman M A and Barker P P 1995 *J. Geophys. Res.* **100** 25701
[58] Malan D J and Collens H 1937 *Proc. R. Soc. London A* **162** 175
[59] Uman M A 1969 *Lightning* (New York: McGraw-Hill)
[60] Orvill R E 1968 *J. Atmos. Sci.* **25** 827
[61] Orvill R E 1968 *J. Atmos. Sci.* **25** 839
[62] Orvill R E 1968 *J. Atmos. Sci.* **25** 852
[63] Kuznetsov N M 1965 *Thermodynamic Functions and Shock Adiabata for High Temperature Air* (Moscow: Mashinostroenie) (in Russian)
[64] Orvill R E 1968 *J. Geophys. Res.* **73** 6999
[65] Orvill R E 1977 in *Lightning*, vol 1, R Golde (ed) (New York: Academic Press) p 281
[66] *Positive Discharges in Air Gaps at Les Renardieres – 1975* 1977 *Electra* **53** 31

Chapter 4

Physical processes in a lightning discharge

Here we shall discuss the basic phenomena occurring in a lightning discharge: a descending negative leader, an ascending positive leader, the return strokes of the first and subsequent components, a dart leader, and some others. Lightning may travel not only from a cloud towards the earth, or from a grounded object towards a cloud, but it may also start from a body isolated from the earth – a plane, a rocket, etc. About 90% of all descending discharges are negative and about as many ascending discharges are positive. For this reason, an ascending leader is said to be positive. Available experimental data on lightning as such are of little use in our attempts to explain the mechanisms underlying the above processes. There are very few observations that might shed light on their physical nature. So, one has to resort to speculations, invoking both theory and experimental data on a long laboratory spark, which relate primarily to a positive leader. Since this process is most simple (to the extent a lightning process may be considered simple), we shall begin with the discussion of an ascending positive leader.

4.1 An ascending positive leader

4.1.1 The origin

The lightnings people observe most frequently are descending discharges, which originate among storm clouds and strike the earth or objects located on its surface. However, constructions over 200 m high and those built in mountainous regions suffer mostly from ascending lightnings. These are of nearly as much interest to the physicist as the seemingly common, descending discharges. An ascending leader is initiated by a charge induced by the electric field of a storm cloud in a conducting vertically extending grounded object. If a metal conductor of height h with a characteristic radius of the rounded top $r \ll h$ is fixed on the earth and then affected by a vertical external field E_0, a field $E_1 \approx E_0 h/r \gg E_0$ is created by the induced charge at the conductor top (see section 2.2.7). This field rapidly decreases in air (for a distance of several r values), creating a potential difference between the conductor end and the adjacent space, $\Delta U \approx E_0 h$. When the cloud bottom is charged negatively and the vector E_0 is directed from the earth up to the cloud, the grounded conductor becomes positively charged, since the field makes some of the negative charges leave the metal to go down to the earth.

No stringent conditions are necessary for the field E_1 to initiate the air ionization (at sea level $E_1 \approx E_i \approx 30\,\text{kV/cm}$) or for a corona discharge to arise at the pointed parts of a high structure (it is necessary to have $E_1 \approx 40–31\,\text{kV/cm}$ for $r = 1–10\,\text{cm}$). The conditions for a leader to be initiated in the streamer corona stem are much more rigorous. The energy estimations made in section 2.6 show that there is no chance for a leader to arise if the leader tip potential U_t, or, more exactly, its excess over the external potential at the tip, $\Delta U = U_t - U_0$, is less than $\Delta U_{t_{\min}} \approx 300–400\,\text{kV}$. This estimate is supported by experiments with leaders, whose streamer zones have no contact with the electrode of opposite sign at the initial moment of time. Therefore, for the desired potential difference $\Delta U_{t_{\min}}$ to be produced, the structure must have, at least, $h \approx \Delta U_{t_{\min}}/E_0 \approx 20–30\,\text{m}$ if the average field of the storm cloud at the site of the grounded object is $\sim 150\,\text{V/cm}$.

On the other hand, even if a leader is produced at such a low potential, $\Delta U_{t_{\min}}$, it can hardly travel for a large distance. The leader current will be too low to heat the channel to a sufficiently high temperature. As a result, the channel resistance will be too high so that a very strong field will be required to support the current in the channel. The channel field E_c is, however, limited by the external field E_0. Indeed, a grounded body of height h, from which a positive leader has started, possesses zero potential. Having covered the distance L, the leader tip acquires the potential $U_t = -E_c L$. Here, the potential of the unperturbed external fields is $U_0 = -E_0(L + h)$, and we have

$$\Delta U_t = U_t - U_0 = \Delta U_i + (E_0 - E_c)L, \qquad \Delta U_i = E_0 h. \qquad (4.1)$$

For a leader to develop from the initial threshold conditions, the potential difference ΔU_t should not decrease relative to the initial value of ΔU_i. For this, the average channel field E_c must be lower than the external field E_0. However, a mature channel possesses a falling current–voltage characteristic $E_c(i)$. A decrease in E_c to $\sim 100\,\text{V/cm}$ requires a channel current higher than 1 A. We discussed this issue in sections 2.5.2 and 2.6. With the approximation accepted there ($E_c \approx b/i$ and $b = 300\,\text{VA/cm}$), the leader current is to exceed $i_{\min} = b/E_0 \approx 2\,\text{A}$ at $E_0 \approx 150\,\text{V/cm}$.

Let us see how large the potential difference ΔU_t should be to make the current exceed $i_{\min}$. In chapter 2, we derived formula (2.35) relating the channel current behind the leader tip to the tip potential U_t and the leader velocity v_L. That formula was applicable to the laboratory conditions considered in that chapter, when a leader travelled through the rapidly decreasing field of a high-voltage electrode. Having covered a distance of only a few radii of the electrode curvature, usually very small, the leader tip found itself in a space with a nearly zero potential, $U_0 \ll U_t$. The neglect of the external field was justifiable in that case. An ascending leader is quite another matter: the potential difference $\Delta U_t = U_t - U_0$ continuously increases with the leader velocity, as its tip approaches a charged cloud, since the tip enters a region of an ever increasing external field. Hence, we have $|U_t| \ll |U_0|$ and the value of ΔU_t largely determined by U_0. Therefore, the approximate formula of (2.35) must be rewritten in its general form, with U_t replaced by ΔU_t:

$$i = \frac{2\pi\varepsilon_0 \Delta U_t v_L}{\ln(L/R)}, \qquad \Delta U_t = U_t - U_0 \qquad (4.2)$$

where R is the effective radius of the leader charge cover.

The available leader theory fails to provide a clear and convincing physical expression to describe the relationship between v_L and ΔU_t. So we shall further use the empirical relation suggested in section 2.6:

$$v_L = a(\Delta U_t)^{1/2}, \qquad a = 15\,\text{m/s}\,\text{V}^{1/2}. \qquad (4.3)$$

This relation was derived from experimental data on rather short gaps, in which the tip potential could be taken to be identical to that of a high voltage electrode.† In accordance with (4.2), expression (4.3) corresponds to the relation $v_L \approx i^{1/3}$ also supported by some laboratory experiments. Now, using the value of i_{min}, we shall find ΔU_i which provides the leader viability:

$$\Delta U_{i_{min}} = \left[\frac{b}{E_0} \frac{\ln(L/R)}{2\pi\varepsilon_0 a} \right]^{2/3}. \qquad (4.4)$$

Assuming $L \approx 10\,\text{m}$ and $R \approx 1\,\text{m}$ for a still-short initial leader at $E_0 = 150\,\text{V/cm}$, we obtain $\Delta U_{i_{min}} \approx 3.1\,\text{MV}$ and $h_{min} = 210\,\text{m}$. The result of this simple estimation agrees with that of lightning observations. In a flat country, ascending lightnings make up an appreciable fraction of the total number of strikes affecting grounded objects of about that height. The continuous ascending leader of a triggered lightning (initiated from a grounded wire raised by a rocket above the earth) is also excited at about 200 m. Note that the value of E_0 used in the calculations is somewhat larger than those measured at the earth surface. The storm cloud field near the earth is always attenuated by the space charge introduced in the air by corona discharges from thin conductors of small height, such as tree branches, shrubs, grass, constructions, etc. Some measurements show that the field at the earth is half that at a height of 10–20 m.

---

† The streamer theory has been advanced further. Note, for comparison, that the streamer velocity is $V_s \approx \Delta U_t$ from formulae (2.6) and (2.8) with ΔU_t instead of U_t and $E_m = \text{const}$.

4.1.2 Leader development and current

Two main leader parameters are accessible to measurement: its velocity and the current through the channel base contacting a grounded object. The current is due to the charge pumped by an electric field into the growing leader. If the field in the leader channel is lower than the external field (otherwise the leader is non-viable), the difference between the tip potential and the unperturbed potential at its site becomes larger as the leader becomes longer (see expression (4.1)). According to (4.2) and (4.3), the leader current, proportional to $i \sim \Delta U_t^{3/2}$, also rises. The current rise becomes more rapid as the leader becomes longer, especially when the leader reaches the region of a very high cloud field. The rising current heats the channel more, so that its linear resistance and field drop. With time, the channel becomes a nearly perfect conductor. Grounded at its base, the channel possesses the same potential $U(t, x)$ everywhere along its length, including the tip, which is low relative to the absolute external potential $|U_0(x)|$. In this case, the value of $\Delta U = U(t, x) - U_0(x) \approx -U_0(x)$ varies only slightly with time at every point x along the channel.

The linear leader capacitance C_1, given by formula (2.8) with length L and cover radius R instead of l and r, also varies very little. Indeed, the cover radius behind the tip is about the same as the streamer zone radius which, according to (2.39), is $R = \Delta U_t / 2E_s$, where $E_s \approx 5\,\text{kV/cm}$ is the streamer zone field under normal conditions. The height of the charged region centre in the cloud, $H \approx 3\,\text{km}$, is much greater than that of the leader starting point, $h \approx 200\,\text{m}$. Suppose the leader length is greater than h but smaller than H by such a value that the cloud field non-uniformity along the channel can be neglected. We then have $\Delta U_t \approx |U_0(L)| \approx |E_0 L|$, and the value of $L/R \approx 2E_s/E_0$ under the logarithm in C_1 of (4.2) and (4.4) is independent of time. If this relation does change, which happens when a leader rises so high that it enters the region of a rapidly increasing external field, the logarithm changes much more slowly. Thus, the linear charge $\tau(x) \approx C_1 \Delta U$ remains nearly constant in time at every leader point. But if there is no charge redistribution along the channel, the current in it, $i(t, x)$, does not change along its length but changes only in time. Entering the channel through its grounded base, the current supplies charge only to the front leader portion. The current in the base is the same as in the channel right behind the tip. It is defined by formula (4.2) with $\Delta U_t \approx |U_0(L)|$, close to the unperturbed potential of the cloud charge at the tip site. Similarly, the leader velocity can be found from (4.3).

Therefore, the velocity and current of a fairly long leader (long relative to the start height), which develops in the average field E_0, are described as

$$v_L = a(\Delta U_t)^{1/2}, \quad i = \frac{2\pi\varepsilon_0 a(\Delta U_t)^{3/2}}{\ln(2\sqrt{3}E_s/eE_0)}, \quad \Delta U_t = E_0 L, \quad e = 2.72\ldots \quad (4.5)$$

The numerical factor $\sqrt{3}/e$ in the second expression of (4.5) has resulted from a more rigorous calculation. This result is obtained if the linear charge $\tau(x)$ is calculated directly from external field E_0 with $\tau(x) = \text{const } x$, as in section 3.6.2, rather than from average linear capacitance C_1 and $\Delta U(x)$, as was done in the derivation of (4.2). The current is found from $i = dQ/dt$, where Q is the net charge of the conductor (the integral of $\tau(x)$ in x).

It follows from (4.5) that the current rises rapidly with time as the leader develops, whereas the velocity increases much more slowly: $v_L = dL/dt \approx L^{1/2} \approx t$ and $i \approx L^{3/2} \approx t^3$ at $E_0 = \text{const}$. In stronger external fields, the leader current also rises with E_0 much faster than the velocity. Numerically, a leader with $L = 500\,\text{m}$ has $i = 4.5\,\text{A}$ and $v_L = 4 \times 10^4\,\text{cm/s}$ in an average field $E_0 = 150\,\text{V/cm}$, i.e., about the same values as for an extremely long laboratory spark. An increase of L to $2000\,\text{m}$ and E_0 to $300\,\text{V/cm}$ gives the typical lightning parameters: $i = 120\,\text{A}$ and $v_L = 12 \times 10^5\,\text{cm/s}$. A rapid rise of the leader current and a much slower increase of its velocity were inevitably registered in observations of both natural and triggered lightnings [1, 2]. These estimations reasonably agree with measurements.

In contrast to (4.2) and (4.3), expression (4.5) ignores the voltage drop across the leader channel because $U_t \ll |U_0(L)|$. It is easy to see the validity of this assumption in the next approximation using the derived formulae. With the voltage–current characteristic $E \approx [\ ]i^{-1}$, the voltage drop across the channel decreases as $U_L \approx E_c L \approx L^{1/2} \approx t^{-1}$ with the leader development. At the tip site, on the contrary, $|U_0(L)| = |E_0|L$ grows even faster than $L \approx t^2$ if one takes into account the increase of the average external field along the channel during its travel up to the cloud. Note that E_c and U_L do not drop to zero in reality but only decrease to a certain limit, because the field $E_c(i)$ in a very heated channel with high current is stabilized due to the greater plasma energy loss for radiation (the current–voltage characteristic should be expressed as $E_c = c + b/i$ rather than as $E_c = b/i$). This issue, however, is of no importance to an ascending leader, since its current becomes very high only when the tip reaches the region with $|U_0(L)| \gg U_L$.

If one desires to refine these simple results by taking account of the voltage drop, charge redistribution, and current variation along the channel, one should regard it as a long line, as was done with the streamer in section 2.2.3. The distributions of potential $U(t, x)$, charge per unit length $\tau(x, t)$, and current $i(x, t)$ along the line can be described by equations similar to (2.13) and (2.14):

$$\frac{\partial \tau}{\partial t} + \frac{\partial i}{\partial x} = 0, \qquad -\frac{\partial U}{\partial x} = E_c(i), \qquad i(L) = \tau(L)v_L \qquad (4.6)$$

where E_c is the longitudinal channel field expressed through current $i(x, t)$ from the current–voltage characteristic (the field in (2.13) was expressed through current and linear resistance, $E_c = iR_1$). The leader velocity v_L is given, for example, by formula (4.3): $dL/dt = v_L$.

Equations (4.6) must involve an electrostatic relation between charges and potentials. In a simple approximation, expressions (2.13) and (4.2) were allowed to contain a local relation, $\tau(x) = C_1[U(x) - U_0(x)]$, through linear capacitance C_1. This approximation was shown by many calculations to be quite acceptable to the case of a uniform or weakly non-uniform external field, but it appears insufficiently rigorous for a strongly non-uniform field, which a leader crosses on entering a storm cloud. In fact, the potential at every point along the channel length is also created by charges located at adjacent channel sites. To simplify the non-local relation, the leader charge can be assumed to be concentrated on a cylindrical surface with an effective cover radius R; then the desired relation takes the form

$$\Delta U(x, t) = U(x, t) - U_0(x) = \frac{1}{4\pi\varepsilon_0} \int_0^L \frac{\tau(z, t)\,dz}{[(z - x)^2 + R^2]^{1/2}}. \qquad (4.7)$$

The boundary conditions for the set of integral differential equations (4.6) and (4.7) are described by the third equality in (4.6) and $U(0, t) = 0$, since the leader base is grounded. Practically, it is convenient to subdivide the channel into N fragments and consider the charge density in each fragment to be dependent only on time, thus replacing the integral equation of (4.7) by a set of linear algebraic equations. Each of them will relate the potential $U(x_k)$ at the middle point x_k of the kth fragment to the intrinsic and all other linear charges. After integrating (4.7), one can easily see that radius R enters logarithmically the factors of the set of equations (compare with (4.2)), thereby justifying the use of linear leader charge τ instead of its cover space charge. The set of algebraic equations for $U(x_k)$ and $\tau(x_k)$ is solved in time at each step, and the progress is made by using equations (4.6). We are presenting the result of this solution.

As the leader tip approaches the cloud, the external field at the tip site becomes stronger and the ever increasing portion of the channel finds itself in a strongly non-uniform field. Since the velocity and current are largely defined by the potential $U_0(L)$ at the tip site, formulae (4.5), in which E_0 is an average field, remain valid. In a simple model of a cloud with a spherical unipolar charged region, the potential distribution in the space free from charges is the same as for a point charge. If H is the height of the spherical charge centre, Q_c, the potential at height x at the point displaced from the vertical charge axis for distance r (with the account of the mirror reflection by the earth's plane) is

$$U_0(x) = \frac{Q_c}{4\pi\varepsilon_0} \left\{ \frac{1}{[(H - x)^2 + r^2]^{1/2}} - \frac{1}{[(H + x)^2 + r^2]^{1/2}} \right\}. \qquad (4.8)$$

Figure 4.1 presents the parameter calculations for an ascending leader propagating in such a non-uniform field. The calculations were made from

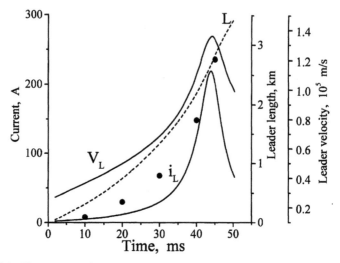

Figure 4.1. The propagation of an ascending leader from a grounded object in a negative cloud field. • Indicate current i_L calculated from (4.5); $Q_c = 5\,C$, $H = 3\,km$, $r = 0.5\,km$.

the set of equations (4.6) and (4.7), as described above. The current at the channel base is defined by the total charge Q and the velocity by expression (4.5):

$$i = \frac{\mathrm{d}Q}{\mathrm{d}t} = v_L \frac{\mathrm{d}Q}{\mathrm{d}L}, \qquad Q(L) = \int_h^L \tau(x)\,\mathrm{d}x. \qquad (4.9)$$

For comparison, the current was also calculated from (4.5). The results show the good accuracy of this simple formula, so the use of average linear capacitance C_1 can be considered justifiable in the calculation of $\tau(L) = C_1 \Delta U_0(L)$ and in the case of a sharply non-uniform field.

4.1.3 Penetration into the cloud and halt

There are two questions to be answered here: how high the maximum leader current is and where the leader halts. To answer the first question, one should keep in mind that the cloud charge is concentrated in a certain volume but not at a point. Suppose it is a sphere of radius R_c with the centre at height H in (4.8). Measurements made during flights through storm clouds indicate that R_c is most likely to be by an order of magnitude smaller than H. The maximum potential at the centre of a uniformly charged sphere is by a factor of 1.5 higher than on its surface and equals $U_{0\,max} = 3Q_C/8\pi\varepsilon_0 R_c$. Penetrating into the charged region, an ascending leader acquires a considerable velocity and a very high current. To illustrate, calculated from (4.8) for $H = 3\,\text{km}$, the field near the earth under the charge centre, $E_0 = -Q_c/2\pi\varepsilon_0 H^2 = 150\,\text{V/cm}$, is created by charge $Q_c = -7.4\,\text{C}$ and $|U_{0\,\text{max}}| = 340\,\text{MV}$ for $R_c = 300\,\text{m}$. From (4.5), a leader that has reached the charged region centre acquires the velocity $v_L = 2.8 \times 10^5\,\text{m/s}$ and the current $i_{\text{max}} = 5\,\text{kA}$ (the field at the sphere boundary is $E_{0\,\text{max}} = 7.5\,\text{kV/cm}$, decreasing to zero towards the centre; in the current estimation, the logarithm was taken to be 1). The lifetime of this high current is short, about $R_c/v_{L\,\text{max}} \sim 10^{-1}\,\text{s}$, with the total duration of the leader ascent of about $3 \times 10^{-2}\,\text{s}$ (these estimations ignore the effect of air density, which is 1.5 times lower than normal at a height of $3\,\text{km}$). Maximum currents of the kiloampere scale were registered during observations of ascending lightnings.

On its way up through the charged region, the leader enters an area of reciprocal external field at height $x > H$. The potential difference ΔU_t is, at first, positive but decreases as the leader elongates. Its velocity and current now decrease with time, but this process has its limits. There is a region of positive charge of nearly the same value high above the negative charge region. Representing it as a sphere with the centre at height $H + D$ and taking the mirror reflection effect into account, as in (4.8), we can find the potential of the dipole thus formed:

$$U_0(x) = \frac{Q_c}{4\pi\varepsilon_0}\left\{ \frac{1}{[(H-x)^2 + r^2]^{1/2}} - \frac{1}{[(H+x)^2 + r^2]^{1/2}} \right.$$

$$\left. - \frac{1}{[(H+D-x)^2 + r^2]^{1/2}} + \frac{1}{[(H+D+x)^2 + r^2]^{1/2}} \right\}. \tag{4.10}$$

Without allowance for the voltage drop across the channel ($U_t \approx 0$) the leader tip will reach the point x_s, where the absolute potential U_0 drops to $\Delta U_t \approx 400\,\text{kV}$, remaining negative as before. Since $\Delta U_{t\,\text{min}}$ is small relative to huge potentials of charged regions ($|U_0|_{\text{max}} \sim 100\,\text{MV}$), a positive ascending leader halts at a slightly lower height than the zero equipotential surface of the external field. Because of the effect of charges reflected by the earth, the zero potential line lies somewhat lower than the dipole centre. For example, at $D = H$, which corresponds, more or less, to reality, we have $x_s = 1.486H$ exactly on the vertical axis ($r = 0$) instead of $1.5H$, as would be the case with a solitary dipole. With greater radial displacement r, the zero equipotential line comes closer to the earth, slowly at first but then more rapidly at $r > H$ (figure 4.2). This is the reason why ascending leaders taking different vertical paths halt at different heights.

It has just been mentioned that the equipotential line $U_0(x, r) = 0$ corresponds to the maximum height attainable by a single ascending leader. With allowance for the voltage drop across the channel, which may appear appreciable in some situations, ΔU_t drops to the threshold value $\Delta U_{t\,\text{min}}$ below the maximum height. This is supported by numerical

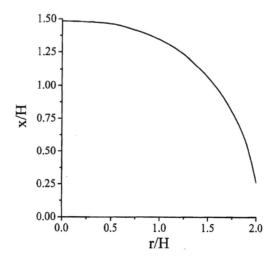

Figure 4.2. The zero potential line of a cloud dipole with the allowance for charges reflected by the earth for $D = H$.

calculations made from the set of equations (4.6) and (4.7) and illustrated in figure 4.3. They also indicate the leader retardation rate. As the leader velocity decreases, the channel current becomes lower, causing the field E_c to rise. The tip potential decreases respectively, together with the potential difference ΔU_t, which limits the current still more, and so on.

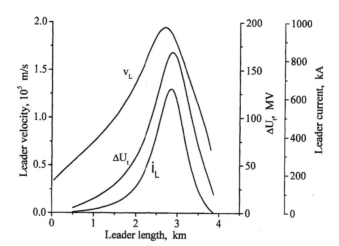

Figure 4.3. Numerical simulation of an ascending leader propagating in a cloud dipole field ($Q_c = 12\,C$, $H = D = 3\,km$, $r = 0.5\,km$), with allowance for the voltage drop across the channel.

When a leader goes beyond the lower cloud charge region, the external field changes its direction along the channel: below the negative charge centre, $x = H$, its vertical component is directed upwards but above the centre it is directed downwards. Correspondingly, the external potential U_0 is non-monotonic and has an extremum at height H (absolute maximum). The leader continues to develop beyond the maximum point, as long as the relation

$$\Delta U_t = U_t(L) - U_0(L) = |U_0(L)| - |U_t(L)| > \Delta U_{t_{min}}$$

is valid. But now, the leader velocity decreases continuously, because $|U_0(L)|$ drops with the leader elongation and $|U_t(L)|$ rises due to the rising channel field. It is clear that a leader can develop successfully in any other direction, since it is capable of propagating in the direction strictly opposite to the external field. The calculations show the leader path along the equipotential line in a zero external field. Here, ΔU_t, i and v_L decrease slowly, only due to the greater voltage drop across the channel; otherwise, the leader would travel for an infinitely long time.

We have focused on this circumstance because it is here that the principal features of a leader process manifest themselves clearly. The external field at the tip site is usually low and cannot affect the instantaneous leader velocity, current and direction of motion. The direction may vary randomly, a fact well known to those making lightning observations. What is important is the voltage U_0 created by this field along the leader path, rather than the field strength. The propagation of a positive leader is provided by the transport of a fairly high positive charge to its streamer zone. The current of many streamers taking the charge out accumulates in the channel, heating it and providing its viability. But for many streamers to be excited off from the leader tip, the latter must possess a high potential relative to the unperturbed potential $\Delta U_t = U_t(L) - U_0(L) \approx |U_0(L)|$. This is indicated by the absence of appreciable discrepancies between the current calculations made straightforwardly from the linear density of induced charge $\tau(x)$ in a strongly non-uniform external field and from formula (4.5) containing only ΔU_t.

Therefore, it is not surprising that the lightning paths exhibit the diversity illustrated in figure 4.4. No random change of the leader path can disturb its viability. A leader can follow any direction: it can move along the external field or in the opposite direction, along the equipotential line, etc. – all ways are open as long as the condition $\Delta U_t > \Delta U_{t_{min}}$ is valid. But the leader acceleration does depend, of course, on its direction of motion. Moving along the field, the leader is accelerated, because the voltage drop is compensated excessively by the increase in $|U_0|$. When the leader moves in the opposite direction, it is decelerated. The maximum acceleration is achieved in the direction of the maximum gradient ΔU_t, and this seems to be the reason for the fact that the main leader branch darts in the direction of the rising field, i.e., towards a charged cloud, a high object, etc. (for details see section 5.6).

Figure 4.4. A photograph of a well-branched lightning with the path bendings.

4.1.4 Leader branching and sign reversal

Leader branches are nearly always visible in photographs of ascending lightnings. Branching may start almost from the channel base or after the leader has covered many hundreds of metres (figure 4.4). The currents of branches are summed up at the branching points, so it is higher at the channel base than in any branch. It is very unlikely that branches would start simultaneously and that the potential differences ΔU_t at their tips would be the same at any moment of time. Rather, the values of ΔU_t are distributed randomly. An abrupt decrease or even an entire cut-off of current in one of the branches does not at all mean that a similar thing has happened in another branch or in the base. Therefore, at least one of the 'main' branches will have a relatively high current and, hence, a greater probability to go up very high and even to reach the maximum leader height x_s than a single leader does. This event is stimulated by the decreasing voltage drop along the branching leader 'stem', where the total branch current has accumulated and where the field is low (in accordance with the current–voltage characteristic), especially if the stem is long and branching occurs at different heights.

Branching can send the leader up above the zero equipotential surface, where its sign reversal occurs. Imagine the situation, in which a branch, running up far from the charged cloud region, has reached the maximum height $x_s(r)$ and stopped. The channel plasma cannot decay immediately but persists for some time. During this time, another, luckier branch has approached the negative cloud bottom and even partly penetrated it. In this region, the cloud has potential $U_{0\,max}$, so that the branch portion that has entered it acquires the positive charge

$$C_1 R_c |U_0|_{max} \approx \frac{2\pi\varepsilon_0 R_c}{\ln(L/R)} \frac{|Q_c|}{4\pi\varepsilon_0 R_c} \approx \frac{|Q_c|}{2\ln(L/R)} \approx \delta|Q_c|$$

which may be as large as $\delta \approx 10\%$ of the negative cloud charge. Due to the partial compensation of the lower cloud charge, with the upper charge being constant, the zero equipotential surface will become lower by the length Δx, which is about the same percentage of $x_s - H$, so that $\Delta x \approx \delta(x_s - H)$. As a result, the upper portion of the first halted branch (from the tip down to the new zero equipotential surface) will be in a field directed downwards. The new external potential at the tip site, $U_0' \approx |dU_0/dx|_{x_s} \Delta x$, will become positive and the potential difference $\Delta U_t = U_t(L) - U_0'(L)$ will be negative. For the example given in the previous section with $x_s - H \approx 1.5$ km at $\delta \approx 0.1$, we have $\Delta x \approx 150$ m; from formula (4.10) with $r \ll H$, we have $|dU_0/dx|_{x_s} \approx 600$ V/cm, so that eventually $U_0'(L) \approx 9$ MV. Even if the branch penetrating the cloud charge misses its centre to enter a region with a potential several times lower than $U_{0\,max}$ (as a result, $U_0'(L)$ will be reduced as much), this will still be sufficient to revive the first leader branch.

Therefore, the halted leader has a chance to revive and move on up to the upper positive cloud charge but as a negative leader this time. The leader position at the point of the first stop is unstable. Even a slight perturbation, such as a decrease in the lower cloud charge (in the example presented, due to the penetration of another branch) may stimulate its further growth with the opposite sign. As the leader develops, it will penetrate into an increasingly higher field of the upper charge and become accelerated. Having passed the upper charge centre, $H + D$, it will be retarded and stop, for good this time, at a height $H_{max} > H + D$, where the potential of (4.10) will drop to a relatively low value of $\Delta U_{t\,min}$. The height H_{max} may be 10–20 km or higher if one accounts for the air density decrease. The currents flow in different directions in different portions of this leader. Above the equipotential surface, the current flows downwards, as in a negative leader. In the lower leader portion which serves as a stem for many positive branches, the current remains directed upwards. The observer, who registers the current at the earth, may not suspect the sign reversal occurring up in the clouds. The channel field is established in accordance with the current. It reverses in the upper channel portion, thereby reducing the total voltage drop across the branch that went far up and stimulating its further development.

4.2   Lightning excited by an isolated object

Like a high grounded body, a large object isolated from the earth can become
a source of lightning in a high electric field of a storm cloud. Discharges can
be induced by fields extending not only between the earth and a charged
cloud but also between oppositely charged clouds. A lightning discharge
can be excited by a large aircraft, rocket or spacecraft when it travels through
the troposphere, and this is a serious hazard to its flight. Therefore, this
phenomenon is of primary practical importance.

4.2.1   A binary leader

In contrast to an ascending leader starting from a high grounded body
(section 4.1), an isolated body produces two leaders, one going along the
external field vector and the other in the opposite direction. The physical
reason for the excitation of two leaders is the same. The external field induces
charge in the conductor, so that a large difference between its potential and
the external potential arises at the conductor end. If the body is extended
along the field, the electrical strength at its end increases abruptly. In contrast
to the situation with a grounded conductor, the opposite charge does not
flow down to the earth but accumulates at the other end, polarizing the
isolated body. A grounded conductor in an external field possesses the earth's
potential, while an isolated conductor acquires a potential corresponding to
an average external potential along its length. Large differences between the
body's and external potentials (of opposite signs) now arise at the ends of the
body, and both ends are capable of exciting leaders of the respective signs. A
long conductor absolutely symmetrical relative to its average cross section
transversal to the uniform field acquires potential U equal exactly to the
external field at the body's centre. The distribution of unlike charges in
each of its halves is identical to the charge distribution in a grounded
conductor of the same size and shape as the isolated conductor half.

The process described here can be easily reproduced in laboratory
conditions. Figure 4.5(*a*) shows streak pictures of leaders which have started
from a rod of 50 cm in length, suspended by thin plastic threads in a 3-m gap
in a uniform field. One can see all characteristic features of a positive leader
propagating continuously to the upper negative plane and those of a stepwise
negative leader travelling down towards a plane anode. Generally, leaders
arise at different moments of time because of the threshold field difference
for the excitation of positive and negative initial streamer flashes or due to
the difference in the curvature radii of the rod ends. The leaders may have
different velocities because the same voltage drop ΔU_t creates streamer
zones of different sizes at the positive and the negative ends. The instantaneous currents at the growing channel ends may also differ. But on average,
every leader transports the same charge, since the net charge remains to be

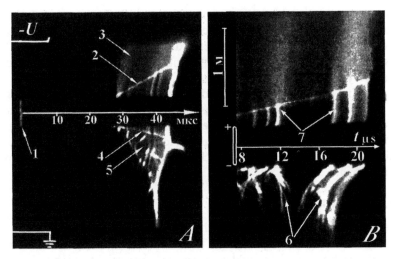

Figure 4.5. Streak photographs of leaders from the ends of a metallic rod placed in a uniform electric field: (*a*) general view; (*b*) fast streak photograph demonstrating the relationship between the positive and the negative leaders; (1) rod, (2), (3) tip and streamer zone of positive leader, (4), (5) tip and streamer zone of negative leader, (6), (7) negative and positive leader flashes.

zero in a system isolated from the voltage source. The discharges appear to be interrelated. Any fluctuation – say, a flash – of one leader appreciably activates the other: the space charge (e.g. positive) incorporated in front of the rod stimulates the accumulation of negative charge across the conductor, thereby enhancing the field at its negative end. The high-speed streak pictures in figure 4.5(*b*), resolving individual streamer flashes, show an activation of the positive leader channel following a negative leader flash.

The conditions for the start of leaders from a long isolated conducting body are the same as from a grounded conductor, and they are also defined by expression (4.4). But now, when estimating the threshold field E_0 from the value of $\Delta U_{i_{min}} = E_0 d$, one should keep in mind that d is a half length of an isolated leader. For a field capable of exciting a discharge from a conductor of length $2d$, we find

$$E_0 = \frac{1}{d^{3/5}} \left[\frac{b \ln(L/R)}{2\pi\varepsilon_0 a} \right]^{2/5}. \tag{4.11}$$

As in the illustration in section 4.1, we take the ratio of the channel length of a young leader to the equivalent charge cover radius to be $L/R \approx 10$. Then we have $E_0 = 440\,\text{V/cm}$ for an aircraft of length $2d = 70\,\text{m}$. This estimate describes the external field component along the aircraft axis. But an aircraft often flies at an angle to the field vector, so that the threshold external field may be several times higher. Fortunately, the lightning excitation threshold is not very low, otherwise airline companies would suffer tremendous losses from lightning damage. On the other hand, fields of this scale are not very rare: much higher fields were registered during airborne cloud surveys. For this reason, the problem of lightning protection in aviation is regarded as being very serious.

Having started from an isolated body, each leader develops as long as the external field permits. This process is basically the same as that discussed in section 4.1 for an ascending leader. Below, we shall consider the specific behaviour of two differently charged leaders developing simultaneously. This specificity becomes especially clear in a non-uniform field typical of a storm cloud.

4.2.2 Binary leader development

The principal features and quantitative characteristics of a binary leader can be understood from a simple model. The x-coordinate will be taken along the leaders. The leader paths should not necessarily be straight lines but they may have various bends, as is the case in reality. Denote the external field potential along the leader lines as $U_0(x)$ and their tip coordinates as x_1 and x_2. In figure 4.6 $U_0(x)$ corresponds to the field of a negatively charged cloud. The leaders were excited by a conducting body somewhere half way between the cloud and the earth. The x-axis is directed upwards, the leader with the subscript 1 travels downwards and the one with the subscript 2 upwards.

Let us neglect the voltage drop across the leader channels, ascribing the same potential U to the channels and the initiating body. The whole system now represents a single conductor. In the satisfactory approximation above, in which the capacitance per unit length C_1 at every moment of time was

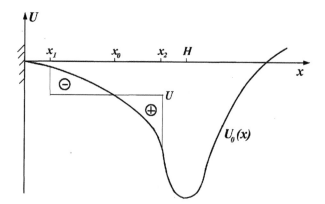

Figure 4.6. A schematic diagram of a binary leader channel in a cloud dipole field. x_1: descending leader tip coordinate; x_2: ascending leader tip coordinate; x_0: position of zero charge point.

assumed to be the same along the conductor length, the general condition for an uncharged conductor is

$$\int_{x_1}^{x_2} C_1[U - U_0(x)]\,dx = 0, \qquad U = \frac{1}{x_2 - x_1}\int_{x_1}^{x_2} U_0(x)\,dx. \qquad (4.12)$$

The condition of (4.12) defines the conductor potential U, which generally varies in time during the leader development (it is constant only if the electric field and the binary leader are symmetrical relative to the centre of the initiating body, which is also symmetrical). The conductor potential is equal to the average external potential along its length. The leader velocity can be calculated from (4.2). It was pointed out above that the available theory cannot provide a clear physical expression for the velocity of even a relatively simple, continuous positive leader, let alone a negative stepwise one. It is this circumstance which makes one resort to the empirical formula (4.2) derived from results of laboratory experiments with currents up to 100 A and justifiable, to some extent, for positive leaders. No similar measurements for negative leaders are available, and this is especially true of natural lightning observations. So, one has to rely on close experimental data on breakdown voltages in superlong gaps at the sign reversal of the high voltage electrode, as well as on the moderate velocity differences between positive and negative lightning leaders. The deviations of their measured values usually overlap. These facts provide good grounds for extending expression (4.2), as a first approximation, to negative leaders. In the latter case, we mean the average velocity neglecting the instantaneous effects of stepwise development. This approximation is the more so justifiable that the direct dependence of the leader velocity on the potential difference at the tip, ΔU_t, raise no doubt and that the variation of the factor a or of the power index in (4.2) cannot change the picture qualitatively.

Thus, with the account of the x-axis directions and velocities, as well as the signs of ΔU at the leader tips, the equations for the leader development can be written as

$$\frac{dx_1}{dt} = -a[U - U_0(x_1)]^{1/2}, \qquad \frac{dx_2}{dt} = a[U_0(x_2) - U]^{1/2}. \qquad (4.13)$$

Together with (4.12), expressions (4.13) describe the evolution of the two leaders starting from the body ends, whose coordinates x_{10} and x_{20} are given as the initial conditions for equations (4.12) and (4.13). The sign reversal point of the conductor, $x_0(t)$, defined by the equation $U(t) = U_0(x_0)$ is displaced, during the leader propagation, in accordance with the nature and degree of field non-uniformity along the channels. Having solved the equations, one can find the currents at the leader tips from (4.3) with $L = x_2 - x_1$. Generally, they differ quantitatively from one another and from the current in other channel cross sections, including the sign reversal at point x_0, through which the total charge flows during the polarization. The current $i(x_0) = i_0$ is defined as

$$i_0 = \frac{dQ_2}{dt} = -\frac{dQ_1}{dt}, \qquad Q_1 = C_1 \int_{x_1}^{x_0} [U - U_0(x)]\, dx. \qquad (4.14)$$

This current is used for changing the charge of the old leader portions, increasing or decreasing them as $\Delta U(x, t)$, and for supplying charge to its new portions. This leads to the current variation along the channels, which can be found by solving the problem.

For some simple distributions of $U_0(x)$, the division of equations (4.13) by one another

$$\frac{dx_2}{dx_1} = -\frac{[U_0(x_2) - U]^{1/2}}{[U - U_0(x)]^{1/2}} \qquad (4.15)$$

allows the functional relationship between x_2 and x_1 to be found from squaring, after which finding the final result $x_1(t)$, $x_2(t)$ reduces to squaring, too. This becomes possible if the cloud field is approximated by the point charge field $U_0(x) \sim |x|^{-1}$, and if a new variable $z = x_2/x_1$ is introduced. The resultant formulas allow an analytical treatment of some characteristic relationships. To avoid cumbersome derivations, we invite the reader to do this independently, while we, instead, shall present some numerical calculations for several variants.

The calculations prove to be quite simple in integrating the set of equations (4.13) and (4.14) as well as in the case of a more rigorous approach to the problem, when the charge distribution along the conductor length is found from an equation similar to (4.7). Figure 4.7 demonstrates the propagation of vertical leaders in the field of a cloud dipole (with the allowance for the earth's effect). The calculation was made using an equation similar to (4.7). The initiating vertical body is located between the lower negative charge of the dipole and the earth, being displaced horizontally by $r = 500$ m from the charge line. As the ascending leader moves up, its tip approaches the bottom charge centre and enters a region of an ever increasing field. The descending leader moves more slowly towards a weaker field. The external field potential approaches zero at the earth but increases rapidly near the charged cloud. As a result, the negative potential of the conductor made up of the leader channels, U, rises with time, with the sign reversal point x_0 going up closer to the cloud. At the initial moment of time, the potential is $U = -27$ MV and the point is at an altitude $x_0 = 1603$ m. When the ascending leader reaches the charged centre 17 ms later, we have the altitude $x_0 = 2040$ m and $U = -64$ MV. The absolute potential rise stimulates the descending leader, increasing its velocity by a factor of three during this time in spite of its propagation through an ever decreasing external field. The calculations made with (4.13) and (4.14) have yielded similar results.

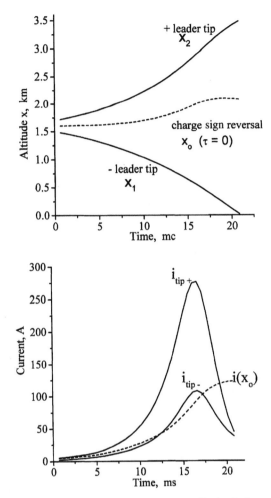

Figure 4.7. The propagation of leaders from a metallic body located between the cloud and the earth ($Q_c = -10\,\text{C}$, $H = D = 3\,\text{km}$, $r = 0.5\,\text{km}$).

Figure 4.8 illustrates the propagation of one of the leaders in a zero external field and refers to the situation when the descending leader has suddenly changed its direction for some reason at a certain height to follow the equipotential surface, i.e. along the zero field. The calculation was made with (4.13) and (4.14). Similar to the first variant, this situation exhibits a remarkable property of a binary leader. The leader developing along a rising field sustains the other leader, which has travelled in less favourable conditions, allowing it to move with a certain acceleration even in a zero field.

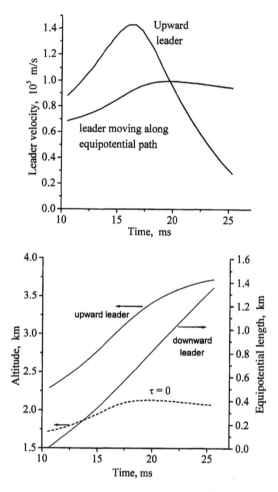

Figure 4.8. The development of a leader pair from a metallic body at 1.5 km above the earth in a cloud dipole field ($H = 3$ km, $D = 3$ km, $Q_c = -10$ C). At the moment of time $\sim$10 ms and 1 km altitude, the descending leader turned to follow an equipotential path: (top) leader velocities; (bottom) position of the zero charge point ($\tau = 0$), the altitude of the ascending leader tip and the length of the portion along the equipotential path.

Where an isolated conducting body may initiate a lightning discharge depends, to some extent, on a mere chance. A leader may start under a storm cloud, as in the illustrations just described, or inside a cloud at the height of the lower or upper charges or somewhere between them. These variants differ considerably in the polarization charge distribution along the conductor and, hence, in the leader propagation conditions. A situation may arise when the positive leader penetrates into the field of the negative lower cloud, thereby transporting a positive charge to the earth, which must apparently be 'attracted' by the negative cloud. (In the two situations above, it was the negative leader that travelled to the earth, 'naturally' extracting a negative charge from the cloud). This exotic situation arises when two leaders are excited from a body located somewhat above the negative charge centre. The negative leader then goes up to a positive cloud. The strong field created by the cloud dipole induces a large negative charge in the ascending leader, displacing the positive charge down. The average external potential between the two leader tips, $U < 0$, appears to be 'more positive' than the external potential of the lower tip, $U_0(x_1) < 0$, so that $\Delta U_{t_1} = U - U_0(x_1) > 0$. This is the reason why the descending leader is positive. With time, when the ascending tip comes closer to the positive charge, the potential U of the binary system does become slightly positive (3–3.5 MV), making up several percent of $|U_0|_{max}$. Then it persists as such, sustaining the descending leader travel to the earth. By the moment of contact with the earth, the positive charge is distributed along the channel in about the same way as in a grounded conductor in a negative cloud field, being mainly concentrated at the height of this charge. For this reason, the return stroke current, which only slightly contributes to the charge after the contact, is weak. The return stroke can be said to make no contribution to the charge redistribution, since the channel potential should be corrected only slightly (as compared with $|U_0|_{max} \approx 100\,\text{MV}$), by reducing it from 3 MV to zero. This reduction enhances, though only slightly, the ascending leader, which travels on until it stops high above the positive charge of the storm cloud.

When an isolated body initiates two oppositely directed leaders, it does not always happen that the descending positive leader reaches the earth. For the contact with the earth to take place, the potential U of the conductor made up of the two leaders must become positive at a certain moment. Otherwise, the descending leader will stop at the point x_{1s}, where the negative leader tip potential U_{t_1} will be by a small value of $\Delta U_{t_{min}}$ higher than the negative potential of the external field (assuming $\Delta U_{t_{min}} = 0$, when U is equal to $U_0(x_{1s})$). The condition for the average conductor potential to be positive at the moment of contact with the earth is described by the inequality

$$\int_0^{x_2} U_0(x)\,\mathrm{d}x > 0 \qquad (4.16)$$

which follows from (4.12). Here, $U_0(x)$ is the cloud dipole potential given by (4.10) with the allowance for its reflection from the earth, and x_2 is the altitude the ascending leader tip has reached by that moment. In principle, there are no reasons for this inequality to be violated, since the integral of (4.16) in the limit $x_2 = \infty$ is necessarily positive (and equal to $-2\ln(1 + D/H)Q_c/4\pi\varepsilon_0$ at $r \ll H$) for the negative lower cloud charge ($Q_c < 0$). This means that the descending positive leader has a chance to reach the earth – this only requires that the ascending leader should reach a sufficient altitude. If the horizontal channel displacement from the vertical line crossing the centres of the cloud charges, $r \ll H$, we obtain $x_2 > 2.47H$ from (4.16) and (4.10) at $D = H$. The ascending leader must cover a distance of about $0.5H$ above the upper positive charge centre.

4.3   The descending leader of the first lightning component

4.3.1   The origin in the clouds

Although lightning observers are familiar with the propagation of a descending (negative) stepwise leader, the conditions and the mechanisms of its origin are literally foggy. No one has ever observed the lightning start or its development in the clouds. Its origin cannot be totally reproduced in laboratory conditions, although negative stepwise leaders have been produced experimentally (section 2.7). But the conditions for their initiation by a high-voltage metallic electrode connected to the condenser of a impulse generator have little in common with what actually occurs in the clouds – a cloud is not a condenser winding and, of course, not a conductor. The negative cloud charge is scattered throughout the dielectric gas on small hydrometeors. It is very hard to perceive how the charges, fixed to particles with low mobility and dispersed in a huge volume, can come together to form a plasma channel in a matter of a few milliseconds.

In our terrestrial practice, we encounter events somewhat similar to the spark initiation in the clouds. Investigation of what has caused an explosion or a fire in industrial premises containing an abundance of electrostatic dust particles or droplets can provide evidence for a spark discharge arising in a medium with a dispersed charge. Lately, there have been reports of studies with gas jet generators ejecting into the atmosphere miniature electrically charged clouds [3, 4]. Sometimes, extended bright structures of about 10 cm in size were observed along a charged spray boundary; on some occasions, they were observed to form spark channels of about 1 m in length. Unfortunately, no measurements could be made of the field at the discharge start, so the fact of discharge excitation was only stated. Therefore, one can do nothing more than just make conjectures about the excitation mechanisms of lightning in the clouds and of sparks in laboratory sprays.

Speculations concerning these mechanisms (the only type of conclusion we can draw today) have to be arrived at via the process of elimination. A cloud medium cannot be considered as being conductive when we speak of current supply to the leader channel. Common charges are not transported directly to the leader, nor do they leave the cloud by themselves during the fast leader process. Therefore, the cloud charges play a different role – they are the source of electric field which ionizes the air molecules, producing the initial plasma, and then sustains the leader process. To fulfil the first task, the field somewhere in the charged region is to exceed the ionization threshold ($E_i \approx 20$–$25\,$kV/cm at the height of the cloud charge) or the cloud is to contain inclusions enhancing the field locally via the polarization charge. It seems that neither mechanism should be discarded entirely, although cloud probing rarely registered fields exceeding several kilovolts per centimetre. These results do not testify to the absence of higher fields, because most of the measurements concerned fields averaged over lengths of several dozens of metres. No measurements were made at the moment of lightning initiation, because the probability of a detector registering a field at the right place at the right moment is extremely low. On the other hand, the conditions necessary for the excitation of a leader process in a cloud are quite rare; otherwise, the number of lightning strikes per square kilometre of the earth's surface would greatly exceed 2–5 per storm season.

Let us estimate the volume to be occupied by a cloud charge capable of creating an ionization field. It was mentioned above that the field E_0 at the earth was often found to be $100\,$V/cm during thunderstorms. This value should not be considered to be the cloud dipole field, since the near-earth charge provided by microcoronas from various pointed objects attenuates the cloud field at the earth. A similar value is obtained from a small positive charge supposed to lie under the principal negative charge [5]. Taking, for estimations, the intrinsic dipole fields E_0 to be $200\,$V/cm and the heights of the lower (negative) and the upper (positive) charges to be $x = H = 3\,$km and $H + D = 6\,$km, respectively, we find, from (4.10), the dipole charges $Q_c = 13.3\,$C. These values will serve as guidelines in further numerical calculations. The charge Q_c can create field $E_i \approx 25\,$kV/cm at its boundary if it is distributed throughout a sphere of radius $R_c = 220\,$m. Measurements show that the charged region is, in reality, 2–3 times larger, but one should not discard the possibility of a short accidental charge concentration in a smaller volume due to the action of some flows in the clouds.

More probable is the situation when a macroscopically averaged maximum field of cloud charge is several times lower than E_i and local fields, enhanced to $E_i \approx 25\,$kV/cm, arise near polarized macroparticles. Note that the maximum field near a metallic ball polarized in an external field E is $E_{max} = 3E$. Similarly enhanced is the external field of a spherical water droplet, since water possesses a very high dielectric permittivity $\varepsilon = 80$ and $E_{max} = 3E\varepsilon/(2 + \varepsilon)$. Therefore, if charge Q_c is concentrated in a sphere of $\sqrt{3}$ times larger radius, $R_c = 380\,$m, the field three-fold enhanced by polarization can achieve the ionization threshold. Following the ionization onset, streamers may be produced around large droplets, giving rise to a possible leader, because streamers may be branched and extended in an average field of $\sim 10\,$kV/cm.

Leaving aside the mechanisms of ionizing fields and leader origin, because they are still poorly understood, we shall take for granted only the mere fact that a leader does occur. At its start, a descending leader is devoid of the possibility of taking the charge it needs away from the cloud.

Observations show that this charge is quite large: an average negative leader transports to the earth a charge $Q_L \approx -5\,C$ and, sometimes, it is as large as $-20\,C$ [1], a value close to the evaluations of Q_c for the storm cloud. But if the cloud charge remains 'intact', the only thing that can provide the charge balance is the ascending leader of opposite sign, which is to develop simultaneously with the descending leader. This idea was suggested in [6], which presented a qualitative distribution of the charge induced along a vertical conductor made up of two leaders prior to and following its contact with the earth. What happens is principally the same as in the excitation of two leaders by a conducting body isolated from the earth and is affected by an external field (section 4.2). This process is independent of the descending leader sign; therefore, one should not think that a negative cloud can produce only a negative leader while a positive cloud always produces a positive one. In any case, two oppositely charged leaders are produced simultaneously, and which of them will travel to the earth depends on the charge position in the cloud and on the leader starting point.

A binary leader is most likely to be initiated near the external boundary of the charged region, because the field there is highest. The field at the centre of an isolated charged sphere is zero. In the case of a uniform charge distribution throughout its volume, the field rises along the radius as $E \sim r$ but decreases from the outside as $e \sim r^{-2}$ with the maximum $E_{max} = Q_c/4\pi\varepsilon_0 R_c^2$ at the boundary. For a dipole configuration of real charges, the field does not practically vary across the boundary surface of the charged region. For the above values of D, H and R_c, the field at the upper point of the lower sphere is about 5% higher than at the lower point. The probability of a binary leader being initiated at either point is nearly the same. However, the final result of the binary leader development will differ radically, and this circumstance was essentially demonstrated in section 4.2. If both leaders are initiated at the bottom edge of the lower negative charge, the negative leader will go down and the positive one will go up. The negative leader has a real chance to reach the earth with a high negative potential equal to that of cloud charges averaged over the whole conductor length. The conductor is mostly in the region of high negative potential, nowhere entering the positive potential domain. The closing of this highly charged channel to the earth leads to the wave processes of charging and charge exchange (the return stroke) involving high current. The latter represents a real hazard. This is what happens in the case of a negative lightning. If a binary leader is initiated at the upper boundary of the lower charge, the positive leader goes down to the earth and the negative one goes up. A positive descending leader can never reach the earth unless it acquires a positive potential. For this, its ascending partner must necessarily go beyond the zero potential point, closer to the upper positive charge of the dipole. Owing to the compensation of positive and negative charges at various sites along the path, the average potential transported down to the earth is quite low. This actually cancels the return stroke current. Positive lightnings with very low currents of about 1 kA present no danger and are quite frequent. The number of their registrations by the observer increases with increasing sensitivity of the detectors used.

4.3.2 Negative leader development and potential transport

The stepwise propagation pattern has been believed by many to be the principal problem for a theoretical description of a negative leader [7]. However, it is of little importance to the leader evolution whether it develops continuously or by relatively short steps. The leader current and velocity are averaged over many steps. Averaged also is the channel energy balance, although the energy release at a distance of several step lengths from the tip has a well defined periodic pulse character. We shall discuss the stepwise effect in section 4.6, following the consideration of the return stroke, since this process is involved in every step as the main component.

The evolution of the descending channel of a binary leader is intimately related to that of its 'twin brother' – the ascending leader. (In this sense, the term 'Siamese twins' would be more appropriate.) A characteristic feature of the twins is the break-off of their potential, which varies but little along their highly conductive channels, from the external potential at the start. In this respect, a lightning leader differs considerably from a laboratory leader starting from an electrode connected to a high-voltage source. Being 'tied up' to the electrode, a laboratory leader with a well conducting channel carries the electrode potential, which may be close to the source emf. Generally, it is lower than the emf by the value of the voltage drop across the external circuit impedance when a discharge current is flowing through it. The underestimation of the principal difference between a laboratory spark initiated from a high-voltage electrode and a natural electrodeless lightning leads to erroneous attempts to derive from observations the voltage drop value across the leader channel. The reasoning is usually as follows. The potential U_1 transported by a lightning leader to the earth can be estimated from the return stroke current and the characteristic channel impedance (section 4.4). The cloud potential U_{0R} can also be estimated (see formula (4.17) below). The leader channel base has the same potential – as if the cloud were an electrode. Therefore, the voltage drop across the leader length, from the cloud to the earth, is $\Delta U_c = |U_{0R}| - |U_1|$, and the average field in the channel is expressed as $E_c = \Delta U_c / L$, where L is the leader length ($L \approx H$, or 30–50% greater with the allowance for the path bendings). Such estimations lead to incredibly large values of $\Delta U_c \approx 100\,\mathrm{MV}$ and $E_c \approx 1\,\mathrm{kV/cm}$. A mature leader channel with $i \approx 100\,\mathrm{A}$ current cannot have such high fields. Its state is very much like that of the quasi-equilibrium hot plasma in an arc, which has a field 1–2 orders of magnitude lower. This follows from theory and from evaluations of fields in superlong laboratory sparks.

The attempts to solve the 'inverse' problem using the expression $|U_{0R}| = \Delta U_c + |U_1|$ to calculate the storm cloud potential have also failed. When one includes in this expression the generally correct values of arc field E_c, one gets unjustifiably low cloud potentials U_{0R} inconsistent with atmospheric probing measurements and other calculations.

The methodological error of both approaches is due to the rigid relation of the base potential of a descending leader to the external field potential at the leader start. In actual reality, the channel potential undergoes a considerable time evolution, being determined by the polarization charge distribution along the binary leader length. When the descending leader approaches the earth, its base potential may differ significantly from the potential created by the cloud charge at the start site at the moment of start.

A simple calculation of the leader development can be made from equations (4.13) and (4.12), but a more rigorous solution can be obtained from equations (4.13) and (4.7), which were used for that purpose in section 4.2. One should also bear in mind that if both leaders start from the boundary of a charged cloud region, at least one of them will enter the charged volume and may even pass through its centre. Then, we have to discard the point model of a cloud dipole and make the next approximation by assuming that charge Q_c is distributed uniformly with the density $3Q_c/4\pi R_c^3$ in a sphere of radius R_c. Inside the sphere, the potential of its intrinsic charge is radially symmetrical and is equal at point r to

$$U_{0s}(r) = U_{0m}\left(1 - \frac{r^2}{3R_c^3}\right), \qquad U_{0m} = \frac{3Q_c}{8\pi\varepsilon_0 R_c}, \qquad r \leqslant R_c. \qquad (4.17)$$

The potentials from the upper dipole charge and from charges reflected by the earth can be found as from point charges. They do not contribute much to U_{0s}. For example, for the centre of a negative sphere with $Q_c = -13.3\,\mathrm{C}$ and $R_c = 500\,\mathrm{m}$, we have $U_{0m} = -360\,\mathrm{MV}$ and at the boundary $U_{0R} = \frac{2}{3}U_{0m} = -240\,\mathrm{MV}$. With all other charges taken into account, we get $U_{0H} = -196\,\mathrm{MV}$ for the bottom edge of the lower sphere at $H = D = 3\,\mathrm{km}$.

Figure 4.9 presents the results of this calculation including those for the charge distribution along the conductor length from an equation similar to (4.7). We have evaluated the development of both leaders along the dipole axis, following the start from the bottom edge of the lower negative

Figure 4.9. (*Opposite*) The model of a descending leader from the lower boundary of the negative dipole charge ($Q_c = -12.5\mathrm{C}$, $H = 3\,\mathrm{km}$, $D = 3\,\mathrm{km}$, $R_c = 0.5\,\mathrm{km}$). Vertical channels have no branches: (top) tip positions of the negative descending leader, x_1, and its positive ascending partner, x_2, with the points of zero potential differences, x_0; (centre) charge distribution along the leader channel; (bottom) potential and velocity of the descending leader.

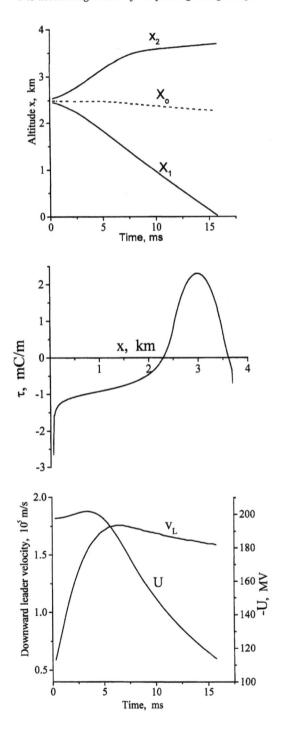

sphere. The dipole potential from (4.10) was used as $U_0(x)$, except for the length in the charged region, where expression (4.17) was employed with $r = |x - H|$. The descending negative leader of a binary system is accelerated quickly after the start. Having covered about 500 m, it travels farther to the earth with a slightly decreasing velocity $v_L \approx (1.6\text{--}1.7) \times 10^5$ m/s, a value close to observations. The leader strikes the earth in 16 ms. By that time, the ascending leader has reached the height $x_2 \approx 3.6$ km. This is far even from the zero potential point located at $x \approx 4.5$ km, let alone from the upper positive charge located at an altitude of 6 km. The descending leader, which started from a site with the local potential -185 MV, transports to the earth nearly half of this value, $U_1 \approx -105$ MV, in spite of the initial assumption of the zero voltage drop across the channel assumed to be a perfect conductor.

The reader should not feel discouraged by the large calculated value of U_1, which is more appropriate to record strong lightnings rather than to a common lightning discharge, especially considering that the cloud parameters taken for the calculation were quite moderate. It will be demonstrated in section 4.3.3 that leader branching, which is a rule rather than an exception, reduces considerably the potential transported down to the earth. It is quite likely, however, that lightnings of record intensities are produced in ordinary clouds rather than in those having a record high charge, but only if the descending leader does not branch (or does so slightly).

The above calculation for an ideal situation with unbranched leaders is interesting and useful for two reasons. First, one should understand the physics of a simple observable phenomenon before one turns to its complex modifications. The other reason is, probably, more important. Practical lightning protection requires the knowledge of both typical average lightning parameters and their record high values. It is the latter that become more important in designing prospective measures for especially valuable constructions and objects. As was pointed out above, the case of an unbranched leader just discussed is likely to be one of the rare but most hazardous phenomena.

The potential U_1 transported by a lightning leader to the earth is an important parameter for practical lightning protection. The return stroke current (section 4.4), the most destructive force of lightning, is proportional to U_1. The nature of U_1 becomes clear from the above conception of descending leader development in a binary leader process. Ideally, potential U_1 is that of a perfect conductor, made up of two leader channels, at the moment of its contact with the earth. But the ascending and descending leaders develop differently, because their paths cross regions possessing different distributions of cloud potentials $U_0(x)$. The descending leader travels nearly without retardation because the potential difference at its tip, $\Delta U_{t_1} = U - U_0(x_1)$, remains almost constant (a decrease in $|U|$ is largely compensated by $|U_0(x)|$ decreasing towards the earth). The ascending

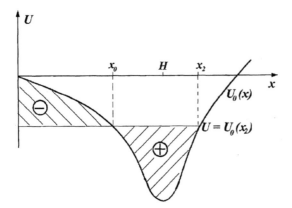

Figure 4.10. Estimation of the potential transported by a negative leader to the earth.

positive leader moves 'against' the field and soon enters a region of a rapidly rising external potential; as a result, $\Delta U_{t_2} = U - U_0(x_2)$ becomes relatively low soon after the start. This leads to a lower velocity of the ascending leader, which now goes up 'unwillingly', being affected by its more active twin which moves faster, pumping its charge into it. For this reason, just before the descending leader contacts the earth, the total potential of the system nearly coincides with the external field potential $U_0(x_2)$ at the site of the ascending leader tip ($\Delta U_{t_2} = U - U_0(x_2) \ll |U|$).

This circumstance makes it possible to determine the transported potential $U_1 = U$ just from the condition $U = U_0(x_2)$ at $x_1 = 0$. The condition has a clear geometrical interpretation (figure 4.10). The shaded regions between the external potential curve $U_0(x)$ and the horizontal line intercepting it must be identical on both sides of the left-hand interception point (the point of the conductor sign reversal, x_0). This results from the net polarization charges of both signs being identical; they are proportional to the shaded regions (see formula (4.12)). This approach can be used to find U_1 in different charge distribution models and for different horizontal deviations of the vertical leader path from the dipole axis. In a simple case when both leaders propagate along the dipole axis, formulae (4.12) and (4.10) with $D = H$ and $r = 0$, together with expression (4.17), yield a dimensionless equality for finding point x_2 and then U_1:

$$\frac{\xi_2}{\xi_2 - 1} = \ln \frac{4e^{8/3}(\xi_2 - 1)}{\kappa^2(\xi_2 + 1)(4 - \xi_2^2)}, \qquad \xi_2 = \frac{x_2}{H}, \qquad \kappa = \frac{R_c}{H}, \qquad U_1 = U_0(x_2).$$

$$(4.18)$$

For the variant shown in figure 4.9 with $\kappa = \frac{1}{6}$, expressions (4.18) give $\xi_2 = 1.27$ and $U/U_{0R} = 0.63$, in a fairly good agreement with the calculations of the leader evolution (note that $U_{0R} \approx U_{0H}$ is the external field potential at the bottom edge of the lower cloud charge that has triggered both leaders).

A negative descending leader can start from any point on the lower hemisphere of the bottom negative charge of the cloud. The location of this point is quite likely to be a matter of chance, since the field across the surface is nearly uniform. Depending on the location of the starting point, the ascending twin crosses the charged region along chords of different lengths, and this, along with the other factors, affects the potential transported to the earth. The maximum potential U_1 at the moment of contact with the earth is characteristic of a leader that has started from the lowermost point of the charged sphere, when the paths of both twins cross the regions of maximum potentials and the ascending leader path in it is longest. The value of $U_{1\,max}$ is about 60% of the external potential U_{0R} at the start and 40% of the maximum U_{0m} value at the centre of the negative cloud charge. But even if the descending leader is initiated near a lateral point of the hemisphere located at a maximum distance from the dipole axis, it transports a considerable potential found from calculations to be $0.65U_{1\,max} \approx 0.4U_{0R}$. Therefore, an unbranched negative leader transports to the earth a high potential, $(0.6–0.4)U_{0R}$, no matter where it has started from the lower hemisphere.

4.3.3   The branching effect

Measurements of return stroke current show that a descending negative lightning rarely transports to the earth a potential as high as 100 MB ($I_M = U_1/Z$, where Z is the channel impedance; see section 4.4.2). The reason for this is not the supposedly lower potentials of most clouds. The value of $\sim100\,MV$ is characteristic of cloud charges moderate in size and density. The reason is most likely to be the leader branching, since an unbranched leader is an exception rather than the rule. Numerous downward branches of a descending leader can be well seen in photographs. Although ascending leaders are screened by the clouds, their branching can be registered by radio-engineering instruments [8–10]. However, the potential U_1 is affected by the branches of a descending negative leader rather than of its positive ascending twin brother.

Let us make sure first that the branches of an ascending leader do not change the situation much. In the limit of a very intensive branching, the negative cloud bottom, pierced by numerous conductive channels, is electrostatically identical to a continuous conductive sphere of capacitance $C_c = 4\pi\varepsilon_0 R_c$, whose charge has been pushed out on to the surface. The net charge of a system made up of a sphere and a negative leader attached to it remains equal to the initial charge $-Q_c$. With the neglect of the voltage drop across the descending channel, the binary system possesses the same potential U along its length. At the moment of contact with the earth, when the leader capacitance C_L corresponds to the length $L \approx H$, this potential is

$$U \approx \frac{Q_c}{C_c + C_L} = \frac{C_c U_{0R}}{C_c + C_L} \approx \frac{U_{0R}}{1 + L/2R_c \ln(L/R)} \qquad (4.19)$$

where U_{0R} would be the boundary potential of a charged sphere, were it isolated (the small capacitance gain due to charges induced in the earth are ignored). For $L/R \approx 100$ and $H/R_c = 6$, as in the previous numerical illustration, the potential $U_1 \approx 0.6U_{0R}$ is nearly the same as that transported to the earth in the absence of ascending leader branching. The potential of the cloud–leader system drops because of the outflow of some of the cloud charge to the new capacitance of the descending leader just produced. A similar effect has been observed in long laboratory sparks. The capacitance of an extremely long spark is often only one order of magnitude smaller than the output capacitance of a impulse voltage generator, connected directly to a gap without a large damping resistor. The charge inflow into the leader is quite appreciable and reveals itself as a voltage drop across the gap.

A branched descending leader possesses a larger capacitance than an unbranched one; it takes away a higher charge from the cloud and decreases the potential more. To estimate this effect, let us represent a branched leader as a bunch of n identical conductors of radius R and length L, spaced at distance d ($L > d \gg R$). Supplied by the same power source, they possess the same potential U and linear charge τ. The potential at the centre of any of these conductors is found by summing the potentials of all charges of all conductors, including the intrinsic potential. Integration with the neglect of the small effect of the earth yields

$$U = \frac{\tau}{2\pi\varepsilon_0}\left[\ln\frac{L}{R} + (n-1)\ln\frac{L}{d}\right] = \frac{\tau}{2\pi\varepsilon_0}\ln\left[\left(\frac{L}{R}\right)\left(\frac{L}{d}\right)^{n-1}\right]$$

The total capacitance of the n conductors, $C_{tn} = n\tau L/U$, is larger than that of a single isolated conductor, but this gain is less than n-fold:

$$\frac{C_{tn}}{C_{t1}} = \frac{n\ln(L/R)}{\ln[(L/R)(L/d)^{n-1}]}$$

The reduction in the potential transported to the earth roughly follows the distribution of the cloud charge Q_c between the capacitances of the charged cloud cell, C_c, and of the leader, C_{tn}, described by the first equality of (4.19). For $n = 10$ branches separated at distances $d = L/3$ and $L/R \approx 100$ derived from photographs, the capacitance is $C_{t10} \approx 3.2C_{t1}$. This well-branched leader will transport to the earth potential $U_1 \approx 0.3U_{0R}$. In view of the real length of a leader (especially, a well-branched one) which is about 1.5 times longer than the charge height H, i.e. $L \approx 1.5H$, the potential decreases to $0.2U_{0R}$, in good agreement with the data on negative lightning currents. The number of branches and their lengths vary randomly with the lightning. The potential U_1 determining the return stroke current vary together with them. This variation is likely to produce a wide range of current amplitudes. The variation of storm cloud charges seems to be less significant.

There is another source of reduction in the potential transported by a negative leader to the earth. In more complex models than the vertical dipole variant, the reduction is due to a low positive charge assumed to be present at the very bottom of a cloud [5]. Calculations show that if a positive $4C$ charge of 0.25 km radius with the centre at 2 km above the earth is added to the above dipole with $Q_c = \pm 13.3C$, $R_c = 0.5$ km and $H = D = 3$ km, the negative leader initiated from the bottom edge of the negative charge will transport half of the potential to the earth.

4.3.4 Specificity of a descending positive leader

Positive leaders do not occur very frequently. Statistics indicate that in Europe their number is 10 times smaller than that of negative ones. But it is quite likely that their actual number is larger than the number of their registrations. It was pointed out in section 4.3.1 that a descending positive leader does not carry high potential to the earth and that its return stroke current is low. For this reason, the electromagnetic field of a positive lightning discharge can be detected at a much shorter distance than that of a negative discharge and, probably, not all of them are registered.

If the bottom charge of a cloud dipole is negative, a positive descending leader may start either from the upper negative hemisphere or from the bottom hemisphere of the upper positive charge. The leader will reach the earth, transporting to it a positive potential, provided the condition of (4.16) is met. With a small deflection of the leader vertical axis from the dipole axis ($r \ll H$), the transported potential found from (4.12) and (4.10) with $x = 0$ will be

$$U = -\frac{Q_c}{4\pi\varepsilon_0 x_2}\ln\frac{H_2^2(x_2 + H_1)(x_2 - H_2)}{H_1^2(x_2 + H_2)(x_2 - H_1)}, \qquad Q_c < 0 \qquad (4.20)$$

where H_1 and H_2 are the heights of the bottom and top charge centres and x_2 is the ascending leader height at the moment the descending leader contacts the earth. We mentioned at the end of section 4.2.2 that an ascending leader must go up at least to $x_2 = 2.47H_1$ at $H_2 = 2H_1$; then we have $U = 0$. At $x_2 \approx 4H_1$, the function $U(x_2)$ crosses the smooth maximum, $U_{max} \approx -Q_c/20\pi\varepsilon_0 H_1 \approx 8$ MV, if $Q_c = -13.3C$ and $H_1 = 3$ km, as in the previous examples. Even the maximum potential transported to the earth is small. This means that the return stroke current of a descending positive leader travelling along the dipole axis will be low. The potential and the current will be still lower, with the real voltage drop U_c across a channel of total length $4H \approx 10$ km taken into account. Even for the channel field $E \approx 10$ V/cm, the value of $U_c \approx 10$ MV is comparable with U_{max}. This lightning is so weak that it has little chance of being registered and included in the statistics.

Vertical channels demonstrate maximum positive potentials transported to the earth. They go through more or less identical regions of negative (at the bottom) and positive (at the top) external potentials, and the respective contributions to the integral of (4.12) are mutually compensated. Positive lightnings, however, can possess very high currents. With the foregoing taken into account, one can suggest at least two reasons for this. One is a favourable random deviation of the channel path from the vertical line. Suppose the ascending leader of a binary system, starting from the upper positive charge point closest to the earth, $x_0 = H_2 - R_c$, moves up vertically, while the other leader, having descended to the zero potential point between the charges, turns aside and goes along the zero equipotential line. After it has deviated for a large distance r from the dipole axis, it turns down vertically to contact the earth this time. In this case, the descending leader misses the region of high negative potential, and positive contribution to the integral of (4.12) remains uncompensated. Calculations with formulae (4.12), (4.10) and (4.17) made at $H_1 = 3$ km, $H_2 = 6$ km, $R_c = 0.5$ km, and $r = 1$ km show that the descending leader will transport to the earth a potential 4.3 times greater than that to be transported along the dipole axis.

Another principal possibility is the deviation of the dipole axis itself from the vertical line, with the vertical leader path preserved. The centres of the top and bottom charges can be shifted from the same vertical line because of the difference in the wind forces at different heights. Then the leader that has started up vertically from the top charged region passes through the region of high positive potential, while its twin, descending vertically, will appear to be shifted aside relative to the bottom charge and go through the region of low negative potential. The effect will be the same as in the first case. Quantitatively, it may even appear to be stronger, since the length and capacitance of the descending leader are smaller due to the lack of an extended path along the zero potential line.

4.3.5 A counterleader

The descending lightning leader does not reach the earth or a grounded body, because it is captured by the ascending leader developing in the electric field of cloud and earth-reflected charges. This field is enhanced by the charge of the descending leader approaching the earth. This can also happen in laboratory conditions, especially if the descending leader is negative. Then the counterleader is positive and requires a lower field for its development. Streak pictures of laboratory sparks clearly show the counterleader start

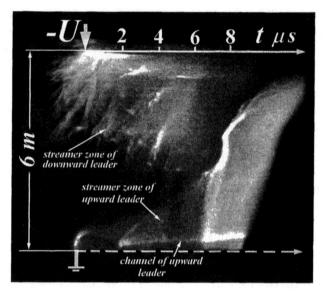

Figure 4.11. A streak photograph of a long spark with a counterleader coming from a grounded electrode.

and motion towards the descending leader (figure 4.11). The altitude at which their encounter occurs depends on the descending leader sign and charge. The length of the counterleader at the moment of their contact is important for lightning protection practice, because it defines the number of strikes at bodies of different heights and, to some extent, the current rise parameters of the return stroke from the affected body.

Let us estimate the altitude z, which the descending leader tip is to reach to be able to create a field at the earth high enough to produce a viable counterleader. The latter does not differ from any other ascending leader, and its development from a body of height d requires that the near-terrestrial field should exceed the value of E_0 from formula (4.11). For the height $d = 30\,\mathrm{m}$ characteristic of industrial premises, the field must be $E_0 \approx 480\,\mathrm{V/cm}$. If the cloud field is $\sim 100\,\mathrm{V/cm}$, the field $\Delta E = 380\,\mathrm{V/cm}$ must be created by the descending leader with charge. The main contribution to the near-terrestrial field is made by the charge concentrated at the leader channel bottom. Therefore, the calculation of the field ΔE under a very long vertical conductor should utilize the constant value of τ averaged over this bottom of length $\sim z$, rather than the linear density of the non-uniform charge $\tau(x)$. With the charge reflected by the earth, we have

$$\Delta E(z) \approx \frac{\tau}{2\pi\varepsilon_0} \int_z^\infty \frac{\mathrm{d}x}{x^2} = \frac{\tau}{2\pi\varepsilon_0 z} \approx \frac{U - U_0}{z \ln(L/R)} \approx \frac{U}{z \ln(H/R)} \tag{4.21}$$

where U is the channel potential and $L \approx H$ is its length, which is about the cloud height at the tip height $z \ll H$. Here, we have used the conventional expression $\tau = C_1(U - U_0)$ with average linear capacitance and accounted for the near-terrestrial potential of the cloud, $|U_0| \ll |U|$. For an unbranched descending leader carrying high potential $U \approx 50$ MV, we obtain $z \approx 260$ m at $\ln(H/R) \approx 5$.

The counterleader arises at the last stage of the descending leader development, i.e., near the earth. Its velocity is not high and is equal to $v_{L_1} \approx 2 \times 10^4$ m/s from the first formula of (4.5), because the potential difference on the leader tip is quite low, $\Delta U \approx E_0 d \approx 1.5$ MV. The descending leader has an order of magnitude higher velocity. For this reason, the counterleader acquires the length $L_1 \approx (v_{L_1}/v_L)z \approx 25$ m by the moment of encounter. This is a large value, since the length L_1 is summed with the body's height d, so that the total height of the grounded conductor becomes nearly doubled. This affects the frequency of the body's damage by lightning strikes.

It follows from formulae (4.21) and (4.11) that the height z, to which the leader descends before it can initiate a counterleader, is greater for higher premises, from which the counterleader starts, $z \sim d^{3/5}$, although this dependence is not very stringent. It is important that as the altitude of a body and z become greater, the counterleader has more time for its acceleration and can acquire a longer length. It is important for applications that it is not only the length L_1 which increases but also the L_1/d ratio.

The simple estimation obtained from (4.21) and (4.11) can be refined by accounting for the $\tau(x)$ non-uniformity in the integral of (4.21) arising from the proportionality $\tau \sim U - U_0(x)$ and by rejecting the approximation of constant linear capacitance C_1. In the latter case, $\tau(x)$ should be found from equation (4.7). Calculations show that the two corrections are rather small, so the estimations above can be considered to be satisfactory.

4.4 Return stroke

4.4.1 The basic mechanism

A return stroke, or the process of lightning channel discharging, begins at the moment the cloud–earth gap is closed by a descending leader. After the contact with the earth or a grounded body, the leader channel (it will be taken to be negative for definiteness) must acquire zero potential, since the earth's capacitance is 'infinite'. Zero potential is also acquired by the ascending leader, which is a continuation of its descending twin brother. The grounding of the leader channel carrying a high potential leads to a dramatic charge redistribution along its length. The initial channel distribution prior to the return stroke was $\tau_0 = C_1[U_i - U_0(x)]$. Here and below, the potential transported to the earth, which acts as the initial potential for the return

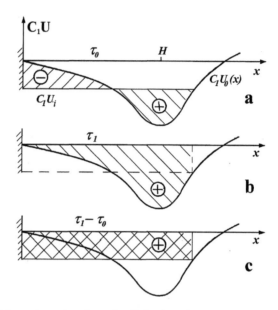

Figure 4.12. Schematic recharging of a lightning channel after the contact of the descending leader with the earth. Shaded regions, charge; (*a*) moment of the leader contact with the earth; (*b*) the return stroke reaching the upper channel end; (*c*) charge change.

stroke, will be denoted as U_i. As before, it will be taken to be constant along both leader lengths, and the voltage drop across the channel will be ignored as an insignificant parameter. We shall assume that the channel is characterized by linear capacitance C_1, which does not vary along its length or in time during the return stroke process. After the whole channel has acquired zero potential, $U = 0$, the linear charge becomes equal to $\tau_1 = -C_1 U_0(x)$. The channel portion belonging to the negative descending leader does not just lose its negative charge but it acquires a positive charge ($U_0 < 0$, $\tau_0 < 0$, $\tau_1 > 0$). Not only does it become discharged but it is also recharged. The twin positive channel high in the cloud acquires a larger positive charge (figure 4.12). The linear charge variation for the return stroke lifetime is $\Delta\tau = \tau_1 - \tau_0 = -C_0 U_i$. At $U_i(x) = \text{const}$, the charge variation is constant along the channel length and has such a value as if a long conductor (a long line) pre-charged to the voltage U_i becomes completely discharged (as if it were $\tau_0 = C_1 U_i$ to become $\tau_1 = 0$).

It has been emphasized that the leader charge is concentrated in its cover. The charge in a non-conducting cover changes due to the charge incorporation from the conductive channel, owing to the streamer corona excitation at the channel surface. This is an exceptionally complicated process, whose rate can be found only from an adequate theoretical treatment.

For this reason, the assumption of capacitance C_1 being constant, which implies a zero-inertia charge variation in the cover with varying channel potential, is quite problematic. But if we discard this seemingly essential assumption, nothing will change qualitatively or even quantitatively. Indeed, suppose the cover charge does not change at all during the time the whole channel acquires zero potential. This is equivalent to the assumption that the channel capacitance is determined, during the return stroke process, by the conductor radius r_c rather than by the cover radius R. Because of the logarithmic dependence of linear capacitance (2.8) on the radius, it decreases to the value C_1' equal to about a half of C_1; for example, we obtain $C_1 \approx 10\,\text{pF/m}$ and $C_1' \approx 4.4\,\text{pF/m}$ at $l = 4000\,\text{m}$, $R = 16\,\text{m}$ and $r_c = 1.5\,\text{cm}$. Then the charge variation during the stroke

$$\Delta\tau = \tau_1 - \tau_0 = [(C_1 - C_1')(U_i - U_0) + C_1'(O - U_0)] - [C_1(U_i - U_0)]$$
$$= -C_1' U_i \qquad (4.22)$$

remains the same in order of magnitude. Consequently, when considering fundamental stroke mechanisms, one can take $C_1' \approx C_1$ and assume the equivalent line to be charged uniformly.

Measurements made at the earth show that a descending leader is discharged with a very high current. For negative lightnings, the current impulse of a return stroke with an amplitude $I_M \sim 10\text{--}100\,\text{kA}$ lasts for 50–100 µs on the 0.5 level. A short bright tip of the return channel well seen in streak photographs runs up for approximately the same time. Its velocity $v_r \approx (0.1\text{--}0.5)c$ is only a few times less than light velocity c. It would be natural to interpret this fact as the propagation of a discharge wave along the channel; this wave is characterized by a decreasing potential and rising current. Due to an intensive energy release, the channel portion close to the wave front, where the potential drops from U_i and a high current is produced, is heated to a high temperature (from 30 000 to 35 000 K, as shown by measurements). This is why the wave front is so bright. The channel behind it is cooled due to expansion and radiation losses, becoming less bright. A return stroke has much in common with the discharge of a common metallic conductor in the form of a long line. The line discharge also has a wave nature, and this process was taken to be a model discharge in shaping the ideas concerning the return lightning stroke.

A lightning channel is discharged much faster than it was charged during its development with the leader velocity $v_L \approx (10^{-3}\text{--}10^{-2})v_r$. But the variations in potential and linear charge during the charging and the discharge are expressed as values of the same order of magnitude: $\tau_0 \sim \Delta\tau$. In agreement with the velocity, the channel is discharged with current $I_M \approx \Delta\tau v_r$ by a factor of $v_r/v_L \approx 10^2\text{--}10^3$ higher than the leader current $i_L \approx \tau_0 v_L \approx 100\,\text{A}$. The linear channel resistance R_1 decreases approximately as much during the leader–stroke transition. This decrease is due to the channel heating by high current. As a result, the plasma conductivity increases and the channel expands, making the conductor cross section larger. In this respect, a lightning discharge certainly differs from a discharge of a common conductive line, whose resistance remains constant (if the skin-effect is ignored). Since resistance is a plasma characteristic, its decrease can be found straightforwardly only if the physical processes occurring in the channel are taken into account. (This situation will be analysed in section 4.4.3.) But this conclusion can be arrived at indirectly from general energy considerations. Over the return stroke lifetime, $t_r \sim H/v_r$, where H is the channel length, the energy dissipated in the channel must be approximately equal to the initial electrical energy $C_1 U_i^2/2$ per unit length:

$$C_1 U_i^2/2 \sim I_M^2 R_1 t_r \sim I_M^2 R_1 H/v_r \sim \Delta \tau H I_M R_1. \qquad (4.23)$$

About as much energy was dissipated in the leader when the capacitance C_1 was charged. If the leader develops in the optimal mode (see section 2.6), to which a natural lightning process is, probably, very close, because Nature usually takes optimal decisions, the voltage drop across the channel is comparable with the excess of the leader tip potential over the external potential. Therefore, the resistances of the channel and the streamer zone are comparable, because the same current flows through them. Therefore, the unit length of the leader dissipates the same energy $C_1 U_i^2/2$ (in order of magnitude), expressed by the leader parameters i_1, v_L and R_{1L} similar to (4.23). This yields $R_1 I_M \sim R_{1L} i_L$, i.e., $R_1/R_{1L} \sim 10^{-2}$–10^{-3}. It is also found that the average electric field in the leader channel and behind the discharge wave in the return stroke, $E_c \approx R_1 I_M \approx R_{1L} i_L$, have the same order of magnitude. This is consistent with the conclusion to be made from a straightforward analysis of the established states in both channels. The situation there is similar to that in a steady state arc. But the channel field E_c in a high current arc does vary but slightly with the current [11].

It follows from the foregoing that if a leader has $i_L \approx 100\,\mathrm{A}$, $E_c \approx 10\,\mathrm{V}/$ cm and $R_{0L} \approx 0.1\,\Omega/\mathrm{cm}$, the return stroke must have $R_0 \approx 10^{-3}$–$10^{-4}\,\Omega/\mathrm{cm}$ in the steady state behind the wave front; the total resistance of a channel of several kilometres in length appears to be $10^2\,\Omega$. This value is comparable with the wave resistance of a long perfectly conducting line in air, Z, whereas the total ohmic resistance of a leader of the same length is two orders of magnitude larger than Z. The ratio of the ohmic resistance of the line portion behind the wave to the wave resistance indicates the degree of the wave attenuation during its travel along the line (section 4.4.2). If the channel resistance were constant and remained on the leader level, the lightning channel discharge wave would attenuate, being unable to cover a considerable channel length. The current through the point of the channel closing on the earth would also attenuate too quickly. Experiments, however, point to the contrary: the visible bright tip has a well-defined front, and a high current is registered at the earth during the whole period of the tip elevation. The transformation of the leader channel during the wave travel decreases its linear resistance considerably, determining the whole return stroke process.

4.4.2 Conclusions from explicit solutions to long line equations

Long line equations with the allowance for the main factor – variation in linear resistance – can be solved only numerically (section 4.4.4). However, the nature of the process and its essential physical characteristics can be understood from the analysis of well-known analytical solutions for simple situations. Their comparison with lightning observations indicate the important points for the formulation and solution of the real problem.

In the absence of transverse charge leakage due to imperfect insulation, a long line is described by the equations

$$-\frac{\partial U}{\partial x} = L_1\frac{\partial i}{\partial t} + R_1 i, \qquad -\frac{\partial i}{\partial x} = C_1\frac{\partial U}{\partial t}. \qquad (4.24)$$

They generalize equations (2.12) by accounting for inductance. The inductance per unit conductor length, L_1, as well as its capacitance C_1, can be assumed to be approximately constant. For an isolated conductor of radius r_c and length $H \gg r_c$, it is

$$L_1 \approx \frac{\mu_0}{2\pi}\ln\frac{H}{r_c} = 0.2\ln\frac{H}{r_c}\,\mu\text{H/m}. \qquad (4.25)$$

Here, the channel length H is about equal to the height of an ascending leader tip. For a perfectly conducting line with $R_1 = 0$, equations (4.24) are re-differentiated to produce a simple wave equation:

$$\frac{\partial^2 U}{\partial x^2} - \frac{1}{v^2}\frac{\partial^2 U}{\partial t^2} = 0, \qquad v = (L_1 C_1)^{-1/2}. \qquad (4.26)$$

If the line is charged to voltage U_i and short-circuited on the earth by its base $x = 0$ at the moment of time $t = 0$, a rectangular wave of complete voltage elimination (from U_i to 0) and an unattenuated current wave of the same shape will propagate with velocity v from the grounding point (figure 4.13):

$$i = -\frac{U_i}{Z}, \qquad Z = \left(\frac{L_1}{C_1}\right)^{1/2}, \qquad \Delta\tau = -C_1 U_i, \qquad x \leqslant vt. \qquad (4.27)$$

While the wave propagates along the line, a detector mounted at its beginning will register direct current. If voltage U_i is low and there is no charge cover around the conductor, the capacitance and inductance are characterized by the same radius r_c in the logarithms of (2.8) and (4.26). In this case, we have

$$v = (\mu_0\varepsilon_0)^{-1/2} = c, \qquad Z = \frac{\ln(H/r_c)}{2\pi}\left(\frac{\mu_0}{\varepsilon_0}\right)^{1/2} = 60\ln\frac{H}{r_c}\,\Omega \qquad (4.28)$$

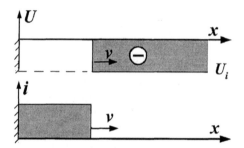

Figure 4.13. Distributions of potential and current during the discharge of a perfectly conducting line.

where c is light velocity. If one describes the capacitance, in contrast to the inductance, with the leader cover radius $R = 16\,\text{m}$, then $r_c = 1.5\,\text{cm}$, $H = 4\,\text{km}$ (as above), $C_1 = 10\,\text{pF/m}$, $L_1 = 2.5\,\mu\text{H/m}$, $v = 0.67\,C$, and $Z = 500\,\Omega$. The wave velocity is now lower than light velocity, but not much. Current of amplitude $I_M = 30\,\text{kA}$, typical of the return stroke of the first negative lightning component, arises at $|U_i| = ZI_M = 15\,\text{MV}$. The values of U_i and v are correct in order of magnitude, but the wave velocity v exceeds several times the observable velocity, and it is impossible to reduce this discrepancy by varying the reactive line parameters. In a line of preset length, C_1 and L_1 vary only slightly (logarithmically) with the conductor radius. What remains to be done is to focus on the only parameter that has not been accounted for – resistance R_1 which is very high in a leader but reduces by 2–3 orders during a return stroke.

Let us discuss the exact solution of equations (4.24) describing the line discharge at $R_1 = \text{const}$:

$$i(x,t) = \frac{U_i\,e^{-at}}{Z}I_0(a\sqrt{t^2 - x^2/v^2}),$$

$$a = \frac{R_1}{2L_1}, \qquad v = (L_1C_1)^{-1/2}, \qquad Z = \left(\frac{L_1}{C_1}\right)^{1/2}, \qquad x \leqslant vt < H$$

(4.29)

where $I_0(z)$ is the Bessel function of a purely imaginary argument jz:

$$I_0(z) \approx 1 + (z/2)^2 \qquad\qquad \text{at } z \ll 1$$

$$I_0(z) \approx e^z(2\pi z)^{-1/2}[1 + O(z^{-1})] \quad \text{at } z \gg 1.$$

(4.30)

The wave has the same velocity as an ideal line, without losses, but the current at the wave front falls exponentially, as it propagates:

$$i_f = -\frac{U_i}{Z}\,e^{-at} = -\frac{U_i}{Z}\exp\left(-\frac{R_1 x_f}{2Z}\right).$$

(4.31)

The attenuation i_f is described by the ratio of the ohmic resistance $R_1 x_f$ of the line behind the wave to the wave resistance. The line base current, which is just the current registered in lightning observations, arises instantaneously (with instantaneous short-circuiting of the line on the earth) and, at the first moment, is determined exclusively by the wave resistance, independent of the value of R_1: $i(0,0) = -U_i/Z$. As the wave moves on towards the cloud, the ohmic resistance the current has to overcome becomes increasingly higher, so the base current decreases. At $at \gg 1$, or at $R_1 x_f/2Z \gg 1$, the current through the base is

$$i(0,t) \approx \frac{U_i}{Z(2\pi at)^{1/2}} = \frac{U_i}{Z(\pi R_1 x_f/Z)^{1/2}}. \tag{4.32}$$

This current decreases much more slowly than at the wave front, because in spite of the negligible front current, the line far behind the wave front is discharged all the same, and all the charge that flows down from it goes through the base.

The wave front propagates at a rate of electromagnetic perturbation. It is independent of the line ohmic resistance but is determined exclusively by its reactive parameters and is close to light velocity. This is a 'precursor' which exists under any conditions, no matter whether the line has a resistance or whether it changes behind the wave front. The precursor carries information about the changes in the line, in our case about the line grounding. If the resistance is zero or, more exactly, has no effect yet because it is much less than the wave resistance ($R_1 x_f \ll Z$), the line is discharged in a resistance-free way, and its initial potential and charge practically vanish right behind the front of the primary electromagnetic signal, the precursor. When the resistance becomes much higher (practically several times higher) than the wave resistance, the charge and potential disappear gradually, and the rate of their reduction decreases as the linear resistance R_1 increases. At $R_1 = 10 \, \Omega/m$ corresponding to the leader channel resistance, the time constant is $a = 2 \, \mu s^{-1}$ and the precursor current decreases, in accordance with (4.31), by an order of magnitude as compared with the initial value of $i(0,0)$ over the period of time $t \sim 1 \, \mu s$, for which the precursor covers only 200 m ($v = 0.67C$). Half way up to the cloud ($x = 1500 \, m$), the front current decreases by a factor of 3×10^6. According to (4.32), the line base current at that moment will be 3×10^5 times higher than at the front. Therefore, the current somewhere behind the precursor will inevitably rise to a much larger value. Let us see where this happens and what will be the velocity of the high current region carrying the charge away from the line.

For the analysis of the relatively late stage in the discharge process with $at \gg 1$, we shall employ the second, asymptotic formula of (4.30) for the Bessel function. For the region $x \ll x_f = vt$ located fairly far from the weak precursor, the root in the argument of I_0 can be expanded. Using formulae (4.29) for a and v, we get

$$i(x, t) \approx -C_1 U_i \left(\frac{\chi}{\pi t}\right)^{1/2} \exp\left(-\frac{x^2}{4\chi t}\right), \qquad \chi = \frac{1}{R_1 C_1}. \qquad (4.33)$$

This expression is the explicit solution to equations (4.22) without the inductance term but with the same boundary and initial conditions. The potential is

$$U(x, t) \approx U_i \frac{2}{\sqrt{\pi}} \int_0^{x/(4\chi t)^{1/2}} \exp(-\xi^2)\, d\xi \equiv U_i \operatorname{erf}\left[\frac{x}{(4\chi t)^{1/2}}\right]. \qquad (4.34)$$

Expressions (4.33) and (4.34) have a clear physical sense, demonstrating the nature of a non-ideal line discharge.

When the perturbation front (the precursor due to the action of inductance) goes far away, the current decreases slowly from the line base to the front. It also varies slowly in time at every point, except for the region close to the front. This is the reason why the inductance effects in the main discharge region are very weak. With the neglect of the inductance term, equations (4.24) transform to equations similar to those for heat conduction or diffusion:

$$\frac{\partial U}{\partial t} = \frac{\partial}{\partial x} \chi \frac{\partial U}{\partial x}, \qquad \chi = \frac{1}{R_1 C_1}. \qquad (4.35)$$

To use an analogy, the potential acts as temperature, current as heat flow, and χ as thermal conductivity (heat diffusion). We did not take $\chi = \mathrm{const}$ out of the derivative deliberately to be able to come back to this equation, also valid at $R_1 \neq \mathrm{const}$.

The process of line discharge is similar to the cooling of a uniformly heated medium, when a low (zero) temperature is maintained at its boundary, beginning with the moment of time $t = 0$. Formulas (4.34) and (4.33) describe the diffusion of the earth potential along the channel (figure 4.14(a)). The current–potential wave, smeared in contrast to the precursor,

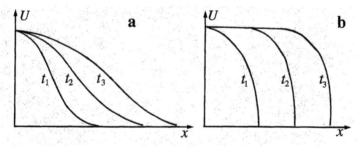

Figure 4.14. The potential wave (a) in linear diffusion with $\chi = \mathrm{const}$ and (b) in nonlinear diffusion with rising χ.

propagates in such a way that its characteristic point x_1, say, where the potential is reduced by half relative to the initial value of U_i, $x_1/(4\chi t)^{1/2} =$ 0.477, obeys the diffusion law $x_1 \approx (\chi t)^{1/2}$ with a decreasing velocity $v_1 \approx \frac{1}{2}(\chi/t)^{1/2} \approx \chi/2x_1$. From expressions (4.33), the current at point x_1 is 20% lower than that at the channel base $x = 0$. Substituting the leader resistance $R_1 \approx 10\,\Omega/\mathrm{cm}$ and $C_1 = 10\,\mathrm{pF/m}$ into the formulae, we get $\chi = 10^{10}\,\mathrm{m}^2/\mathrm{s}$. Over the time $t = 10\,\mu\mathrm{s}$ (at $at = 20$), during which a weak precursor will cover a distance of 2000 m, the half-potential point characterizing the propagation of the line discharge wave will diffuse for 315 m only and will be moving at velocity $v_1 \approx 0.05c$. By that time, the base current $i(0, t)$ will have dropped by a factor of 11 relative to the initial current $i(0, 0)$ (formula (4.32)).

The calculated values of x_1, v_1, and $i(0, t)$ can be brought closer to measurements at a certain stage of the lightning discharge. Instead of the leader resistivity, one should then deal with a lower resistivity averaged over the perturbed region. This makes sense in some evaluations. But the illusion of a satisfactory numerical agreement with measurements in a short stage of the process is destroyed, as soon as we recall one of the important qualitative observations. At the return stroke stage, a bright and well-defined wave front – the channel tip, which becomes smeared only slightly with time – is moving up to the cloud. This indicates that the energy release and, hence, the current rise occur faster than in the solution to (4.33). Clearly we deal with a wave possessing a steep front, at least for powerful lightnings, rather than with diffuse current profiles. This contradiction can be resolved by rejecting the approximation $R_1 = \mathrm{const}$ and by including, in the theoretical treatment, the time evolution of the leader channel and its transformation to a return stroke channel.

Note that the simple and attractive model of an immediate transformation of the leader channel at the wave front to an ideal conductor cannot rectify the situation. This model would take us back to equalities (4.26) and (4.27) describing the wave of immediate voltage removal and sustained current, which propagates with the velocity of an electromagnetic signal close to light velocity. But this possibility was already refused above. It was mentioned in section 4.4.1 that the key to the phenomenon of return stroke should be the analysis of the channel transformation dynamics.

The effect of the gradual resistance reduction during the Joule heat release can be understood from equations (4.35) and (4.24) without the inductance term. It would be justifiable to replace U by the potential variation $\Delta U = U - U_i$, since $U_i = \mathrm{const}$ in our approximation, so we get

$$\frac{\partial \Delta U}{\partial t} = \frac{\partial}{\partial x}\chi\frac{\partial \Delta U}{\partial x}, \qquad i = -C_1\chi\frac{\partial \Delta U}{\partial x}, \qquad \chi = \frac{1}{R_1 C_1}. \qquad (4.36)$$

The resistance decreases while χ increases, as the amount of charge flowing through the particular channel site becomes larger, or with the increase in

ΔU. Consequently, the rear sites of the diffusion wave, where ΔU and diffusion coefficient χ are already higher, propagate faster than the front sites, where ΔU and χ are still low. To supply current to the region close to the discharge wave front (a weak precursor is out of the question now), the potential gradient there must be large because of the small diffusion coefficient. Both circumstances indicate that the wave acquires a sharp front, its profile becomes steeper and convex. In contrast to the gradual asymptotic approximation at $\chi = \text{const}$, the curves $U(x)$ and $i(x)$ for a given moment of time look as if they stick into the abscissa (figure 4.14; the same will be seen from the numerical simulation illustrated in figure 4.17). The effect described here is well known [12]; this is a non-linear heat wave driven, for example, by radiative heat conduction, whose coefficient drops with decreasing temperature T approximately as $\chi \approx T^3$.†

The variant with $R_1 = \text{const}$, for which the solution to (4.33) and (4.34) is valid, probably corresponds to low current lightnings, when the energy release is too small to provide an essential reduction in the former channel resistance. In any case, there are streak pictures of return strokes with unclear wave fronts or those becoming smeared after the propagation for a few hundreds of metres [13, 14]. To obtain conclusive evidence, stroke streak pictures should be analysed at different currents. Regretfully, no simultaneous recordings of currents and stroke waves are available.

One can draw another conclusion from the solution to the set of equations (4.24) at $R_1 = \text{const} \neq 0$, which is important for the analysis of observations and for the formulation of boundary conditions necessary for finding a numerical solution. According to (4.29), when the line closes on the earth instantaneously the discharge current through the closed end also reaches its maximum instantaneously. As mentioned above, the maximum is independent of R_1, being determined exclusively by the wave resistance. Clearly, the same will also be true for any time-variable resistivity, and the only question is how fast the current will decrease after the maximum. However, the current in a real return stroke rises for several microseconds, sometimes for several dozens of microseconds, and this time may become even comparable with the total impulse time. Such a slow current rise may

---

† Equations (4.36) with $\Delta U(x,0) = 0$, $\Delta U(0,t) = -U_i$ and no inductance terms allow self-similar solutions. The simplest of them are (4.33) and (4.34) for $\chi = \text{const}$. The process is self-similar in a more complex approximation for $\chi = b(|U_i|)^n t^\nu$, which corresponds qualitatively to the $R_1 \sim \chi^{-1}$ evolution during the channel transformation. Constants b, n, and ν can be chosen from the analysis of R_1 behaviour (section 4.4.3): $n \approx 1$–2; $\nu \approx 0.5$–1. The wave front follows the relations

$$x_f = \xi[b(|U_i|)^n]^{1/2} t^{(\nu+1)/2},$$

$$v_f = \tfrac{1}{2}(\nu+1)\xi[b(|U_i|)^n]^{1/2} t^{-(1-\nu)/2},$$

where ξ of about 1 is to be found by solving an ordinary differential equation [12].

be only due to the properties of the commutator, whose role is played by the streamer zones of the descending leader and the counterleader. Their contact actually gives rise to the return stroke. The streamer zone field rises, as the streamer zones are reduced and the leader tips come close to each other or as the descending tip approaches the earth with no counterleader formed. The streamers are accelerated to a velocity 10^7 m/s, transporting kiloampere currents even in laboratory conditions [15, 16]. Thus, the rate of current rise and the impulse front duration at the earth, t_f, are determined by processes occurring in the vanishing streamer zone rather than in the former leader channel. Measurements provide indirect evidence for this, showing that the impulse rise time t_f in positive lightnings possessing a longer streamer zone than negative ones, at the same voltage, is several times longer.

4.4.3   Channel transformation in the return stroke

It has been shown above that electromagnetic perturbation propagates along a line with a velocity equal to or somewhat lower than light velocity, independent of the initial resistivity. When the resistance is high, as in the leader channel, the current and the potential variation induced by the perturbation attenuate rapidly. But the precursor is followed by a stronger perturbation propagating at a lower velocity, which reduces the potential considerably, to zero with time. The potential of a negative lightning drops to zero at this channel site due to positive charge pumping; this compensates the initial negative potential there. This process is accompanied by Joule heat release with a linear power $i^2 R_1$, which is high at first since the impulse front of the 'genuine' (not the precursor) current is quite short and the initial (leader) resistance R_1 is relatively high. The processes that follow – the channel heating, its radial gas-dynamic expansion, the shock wave propagation, and the resistivity reduction – have much in common with those in powerful spark discharges in short laboratory gaps. The latter have been extensively studied experimentally, theoretically, and numerically [17–24]. Also, calculations have been made with the initial parameters characteristic of a lightning return stroke, accounting for radiative heat exchange which is especially important in this large-scale phenomenon [22–24]. The stroke channel gas is heated up to 35 000 K. Most of the Joule heat is radiated by the highly heated gas in the ultraviolet spectrum. The emission from this spectral region is absorbed by the adjacent colder air, adding the newly heated gas to the conductive channel.

Such a treatment of the process would take us far from the point of interest, so we shall restrict ourselves to a description of two numerical results for atmospheric air, obtained with a rigorous allowance for radiative heat exchange [23, 24]. In both calculations, the shape and parameters of the current impulse were preset, as is usually done in lightning calculations. Of course, the current behaviour here depends on what happens in the whole

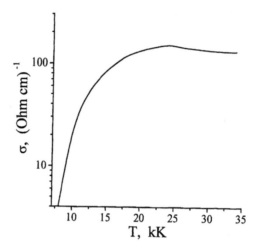

Figure 4.15. The conductivity of thermodynamically equilibrium air at atmospheric pressure.

of the perturbed region. But the formulation of a self-consistent problem requires a combined solution of a set of equations for a long line discharge and extremely cumbersome equations describing the physical evolution of each section along the line. A problem of this complexity has not been approached yet.†

Both calculations for one-dimensional cylindrical geometry were made with the current impulse

$$i(t) = I_M t/t_f \quad \text{at } t \leqslant t_f, \qquad i(t) = I_M \exp[-(t - t_f)/t_p] \quad \text{at } t > t_f$$

possessing a linearly rising front and exponentially decreasing tail. The calculation in [23] was made with moderate parameters $t_f = 5\,\mu\text{s}$, $I_M = 20\,\text{kA}$, and $t_p = 50\,\mu\text{s}$ corresponding to a moderate power lightning. The other calculation [24] was for $t_f = 5\,\mu\text{s}$, $I_M = 100\,\text{kA}$, and $t_p = 100\,\mu\text{s}$ of a very powerful lightning. It is generally believed that the air conductivity σ corresponds to its thermodynamic equilibrium and is determined by temperature (figure 4.15).

Figure 4.16 shows the evolution of pressure, gas density, temperature, and radial velocity distributions behind the shock front for a powerful current impulse [24]. The curves for moderate current impulses are qualitatively similar.

---

† The problem of a short laboratory spark is much simpler. The set should include a simple discharge equation for a capacitor bank as a high-voltage source for a spark gap with the desired resistance and allowance for the circuit inductance. Note that this kind of *LRC* circuit usually registers damped oscillations unobservable in lightning.

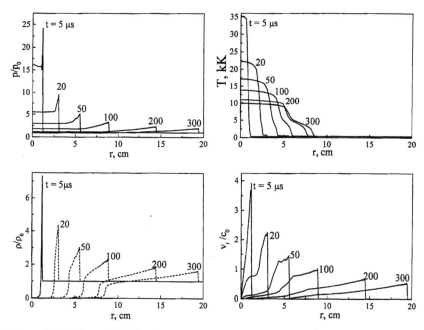

Figure 4.16. The radial distributions of pressure p, density ρ, temperature T, and the velocity v behind the cylindrical shock wave of a return stroke: $I_M = 100\,\text{kA}$, $t_f = 5\,\mu\text{s}$, $t_p = 100\,\mu\text{s}$; $p_0 = 1\,\text{atm}$, ρ_0 and c_0 are the initial presure, air density and sound velocity; $T_0 = 300\,\text{K}$.

The point of primary interest in a return stroke treatment is the behaviour of the integral channel parameter – its linear resistance:

$$R_1 = \left[\int_0^\infty 2\pi r \sigma(r)\,\mathrm{d}r \right]^{-1}. \tag{4.37}$$

Table 4.1 presents, among other parameters, the linear resistance values obtained from $T(r)$ data borrowed from [23, 24]. One can see that the resistivity drops at first for $1\,\mu\text{s}$ but then falls rather slowly. This decrease ceases closer to the pulse tail, and the resistance begins to rise gradually. The dramatic initial drop in R_1 is due to the primary heating of a very thin initial channel by high density current.† As T increases to about $20\,000\,\text{K}$, the conductivity σ rises but remains nearly constant with further temperature

---

† The gas is assumed to be in thermodynamic equilibrium at every moment of time. This assumption is justified by a fast energy exchange (for 10^{-8}–10^{-7} s) between electrons and ions, resulting in a small difference between the gas and electron temperatures. The ionization is of thermal nature: a Maxwellian distribution is established in the electron gas, and the amount of ionizing electrons is defined directly by the electron temperature, rather than by the field. The electron temperature, in turn, is determined by the Joule heat release and energy balance of the gas. Equilibrium ionization is also established rapidly (for details, see [11]).

Table 4.1. The evolution of a return stroke channel

t (μs)	T_{max} (kK)	T_{eff} (kK)	σ_{eff} (Ω/cm)$^{-1}$	p (atm)	r_{eff} (cm)	W_1 (J/cm)	R_1 (Ω/m)
\multicolumn{8}{c}{Current impulse 20 kA with $t_f = 5\,\mu s$, $t_p = 50\,\mu s$, $Q = 1.05C$}							
0.074	8	5	2	>24	0.1	–	1600
3.7	28	18	120	5.5	0.3	8	3.1
11	24	15	170	4	0.7	26	0.38
39	17	12	110	2	1.2	38	0.20
91	14	10	60	1	1.4	46	0.27
\multicolumn{8}{c}{Current impulse 100 kA with $t_f = 5\,\mu s$, $t_p = 100\,\mu s$, $Q = 10C$}							
5	35	25	180	16	0.8	–	0.28
20	22	20	140	5.7	2	–	0.057
50	18	15	90	2.6	3	–	0.039
100	14	12	70	1.8	4	–	0.028
200	12	11	40	1.0	5	~150	0.032
300	10	10	20	1.0	5	–	0.064

Note. T_{max} is the temperature along the channel axis, T_{eff} is the average temperature in the conductive channel, σ_{eff} is an average channel conductivity, p is channel pressure, r_{eff} is the effective radius of the conductive channel, W is the total energy released (no data for the second variant; the given values was estimated as $W \approx i_{max}^2 R_1 t_p$), and Q is the charge transported during the current impulse.

rise. In a strongly ionized plasma, with ions of constant charge $\sigma \sim T^{3/2}$, but doubly charged ions appear with increasing T. Since $\sigma \sim Z_i^{-2}$, where Z_i is the ion charge multiplicity, the two effects compensate each other. The resistance of a highly heated channel decreases with time due to its expansion only. Some time later, however, the pressure at the channel centre drops to atmospheric pressure, and the expansion ceases. The conductive channel cross section is reduced gradually because of the gas cooling caused by thermal radiation. The channel resistance begins to rise slowly because of decreasing r_{eff} and T_{eff}.

The expansion time of the channel becomes longer and its minimal resistance decreases for the stronger current impulses. Physically, the linear resistance is affected by the energy released per unit channel length, W, rather than by the current. This value is not described unambiguously by the current amplitude; what is more important is the amount of the transported charge Q: $W_1 \sim i^2 R_1 t \sim Q i R_1 \sim QE$ and the field does not vary much. The calculations, however, deal with the current impulse but not with W_1. Semi-quantitatively, the time dependence of resistance can be understood using the relations for the shock wave of a powerful cylindrical explosion. The explosion can be considered to be strong as long as energy is released in a thin channel and the pressure of the explosion wave does not fall close to the atmospheric pressure. In this case, the flow is self-similar.

The shock front radius r_s and pressure p in the affected region depend, within the accuracy of numerical factors, on W_1 and t as $r_s \sim (W_1/\rho_0)^{1/4} t^{1/2}$ and $p \sim W_1/r_s^2 \sim (W_1\rho_0)^{1/2} t^{-1}$, where ρ_0 is cold air density. The channel expansion is completed when the pressure drops to a certain value close to atmospheric pressure. This sets the limit to the validity of formulae for self-similar motion. This means that they are still applicable, and the duration of the resistance reduction then is $t \sim (W_1\rho_0)^{1/2}$. It can be shown [12] that for a self-similar cylindrical explosion in the central region with the pressure equalized along the radius (figure 4.16), the internal specific energy depends on r, t and W_1 as $\varepsilon \sim (W_1^{\gamma/2} t^{2-\gamma} r^{-2})^{1/(\gamma-1)}$, where γ is the adiabatic exponent. A point with fixed temperature, e.g., $T \approx 10\,000$ K, can be regarded as the conductive channel boundary, since the plasma conductivity below this point is relatively low. The radius of a point with fixed T and $\varepsilon(T)$ varies with time as $r \sim W_1^{\gamma/4} t^{1-\gamma/2}$, reaching a value proportional to $r_{max} \sim W_1^{1/2}$ by the moment the channel stops expanding, $t \sim W_1^{1/2}$. Therefore, the linear channel resistance drops to a value proportional to $R_{1_{min}} \sim r_{max}^{-2} \sim W^{-1}$, and this occurs for the time $t \sim W^{1/2}$. These relationships are qualitatively consistent with the calculations for the two variants described in table 4.1.

4.4.4 Return stroke as a channel transformation wave

The first substantiated attempt to make a numerical simulation of the lightning return stroke with allowance for the resistance variation was undertaken as far back as the 1970s [25, 26]. The most important features of the process, which are due to an abrupt conductivity rise at the site of intensive Joule heat release, became evident at once. The simulation showed that a weak initial perturbation (precursor) propagating up along the channel at an electromagnetic signal velocity close to light velocity does not change the plasma state and cannot be treated as the return stroke wave front visible in streak photographs. The main wave of current and decreasing potential travels several times slower; its velocity is defined by the transformation of the low conductivity leader to the low resistivity stroke channel. This conclusion was formulated explicitly in [25–28]; it reflects the nature of the lightning return stroke.

Turning to numerical simulation today, we should like to formulate this problem in a simple and clear physical language and to try to outline problems to be solved within this model. An obviously essential aspect of the theory still is the resistivity dynamics of the lightning channel. An exhaustive formulation of this problem would involve a simultaneous solution of equations describing the propagation of a current–voltage wave and the channel dynamics at every point along its length, affected by the ever varying energy release. So we shall restrict the discussion to a simple model, having accepted a probable law for the linear conductivity rise, $G = R_1^{-1}$, and focusing on the qualitative results of the solution.

Let us describe G in the simplest way reflecting the main qualitative features of the channel evolution. It will be assumed that the linear conductivity increases with current. This partly reflects the fact that resistance decreases with increasing charge through a particular channel site. But the resistance is stabilized with time, even though the current continues to flow. In principle, the stable state of a lightning channel hardly differs from that of an arc. The field E in a high current arc only slightly varies with current; in other words, the linear conductivity of the channel is $G_{st} = i/E \sim i$. Assume E to be equal to the field E_L in the lightning leader, whose current is not low on the arc scale; then the conductivity is $G_{st} = i/E_L$. In a mature gas-dynamic process when the shock wave is still strong, the resistance will drop with time. As the shock wave becomes weaker, the decrease in R_1 and the increase in G become slower. These tendencies are described by the relaxation-type formula

$$\frac{dG}{dt} = \frac{i/E_L - G(t)}{T_g} = \frac{G_{st}(i) - G(t)}{T_g} \qquad (4.38)$$

where T_g is the characteristic time of linear conductivity variation (relaxation time). In a simple case with $i = \text{const}$, $T_g = \text{const}$, and $G(0) = 0$, we have $G = G_{st}[1 - \exp(-t/T_g)]$.

Equations (4.24) are solved with the initial conditions $U(x, 0) = U_i$ and $i(x, 0) = 0$, $R_1(x, 0) = R_{1L}$; the reactive parameters are taken to be constant: $C_1 = 10\,\text{pF/m}$ and $L_1 = 2\,\mu\text{H/m}$. The channel does not close on the earth in an instant but does so through the time-decreasing resistance of the commutator (similarly to the real lightning length decreasing through the streamer zone). The accepted values of $R_{com} = R(0)\exp(-\alpha t)$, $R(0) = 10\,\Omega$ and $\alpha = 1\,\mu\text{s}^{-1}$ provide a typical duration of the negative current impulse front $t_f \approx 5\,\mu\text{s}$. The boundary condition at the grounded end of the line raises no doubt: $U(0, t) = i(0, t)R_{com}$. The problem of the far end up in the clouds, $x = H$, is much more complex. Conventionally, it is considered as being open, assuming $i(H, t) = 0$. In reality, the situation is far from being self-evident. When the line gets discharged and its end in the clouds takes zero potential, a high electric field must arise near it due to the voltage difference $\Delta U = -U_0(H)$. This gives impetus to very intensive ionization processes, probably involving high current. This situation will be partly discussed below. Now, we shall assume the upper end to have no current. The results to be presented were obtained for a vertical unbranched channel with the total length $H = 4\,\text{km}$. This is the height the ascending leader tip reaches when the descending leader, which has started from the point closest to the earth in the bottom negative sphere with the centre 3 km high, contacts the earth. It is normal practice to use the following averaged parameters of the leader prior to the contact: $E_L = 10\,\text{V/cm}$, $i_L = 100\,\text{A}$, and $R_{1L} = E_L/i_L = 10\,\Omega/\text{m}$. For a realistic description of the resistivity dynamics (section 4.4.3), the relaxation time should be taken to be $T_g = 40\,\mu\text{s}$, when the current at this channel site rises, and $T_g = 200\,\mu s$, when the current decreases. The model calculation reproduces the distributions of current $i(x, t)$ and potential $U(x, t)$ along the channel; the linear charge is $\tau(x, t) = C_1[U(x, t) - U_0(x)]$. Generally, the external field potential $U_0(x)$ can also vary in time, because we should not discard a possible partial neutralization of the charge in one of the regions of the cloud dipole. The latter point will not be discussed for the time being.

The calculations are presented in figures 4.17–4.21. The precursor travelling with velocity $0.64c$ is damped so fast that this is not shown in the plots after a noticeable break from the principal wave re-charging the channel (we shall term it a discharge wave for simplicity). The wave in figure 4.17 travels along the channel with velocity $v_r \approx 0.4c$, i.e., 1.6 times slower than the precursor. This velocity somewhat decreases as the wave moves up. Its variation can be conveniently followed from the change in the well-defined maximum linear power of the Joule losses $i^2 R_1$ (figure 4.17 (centre)). The wave front power rises abruptly along a 100–200 m length, then it decreases towards the earth, making the channel tip with intensive energy release stand out clearly. It seems that it is this region which is clearly discernible in streak photographs. The linear power proportional to the squared current drops remarkably on the way up the cloud, and the maximum becomes smeared. This is also consistent with observations of radiation intensity [14, 29]. A photometric study has shown that the radiation from the wave front is attenuated and the front loses its clear boundary. The current wave is not attenuated so rapidly (figure 4.17 (top)). For the time of its earth-cloud travel lasting for $34\,\mu s$, the current at the channel base drops from the maximum of 35 kA to 24 kA. This agrees with observations indicating that an average current impulse duration in a negative lightning is close to $75\,\mu s$ on the 0.5 level. The wave front deformation depends on the initial potential U_i delivered by the leader to the earth. The higher the value $|U_i|$, the higher the discharge current. The rate of resistivity decrease at the wave front grows respectively, so the front steepness increases. This is evident from a comparison of figures 4.18 (top) and 4.18 (bottom). At $|U_i| = 50\,\text{MV}$, the current wave goes along the channel practically without elongating the front† while at $|U_i| = 10\,\text{MV}$ it has a lower velocity and a smooth front. Unfortunately, there have been no registrations of current and streak photographs of the return stroke taken simultaneously. A comparison of the relationships between current and wave velocity could provide a good test for the return stroke theory.

As the current impulse amplitude rises, the linear resistance falls more quickly and to a lower level, so the wave is damped more slowly during its

---

†The motion of a high current wave with attenuation but without noticeable distortions facilitates the electromagnetic field calculation necessary in many applied problems of lightning protection and in substantiation of remote current registration methods.

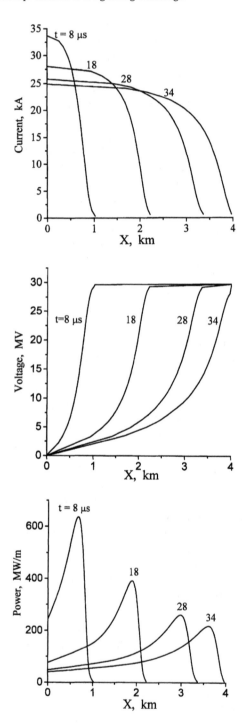

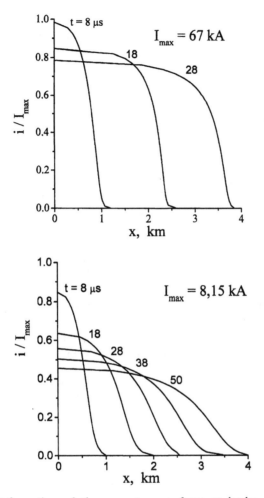

Figure 4.18. Deformation of the current wave front at leader potential (top) $U_i = -50\,\text{MV}$ and (bottom) $-10\,\text{MV}$; for the other parameters, see figure 4.17.

propagation along the channel. There is no damping at a very high current and the impulse front becomes steeper, as was discussed in section 4.4.2 (figure 4.18). Non-linearity is also observed in the current amplitude dependence on the initial potential U_i at the earth. If the commutator were perfect ($R_{\text{com}} = 0$), the current at the earth at the moment of contact would instantly

Figure 4.17. (*Opposite*) Numerical simulation of the return stroke excited by a descending leader with potential $-30\,\text{MV}$: (top) current and (centre) voltage distributions; (bottom) the power of Joule losses. The initial leader resistance, $10\,\Omega/\text{m}$. Steady state field in the channel behind the wave, $10\,\text{V/cm}$.

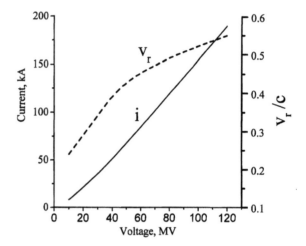

Figure 4.19. Calculated dependencies of the current amplitude and average wave velocity in the return stroke on the leader potential U_i.

take the maximum value $I_M = |U_i|/Z$, independent of the actual channel resistance, and would be $I_M \sim U_i$. With the finite time of R_{com} decrease to zero, the current wave is able to cover some distance along the channel and to include in the circuit the ohmic resistance of this channel portion. For this reason, the current amplitude appears to be lower than U_i/Z and rises somewhat faster than potential U_i (figure 4.19). It is important that the lightning current amplitude I_M is found to be appreciably smaller than its theoretical limit U_i/Z: e.g., $I_M \approx 0,6 U_i/Z$ at $U_i = 30\,\mathrm{MV}$. This is another source of errors in evaluations using the equality $U_i = ZI_M$, in particular, in the calculation of cloud potential from lightning current data.

4.4.5   Arising problems and approaches to their solution

The current at the earth is independent of the boundary condition at the upper channel end, until the reflected wave comes back to the earth with the information about the processes occurring there. Before that moment, the positive charge is pumped into the line from the earth. In virtue of the boundary condition – zero current at the upper end – the current wave is reflected there with the sign reversal. As a result, the current behind the reflected wave, i.e., between its front and the channel end (figure 4.20), decreases (it would drop to zero in the absence of damping). The incoming positive charge now re-charges the line making it positive (an ideal line would be re-charged to $-U_i$). The reflected wave moves faster and is damped more slowly, because the linear resistance in most of the channel has dropped by an order of magnitude or more due to the action of the forward current wave.

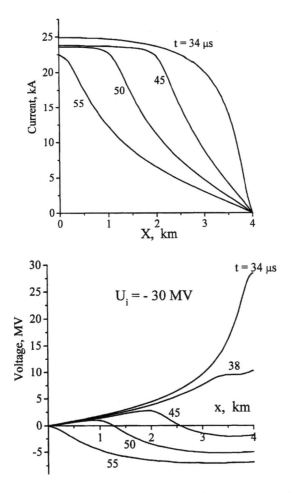

Figure 4.20. Current and potential distributions during the propagation of waves reflected by the cloud end of the channel.

When the reflected wave reaches the earth, delivering a positive potential to it, a new discharge cycle begins. The newly acquired positive charge flowing into the earth is equivalent to the extracted negative charge. The current sign at the grounded end is reversed (figure 4.21). In the absence of dissipation in a distributed system such as a long line, undamped oscillations with a period $T = 4H/v_r$ would arise similar to those in an LC circuit. Nothing of the kind is observed in lightning registrations, nor is there a single change in the current direction. This means that the discharge wave is either not reflected by the upper end of the line or the reflected wave becomes so damped on the way back to the earth that it is unable to manifest

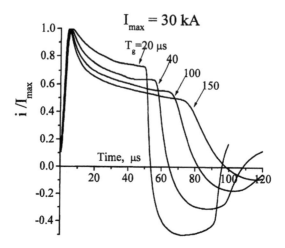

Figure 4.21. Calculated current impulse through the grounded channel end.

itself against the background of other variations in the current. By changing the parameter T_g or the quasi-stationary channel field E_L, one can reduce or even cancel part of the current impulse of opposite sign, but it is impossible to attain a portrait likelihood between the calculated and observable currents. The suppression of the reflected wave by raising the instantaneous values of the channel resistance $R_1(x, t)$ inevitably results in an excessive reduction of the impulse duration at the grounded end of the line. There seems to be no way of avoiding this even by changing the resistivity reduction law.

The first thing that seems to be suitable for rectifying this situation is to question the boundary condition at the upper end. This idea appears reasonable because it generally agrees with lightning current registrations at the earth for the double path time $t \approx 2H/v_r$ while the model solution for $i(0, t)$ remains independent of the boundary condition. It is obvious that the open circuit condition is an excessively rough idealization. It was mentioned in section 4.3.3 that if the negative cloud bottom is filled with a large number of branches which stem from the ascending leader, this region becomes similar to a metallic sphere. Assuming such a 'metallization' of the cloud, it would be more reasonable to consider the upper end to be connected to a lumped capacitance $C_c = 4\pi\varepsilon_0 R_c$, defined by the cloud charge radius R_c, instead of being open. The boundary condition at $x = H$ would have the form $i(H, t) = C_c \, dU/dt$. When the current wave reaches the line end, the delivered current also discharges the negative 'metallized' cloud region. This, however, does not prevent the appearance of the reflected wave. At the first moment of time, the capacitance still preserves its charge and is similar, in accordance with the reflection condition, to a short-circuited end of the line, which generates a reflected current wave of the same sign and amplitude as the incident one. As the capacitance becomes discharged, the reflected wave amplitude decreases and then the sign is reversed. The completely discharged capacitance, incapable of supporting current, eventually becomes equivalent to an open line end. It is clear even without a numerical calculation how much the current changes at the earth after the arrival of a reflected wave of such complexity. It should be emphasized again that nothing of the kind has ever been observed in real lightning.

One can also try to rectify the situation by complicating the boundary condition with the allowance for the final resistivity of the 'metallized' cloud region. The streamer and leader branches filling the cloud possess a resistance at the moment of their generation. A leader branch can hardly be heated as much as a single descending leader. The resistance of the 'metallized' cloud, R_{cl}, is quite likely to be high during the whole return stroke stage. If this is so, the boundary condition should be formally represented as $i(H, t) = C_c \, dU/dt - R_{cl} \, di/dt$. Strictly, it is not only the boundary condition that changes in this case, like in the case of ideal metallization, but also the set of equations. The cloud potential U_0 can no more be considered as being constant in time. The second equation of (4.24) should be re-written as

$$-\frac{\partial i}{\partial x} = \frac{\partial \tau}{\partial t}, \qquad \tau = C_1[U(x,t) - U_0(x,t)]$$

having taken into account the change in U_0 due to the change in the cloud charge Q_c. Then the function $U_0(Q_c)$ must allow for the delay because of the finite rate of the electromagnetic field propagation. The problem becomes extremely complicated. Although radar registrations do indicate the development of a wide network of branches in clouds, there has been no investigation of cloud 'metallization'. The reason for this, no doubt, is the lack of initial data.

One should not discard two other factors unaccounted for by the numerical model. First, a leader channel can practically never be single. Owing to the numerous branches of different lengths developing at different heights, numerous reflected waves will arrive at the earth at different moments of time, creating a sort of 'white noise' with a nearly zero total signal. This will deprive the current of its characteristic bending which is usually created by a single reflected wave at the earth. Second, constant linear capacitance only approximately describes the real re-charging of a lightning leader. We have mentioned above that the cover charge around a leader channel is changed by numerous streamers starting from it. Their velocity decreases rapidly when the streamer tips go away from the channel surface with its high radial field. So, when the voltage at the wave front changes, the charge near the channel changes almost immediately, while its change at the external cover boundary occurs with a delay. In other words, a lightning discharge can proceed for a fairly long time. The quasi-stationary current from the discharge of the cover periphery, having the same direction as the current in the forward wave, can compensate for the reverse current induced by the wave reflected by the earth.

To conclude, the model of a single long line with varying linear resistance allows elucidation of many aspects of the return stroke but cannot claim to give reliable quantitative description of all characteristics of this phenomenon. The much more simplified models of return stroke are usually used calculating electromagnetic field for technical application. A review of this model is given [30].

4.4.6 The return stroke of a positive lightning

Two kinds of current impulse can be distinguished in oscillograms taken at the earth after the arrival of a positive leader. Common impulses are similar to those registered in negative lightnings, although they have a slightly longer duration t_p and less steep fronts. Such impulses can be naturally interpreted as return stroke currents corresponding to the wave discharge of the leader channel, as described above. Sometimes, however, quite different impulses are registered with an order longer duration and an amplitude as large as 200 kA. A closer examination shows that impulses with an 'anomalous' duration cannot be interpreted as resulting from a grounded leader discharge. They appear to result from another process, and we shall offer some suggestions concerning their nature in section 4.5. Here, only common impulses will be discussed.

It was shown in section 4.4.2 that the stroke current front is unrelated to the wave discharge process in the channel but, rather, is associated with an imperfect commutator closing the channel on the earth. The leader streamer zone acting as a commutator possesses a finite resistance and is reduced during a finite period of time. The front steepness is determined by the rate of resistance reduction in this transient link between the channel and the earth. But the streamer zone length of a positive leader at the same voltage is about twice as large as that of a negative leader and takes more time to be reduced. It is quite likely that this is the main reason why, with the 50% probability, the current front duration in positive lightnings, $t_f = 22\,\mu s$, is four times longer than in negative leaders [1]. Approximately the same proportion is characteristic of the maximum pulse steepness.

The duration of the pulse itself, t_p, is primarily determined by the stroke channel length. It was shown in section 4.3 that it is only positive leaders starting from the top positive region of a storm cloud which have a real chance to reach the earth. This region is twice as high as the negative cloud bottom. Hence, the channel length of a positive descending leader is, at least, twice as long. But the vertical positive channel transverses the negative cloud region, delivering a very low potential U_i to the earth, so it is incapable of producing a return stroke with an appreciably high current (section 4.3.6). Only those lightnings, whose positive descending leaders bypass the negative cloud region along a very curved path, can actually be identified in the registrations. The statistics shows that the total length of such a leader, including the path bendings, is 1.3–1.7 times greater than that of a straight leader. Therefore, a positive channel length and its stroke pulse duration appear on average to be three times greater than in a negative leader. As for other characteristics, common positive pulses are the same as the negative ones described above.

4.5 Anomalously large current impulses of positive lightnings

Anomalous impulses of a positive lightning have the duration $t_p \approx 1000\,\mu s$ of the 0.5 amplitude level and the rise time $t_f \approx 100\,\mu s$. The current in some of them is as high as 100 kA or more [1]. Although such lightnings are rare, their effects on industrial objects are so hazardous that they should not be underestimated. A current impulse delivers to the earth a charge $Q \approx 100C$; therefore, as large a charge must be located in the cloud cell where the lightning originated. The potential at the boundary of a charged cloud region of radius, say, $R_c \approx 1\,km$ is $U_{0R} \approx 1000\,MV$, with $1500\,MV$ at its centre. Any attempt to treat a long current impulse as return stroke current inevitably leads to contradictions. Indeed, in order to reduce the near-earth current by half of its maximum value for $1000\,\mu s$, it would be necessary to assume in (4.29) $at_p = 0.7$ and $a = R_1/2L_1 = 700\,s^{-1}$; hence, the average linear resistance behind the wave front of the return stroke would be $R_1 \approx 3.5 \times 10^{-3}\,\Omega/m$. The total resistance of a channel of length $H = 4000\,m$ would be $R_1 H \approx 14\,\Omega$, i.e., 40 times less than the wave resistance. The line would seem to be discharged as an ideal line, i.e., for $20\,\mu s$ instead of $1000\,\mu s$, with the velocity of an electromagnetic signal.

Excessively smooth impulses are sometimes observed in ascending leaders. A positive impulse $I_M \approx 28\,kA$ with $t_p \approx 800\,\mu s$ was registered during the propagation of a negative ascending leader from a 70-m tower on the San Salvatore Mount in Switzerland [31]. This fact in itself is of interest, but its analysis may offer an explanation of 'anomalous' currents of descending positive lightnings. Note, at first, the unusual situation at the start. Since the negative charge of the dipole is located at the cloud bottom, the ascending leader is to be positive rather than negative. Therefore, the dipole axis has either deviated from the normal or the bottom negative charge was neutralized earlier by, say, an intercloud discharge. This situation occurs rarely but it is possible.

We mentioned in section 4.1 that ascending lightnings have no return strokes because their channels are grounded from the very beginning.

However, when the ascending leader penetrates the charged cloud region (positive, in this case), a large potential difference arises between the front end of its grounded channel and the space around it, so the leader current has been found from many registrations to rise to several kiloamperes. This event seems to be triggered by the same mechanism, but its effect is greatly enhanced by the leader hitting the very centre of a large cloud charge of, say, $Q_c \approx 30C$ and radius $R_c \approx 500\,m$, where the potential is as high as $U_0 \approx 500\text{--}800\,MV$. At such voltages, the streamer zone and cover appear much extended. Negative streamers develop until the average field in their streamers drops below $E_s \approx 10\,kV/cm$ under normal conditions (or 1.5 times less at a 5–6 km height [16]). Streamers elongate very quickly when the field is higher. Therefore, a very powerful streamer corona consisting of numerous branched streamers (they are likely to originate not only from the stem but from its numerous branches, too) will fill up a space of size $R \approx U_0/E_s \approx R_c$. The negative charge of the streamer zone will partly neutralize the positive charge of the cloud cell. If the streamers have velocity $v_s \approx 10^6\,m/s$, they will fill the charged cloud region for $t \approx R_c/v_s \approx 10^{-3}\,s$. Since the capacitance of the leader portion inside the cloud, C_L, is comparable with that of the charged cloud region, C_{cl},

$$ C_L \approx \frac{2\pi\varepsilon_0 2R}{\ln(H/R)}, \qquad C_{cl} \approx 4\pi\varepsilon_0 R_c, \qquad C_L \approx 0.5 C_{cl} $$

a charge of opposite sign, comparable with the cloud intrinsic charge, penetrates the cloud. The resultant effect is such that most of the cloud charge would seem to run down to the earth with current $i \approx Q_0/t \approx 30\,kA$ for $t \approx 10^{-3}\,s$. Microscopically, the cloud medium remains non-conductive, as before. Charges do not recombine but neutralize one another on average. The process of current organization reduces to the neutralization of the cloud rather than leader charge.

Returning to long current impulses after the positive leader arrival at the earth, let us imagine that the leader has been developing along a vertical line somewhat away from the axis of a powerful cloud dipole with $Q_c \approx 100C$ or more, $R_c \approx 1\,km$, and $U_{0R} \approx 1000\,MV$. Suppose the leader cover has no contact with the cloud charge boundary but is close to it. All the same, a huge, actually induced charge comparable with Q_c arises in the vicinity of the cloud charges. Note that the arrival of a vertical positive leader does not reveal itself in any way, since its potential is close to zero because of a nearly complete symmetry of charges induced in the lightning channel.

Suppose now that while this leader still preserves conductivity (this period of time is measured in dozens of milliseconds because of the current supply of ~100 A), an intercloud discharge occurs, connecting the lower negative charge of the dipole to another positive charge. Intercloud discharges have been observed to be a much more frequent phenomenon than cloud–earth discharges. So our suggestion is not at all improbable. The charges of opposite signs connected by intercloud leaders will gradually neutralize each other via the same mechanism as the one underlying an ascending leader.† As the neutralization goes on, the earlier induced but now liberated positive charge of the grounded leader will flow down to the ground. This will occur at a lower velocity than the return stroke velocity, in accordance with the neutralization rate of the negative cloud charge. This is a likely explanation for long powerful current impulses in positive lightnings.

4.6 Stepwise behaviour of a negative leader

When discussing the negative leader in section 4.3.2, we put off the consideration of its stepwise behaviour until the reader has become familiar with the return stroke, since a similar phenomenon is the principal event occurring in each step. It would be reasonable, at this point, to turn to the nature and effects of the stepwise leader behaviour. But we should like to warn the reader that there is no clear answer to the question why a negative leader has a stepwise structure while a positive one has not. Nonetheless, some observations of stepwise positive lightning leaders were presented in [32]. This phenomenon has never been observed in laboratory conditions.

4.6.1 The step formation and parameters

The only thing one can rely on today in discussing the nature of leader steps is the results of laboratory experiments with long negative sparks (section 2.7). Natural lightning observations are not informative, except for the step lengths $\Delta x_s \approx 5{-}100$ m [13, 32–37] and the registrations of leader channel flashes occurring at the step frequency. Streak photographs indicate that only the front channel end of 1–2 steps in length shows bright flashes. But weak flashes may appear even along a kilometre length (the vision field of a photocamera does not always cover the whole channel).

Laboratory streak pictures show that a step originates from two secondary twin leaders at the front end of the streamer zone in the main negative leader (in the Russian literature, these are termed bulk leaders). The positive leader moves towards the main leader tip while the negative one develops along the latter (figure 4.22). During the pause between two steps, the secondary negative and the main leaders do not have a high velocity, but the positive leader moves faster for two reasons. First, as the distance to the main leader tip becomes shorter, the difference between the positive tip

---

† The fact that intercloud discharges do neutralize charged regions is supported by electric field measurements, and high currents that flow through lightning channels are indicated by peals of thunder.

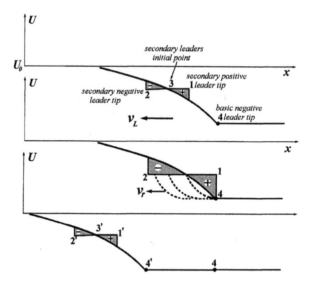

Figure 4.22. The potential distribution for various stages of a step formation. The main leader potential $U(x)$ is counted from the external potential U_0: (top) secondary leaders 1 and 2 are formed at point 3 at the streamer zone end; (centre) the tip of a positive secondary leader has reached tip 4 of the main leader, and a discharge wave has started its travel along the secondary leader channel (dashed line); (bottom) the main leader tip after the step formation has taken a new position, and the process is repeated.

potential U_1 and the external (for the tip) potential $U(x_1)$ at the tip site x_1 increases (figure 4.22 (top)). Second, the streamer zone field of the main leader, $E_s \approx 10\,\text{kV/cm}$, which must support the generated negative streamers, is higher than the field $E_s \approx 5\,\text{kV/cm}$ required for the development of positive leaders. For this reason, the streamers generated by the secondary positive leader tip develop in a fairly strong field, become accelerated and all reach the main leader tip. Since the long channel of the main leader has a capacitance greatly exceeding that of the short secondary leader, it absorbs completely all charges carried by the positive streamers. In other words, the secondary positive leader develops in the final jump mode. We know from section 2.4.3 that this leads to its acceleration. The secondary negative leader, on the contrary, develops in a decreasing field beyond the streamer zone of the main leader, whose streamers stop in space, so it moves much more slowly, similarly to the main leader.

When the tip of the secondary positive leader comes in contact with the main channel, the positive leader experiences the transition to the return stroke. Charge variation waves run along both channels, as described in section 4.4, and their potentials tend to become equalized (figure 4.22 (centre)). But the capacitance of the kilometre length channel is much higher than that of the shorter secondary channel, so their fusion results in establishing a potential only slightly differing from the initial potential of the main leader tip, U_t.

The moment at which the potential U_1 is taken off the secondary leader channel and the latter joins the main leader, manifests the end of the step. The main leader tip 'jumps' over to a new place, the one occupied previously by the tip of the secondary negative leader, delivering to it its potential U_t (figure 4.22 (bottom)). The tremendous potential difference that arises in the vicinity of the newly formed tip at this moment produces a flash of a powerful negative streamer corona, which transforms to the novel streamer zone of the main leader. Then the sequence of events is repeated. The combination of the charge utilized for the short recharging of the secondary positive leader and for charging the secondary negative one, plus the charge incorporated into the new streamer zone, create the step current impulse. (Recall that there is always a local current peak at the streamer tip or in the leader streamer zone, related to the displacement of the charge in this region; see sections 2.2.3 and 2.3.2.) Part of the step impulse creates a current impulse in the main channel and the other part is spent for the formation of a new cover portion. The charge Q_s pumped into the main channel can be evaluated in terms of the mean velocity of the stepwise leader, $v_L \approx 3 \times 10 \, \mathrm{m/s}$, the length of a step $\Delta x_s \approx 30 \, \mathrm{m}$, and the current $i_L \approx 100 \, \mathrm{A}$ averaged over the whole duration of the process. Since the time between two steps is $\Delta t_s \approx \Delta x_s / v_L \approx 10^{-4} \, \mathrm{s}$, the charge is $Q_s \approx i_L \Delta t_s \approx 10^{-2} \, \mathrm{C}$.

4.6.2 Energy effects in the leader channel

The energy pumped by the charge pulse Q_s into the channel can be evaluated if the effect is assumed to be similar to that observed when the small capacitance (of the secondary leader) is added parallel to the large capacitance (of the main leader). While a common potential is being established, there is a dissipation of energy nearly equal to the electrical energy stored by the small capacitance at a voltage equal to the difference between the initial capacitance voltages $U_t - U_1$, where U_1 is the potential of secondary leaders. It was pointed out in section 2.4.1 that the leader tip potential is shared nearly equally between the streamer zone and the space in front of it. Secondary leaders are produced at the streamer zone edge, so we have $U_1 - U_0 \approx \frac{1}{2}(U_t - U_0)$; hence, $U_t - U_1 = \frac{1}{2}(U_t - U_0)$. With the accepted average values of current $i \approx C_1(U_t - U_0)v_L$ and of velocity v_L and assuming $C_1 \approx 10 \, \mathrm{pF/m}$, we find $U_t - U_0 \approx 30 \, \mathrm{MV}$ and $U_t - U_1 \approx 15 \, \mathrm{MV}$. Thus, the step energy is

$$W \approx Q_s(U_t - U_1)/2 \approx 7.5 \times 10^4 \, \mathrm{J}. \qquad (4.39)$$

Of course, not all of this energy is released in the main channel. During the dissipation of charge Q_s, the channel potential rises appreciably. This leads to the radial field enhancement and to the excitation of a streamer corona which pumps some of the charge into the leader cover. This process, the cover ionization in particular, requires much energy. But even if the energy released in the channel is assumed to be $W \approx 10^4$ J, this is still a very large energy.

The power required to support an average current of 100 A in remote channel portions, where the effects of current impulses are averaged and smeared, should be $P_1 \approx 10^5$ W/m. It is removed from the channel by heat conduction and, partly, by radiation. These parameters correspond to the maximum channel temperature $T \approx 10\,000$ K, field $E \approx 10$ V/cm, and resistance $R_1 \approx 10\,\Omega/\text{m}$ taken for the above estimations (section 2.5.2). At this power, the energy released between two flashes per unit channel length will be $W_{\text{lav}} \approx 10$ J/m for the time $\Delta t_s \approx 10^{-4}$ s. Therefore, the single pulse energy W would be sufficient to support a channel 1 km long. In reality, a step pulse is damped at a much shorter length. At a distance of about 1 km, the step effects are smeared and the energy released in the channel becomes totally averaged. But at a short distance from the tip, the energy effect of the step is very strong, as is indicated by the intensive flash. The temperature registered in some measurements was as high as 30 000 K [35], i.e., the same as at the wave front in a return stroke.

Let us evaluate the distance at which the energy effect of an individual step is still essential. When a short step joins a long channel, charge Q_s is pumped into the channel for a short time. Since we are interested in distance and time much larger than the real length and duration of a charge source, let us assume the source to be instantaneous and point-like, as is usually done in physics: a point charge Q_s is introduced at the initial point of the line, $x = 0$, at the initial moment of time $t = 0$. The resistance of not very short channel fragments, $R_1 x$, is higher than the wave resistance, so the inductance effect will be neglected. At an average resistance of $10\,\Omega/\text{m}$, this distance is just about the step length Δx_s. At shorter distances, the instantaneous point source model is invalid, since it implies an infinite initial voltage and energy. They drop to realistic values only if the charge affects a length exceeding Δx_s, at which the source was actually placed. Therefore, with the neglect of inductance (and the precursor), the line charging to potential $U_p(x, t)$ above the background potential is described by equations (4.36). On the assumption of $R_1 = \text{const}$,† they have an exact solution corresponding to heat flow from an instantaneous lumped source:

$$U_p = \frac{2Q_s}{C_1(4\pi\chi t)^{1/2}} \exp(-x^2/4\chi t), \qquad C_1 \int_0^\infty U_p\, dx = Q_s \qquad (4.40)$$

$$i_p = -\frac{1}{R_1}\frac{\partial U_p}{\partial x} = \frac{Q_s x}{(4\pi\chi t)^{1/2} t} \exp(-x^2/2\chi t), \qquad \chi = \frac{1}{R_1 C_1}. \qquad (4.41)$$

† The value of resistance R_1 to be taken for evaluations may be smaller than that in the leader, having in mind a transformation of the channel due to the step current.

The power released by the current step per unit channel length is described as

$$P_1 = i_p^2 R_1 = \frac{Q_s^2 x^2}{4\pi\chi^2 C_1 t^3} \exp(-x^2/2\chi t). \tag{4.42}$$

At the point x, the power reaches a maximum at moment $t_m = x^2/6\chi$, and

$$P_{1_{max}} = \frac{54}{\pi e^3} \frac{Q_s^2 \chi}{x^4 C_1}. \tag{4.43}$$

For the time of the pulse action, the energy released per unit length at point x is

$$W_1 \approx \int_0^\infty i_p^2 R_1 \, dt = \frac{Q_s^2}{\pi C_1 x^2} \approx \frac{W}{\Delta x_s} \left(\frac{\Delta x_s}{x}\right)^2, \qquad x > \Delta x_s \tag{4.44}$$

where W is the total energy injected into the channel by the pulse, with its upper limit given by formula (4.39). The effective duration of energy release from a single step at point x is expressed as

$$\Delta t_W \approx \frac{W_1}{P_{1_{max}}} \approx 2.2t_m = \frac{x^2}{2.7\chi}. \tag{4.45}$$

Consequently, the contribution of charge injection to the energy release at a given channel site decreases in the direction of perturbation propagation as $W_1 \approx x^{-2}$ and is independent of R_1. The latter fact justifies the use of $R_1 = \text{const}$ without reservations concerning the resistance variation during the current impulse passage. The energy pulses released at point x owing to the two subsequent steps superimpose at $x > (2.7\chi\Delta t_s)^{1/2}$; this critical distance follows from the condition $\Delta t_W \approx \Delta t_s$. For example, at the average resistance $R_1 = 10\,\Omega/\text{m}$, with $\chi = 10^{10}\,\text{m}^2/\text{s}$ and step frequency $\Delta t_s = 10^{-4}\,\text{s}$, this happens at a distance $x \approx 1.6\,\text{km}$ in $t_m \approx \Delta t_s/2.2 \approx 45\,\mu\text{s}$, after the pulse arrival here. Thus, the effects of energy release from individual steps are detectable even along an extended lightning path, and this is the cause of observable flashes of almost the whole channel. For a flash to arise, there is no need for a strong energy effect. A short temperature rise of, say, above $1000\,\text{K}$ over $10000\,\text{K}$ would be sufficient for a flash to be detected by modern photographic equipment.

The channel energy is affected by the temperature rise above the average background, rather than by the time separation of the energy pulses between two subsequent steps. In this respect, the impulse effect on the channel during the wave propagation is damped at distances close to the tip. The plasma temperature modulation determining the flash intensity at large distances is due to the imbalance between the energy release and heat removal from the channel during the pauses between the steps. There is no imbalance at a large distance from the tip after the channel development has been established. At $T \approx 10\,000\,\mathrm{K}$, the losses for air plasma radiation are not particularly great but become appreciable at $T \approx 12\text{--}14\,000\,\mathrm{K}$. The Joule heat of current is eliminated from the channel primarily by heat conduction. This process occurs at constant (atmospheric) pressure when the energy release is moderate, as is the case for distances of hundreds of metres from the tip. At $T \approx 10\,000\,\mathrm{K}$, the air heat conductivity is $\lambda \approx 1.5 \times 10^{-2}\,\mathrm{W/cm\,K}$ and the thermal conductivity at pressure $p = 1\,\mathrm{atm}$ is $\chi_{\mathrm{T}} = \lambda/\rho c_{\mathrm{p}} \approx 180\,\mathrm{cm^2/s}$, where ρ is air density and c_{p} is heat capacity. The average conductivity in the channel σ corresponds to a temperature lower than the maximum temperature. To illustrate, at $\sigma \approx 10\,(\Omega \cdot \mathrm{cm})^{-1}$ corresponding to $T = 8000\,\mathrm{K}$, the effective radius of a conductive channel is $r \approx (\pi \sigma R_1)^{-1/2} \approx 0.6\,\mathrm{cm}$ for $R_1 = 10\,\Omega/\mathrm{m}$. The heat is removed from the channel for time $t \sim r^2/2\chi_{\mathrm{T}} \sim 10^{-3}\,\mathrm{s}$, an order of magnitude longer than the pause between the steps. The repeated energy pulses dissipate rather slowly, and the temperature modulation relative to its average value $T \approx 10\,000\,\mathrm{K}$ is not large at long distances x. Indeed, the energy released in the remote channel portions during a pause is $W_{\mathrm{lav}} \approx P_{\mathrm{lav}}\Delta t_{\mathrm{s}} \approx 10\,\mathrm{J/m}$ at an average power $P_{\mathrm{lav}} \approx 10^5\,\mathrm{W/m}$. Even if we assume that all energy of a step is released in the channel and $W/\Delta x_{\mathrm{s}} \approx 2500\,\mathrm{J/m}$ in (4.45), the excess of the pulse release over the average heat removal, which is equal to the average energy release W_{lav}, will be small at $x > \Delta x_{\mathrm{s}}(W_1/W_{\mathrm{lav}})^{1/2} \approx 10\Delta x_{\mathrm{s}} \approx 500\,\mathrm{m}$. With allowance for other energy expenditures, this reduction in the pulse effect will be evident even at shorter distances. This circumstance makes the use of average parameters reasonable in the consideration of the evolution of long stepwise lightning leaders, ignoring the stepwise behaviour effects. In any case, laboratory experiments show that there is no appreciable difference between a positive continuous and a negative stepwise spark discharge as for the velocity, average leader current or breakdown voltage in superlong gaps.

However, even a small excess of the average temperature over its average value may be sufficient for a flash to be registered optically. As for channel portions located at a distance of one or two steps from the tip, the energy pulses and outbursts of temperature and brightness are found to be very strong there. A gas-dynamic expansion of the channel is also possible, as happens in the return stroke (section 4.4), although it occurs on a smaller scale. No doubt, a flash is also produced by a powerful impulse corona giving rise to a new streamer zone of the elongated leader. Photographs show that the transverse dimension of a step flash is about 10 m [38].

4.7 The subsequent components. The M-component

The processes in the lightning channel following the first component are known as subsequent components. Of interest among these are so-called M-components and dart leaders. In the first case, the current impulse registered at the earth has a very smooth front (0.1–1 ms), a similar duration and an amplitude of several hundreds of amperes, sometimes of 1–2 kA. The channel radiation intensity increases abruptly during the impulse, but one can hardly identify in the photographs a structure similar to the impulse front. The current impulse of an M-component is always registered against the background of about 100 A continuous current of the interpulse pause. For a dart leader to arise, this current must necessarily be cut off [39, 40]. A few microseconds after the cut-off, a short high-intensity region – a dart leader tip – runs down to the earth along the previous channel with a velocity of $\sim 10^7$ m/s. The contact of the dart leader with the earth produces a return stroke with its typical characteristics but having a much shorter impulse front than in the first component (less than 1 μs or even 0.1 μs in some impulses). It is hard to say anything definite about the lower limit of the front duration: it is likely to lie beyond the time resolution of the measuring equipment.

The papers published almost simultaneously [41, 42] interpret the subsequent component qualitatively as representing the discharge, into the earth, of an intercloud leader after its contact with the upper end of the preceding grounded but still conductive channel. Here we describe the evolution of an M-component in terms of a numerical simulation.

The model underlying the simulation is as follows (figure 4.23). Initially, there is a grounded plasma channel of length H_1 with zero potential, which was left behind by the preceding lightning component. At time $t = 0$, a leader channel of length H_2 and potential U_i joins it in the clouds (the voltage drop from the leader current and from the intercomponent current is neglected). The short process of the channel commutation through the streamer zone

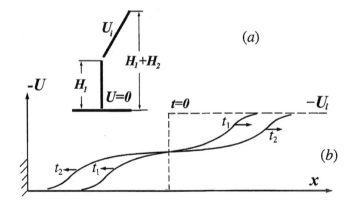

Figure 4.23. The formation of a subsequent component: (*a*) the grounded channel of the previous component and an intercloud leader; (*b*) channel charging-discharging waves.

of the intercloud leader is ignored. At the moment of closure, the leader channel possesses a typical resistance $R_{1L} \approx 10\,\Omega/m$. The resistance R_{1c} of the previous channel depends on the duration of the intercomponent pause. After the return stroke current impulse of the previous component

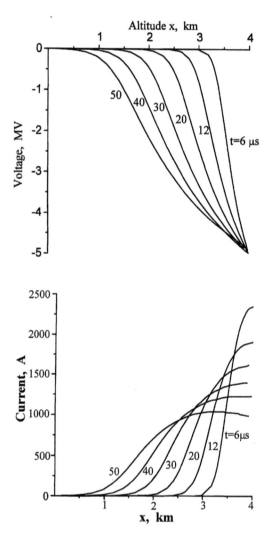

Figure 4.24. Simulation of the M-component on closing an intercloud leader 2 km in length and 10 MV potential on a 4-km grounded channel. The initial linear resistances of the channels $R_1 = 10\,\Omega/m$ and the steady field $E_L = 10\,V/cm$. The waves of potential (this page, top), current (this page, bottom), the power of Joule losses (opposite, top) and the current impulse at the grounded end of the channel (opposite, bottom); for comparison, the latter is also given for $R_1 = 20\,\Omega/m$ and $E_L = 20\,V/cm$ (curve B).

is damped, the channel resistance increases gradually due to the gas cooling. But if the intercomponent current is comparable with the leader current, as is usually the case by the moment the M-component arises, the increased resistance of the grounded channel may be suggested to be limited by the value of $R_{1c} \approx r_{1L}$. The reactive parameters of both lines, which are not very sensitive to the channel plasma state, can be taken to be identical to those of the leader: $C_1 \approx 10\,\mathrm{pF/m}$ and $L_1 \approx 2.7\,\mu\mathrm{H/m}$.

During the intercloud leader discharge into the earth via the preceding channel path, the channel resistances change, as in the return stroke

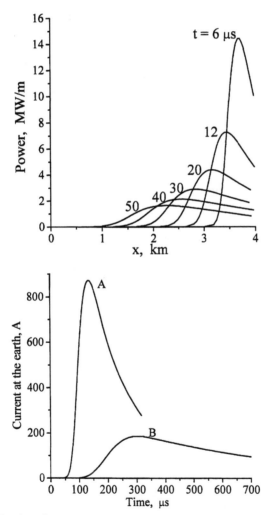

Figure 4.24. Continued.

(section 4.4.4). Suppose that these changes follow the relaxation law expressed by formula (4.38). This formula describes adequately the qualitative tendencies; there are no quantitative results to compare them with.

This process is described by the long line equations of (4.24) with the following initial and boundary conditions:

$$U(x,0) = 0 \quad \text{at } 0 \leqslant x \leqslant H_1,$$

$$U(x,0) = U_i \quad \text{at } H_1 \leqslant x \leqslant H_1 + H_2, \tag{4.46}$$

$$i(x,0) = 0, \qquad U(0,t) = 0, \qquad i(H_1 + H_2, t) = 0.$$

At the site of contact of the two lines, their potentials are identical at $t > 0$. After the contact of the intercloud leader with the grounded channel, current and voltage waves start running along both lines away from the point of contact. As a result, the grounded channel becomes charged while the leader channel is discharged. If the initial parameters of the lines are identical, the initial current at the point of contact is $i = U_i/2Z$, where Z is the wave resistance of the lines. Rapidly attenuated precursors run in both directions at velocities of electromagnetic signal, while the main current and voltage waves propagate via the diffusion mechanism (figure 4.24). These waves spread much stronger than in the return stroke, since the current and voltage are lower here (the more so that the initial voltage amplitude is half the value of U_i). The channel resistance decreases more slowly and the wave fronts become smooth instead of becoming steeper. The initial voltage U_i in the subsequent components seems to be lower on average than in the stepwise leader of the first component because this process involves the increasingly less mature cloud cells with lower charges. If we ignore the weak displacement current induced by the changing charges of the recharged channels, the current impulse at the earth can be registered only after the diffusion wave front has reached the earth. By that moment, the wave has become very diffuse, so the current impulse front appears to be very smooth (figure 4.24 (bottom, page 205)).† The more or less uniform power distribution along the channel is to look as a uniform enhancement of its radiation intensity. The calculations of this distribution and such current impulse characteristics as the front steepness, duration, and amplitude are similar to their observations in M-components. A still better agreement with the measurements can be attained by varying the parameters in the calculations, primarily R_{1L} and quasistationary field E_L in fully transformed channels. These arguments favour the above suggestions concerning the nature of lightning M-components.

---

† The current impulse of a return stroke is of a different form. The current amplitude is registered right after the short-term commutation process when the leader contacts the earth via its reducing streamer zone, which takes a few microseconds.

4.8   Subsequent components. The problem of a dart leader

There is still no clear understanding of the nature of a dart leader, so we shall discuss the scarce experimental data available and suggest a hypothesis based on them. Then we shall consider some possible consequences of this hypothesis and the difficulties that may arise. The dart leader problem remains unsolved but it cannot be put aside because of the importance of the dart leader process.

4.8.1   A streamer in a 'waveguide'?

There are no grounds to believe that the mechanism of dart leader initiation in the clouds is essentially different from that of a M-component. Rather, both processes result from the closure of an intercloud leader on a grounded channel remaining after the passage of the return stroke in the previous lightning component. But the potential wave running along this track to the earth, known as a dart leader, differs radically from that of an M-component. It has a well-defined front identifiable by the intense radiation of dart leader tip travelling to the earth with velocity $v_{dL} \sim 10^7$ m/s, which is at least by an order of magnitude higher than the typical velocities of the first stepwise leader. The ability of a dart leader to travel so fast is especially remarkable because its potential is most likely to be lower than that of a stepwise leader. This is indicated by the return stroke currents, which are on average 2–2.5 times lower ($I_M \approx U_i/Z$). The potential drop from the dart leader tip in the previous, still untransformed channel towards the earth must occur very quickly. This is indicated by a very fast front rise of the return current impulse, t_f. To gain the full current $I_M \approx U/Z$, where U is the potential carried by the dart leader, the return wave must run along a leader section with a rising potential δx and reach the totally charged portion of the channel. Therefore, we have $\Delta x \sim v_r t_f$, and if the return wave velocity is $v_r \approx 10^8$ m/s and $t_f \approx 0.1\,\mu s$, the length of the region with an abrupt potential drop in the dart leader front is $\delta x \approx 10$ m. It is quite possible that this value is actually smaller because the return wave cannot at first gain the full return stroke velocity $v_r \approx 10^8$ m/s. On the other hand, the potential drop region should not be shorter than $\Delta x \approx v_{dL} t_f \approx 1$ m, since the cross section of a channel with total potential U approaches the earth at velocity v_{dL}. Such a steep front of 1–10 m is unattainable not only by a diffusion wave with its potential varying along many hundreds of metres (figure 4.24) but even by an 'ordinary' leader of the first lightning component, in which Δx is determined by the streamer zone length. At the moment of contact with the earth, the latter is measured in dozens of metres at the tip potential of 20–30 MV. This is the reason why the time necessary for the return wave current to grow to its maximum value is dozens of times longer than that for the dart leader.

It follows from the foregoing and the fact that a dart leader travels as fast as a very fast streamer that the former has no streamer zone which would serve as the primary prerequisite for a leader mechanism. It appears that the dart leader, contrary to its name, is essentially not a leader, although it has a charge cover, which it has to acquire under somewhat different circumstances (see below). Nor does it look like a diffusion wave of the M-type. The latter would have an order of magnitude higher velocity and a very diffuse front.

A dart leader looks more like the oldest of the known types of propagating plasma channel – a streamer, whose head represents an ionization wave. The velocity of a dart leader is close to that of a high voltage streamer. The principal reason for a streamer channel being non-viable in air – a rapid loss of conductivity by the cold plasma – is very weak in this case. A dart leader follows the track heated by the preceding component, so that the still-hot track serves as a kind of waveguide to the leader. The high gas temperature greatly retards electron losses. Therefore, the possibility for the region behind the tip to be heated to arc temperatures increases considerably. This provides a stable highly conductive state inherent in a 'classical' leader.

The preheated air pipe serves another, probably more important, function. Its hot and rarefied air is surrounded laterally by cold dense air. Since the rate of ionization due to the field is described by the E/N ratio, the radial expansion of the channel region behind the streamer tip is abruptly retarded as compared with the forward motion of the ionization wave. So the air mass to be heated by the current is reduced, permitting the channel gas to be heated to a higher temperature. The cold air restricts the channel expansion because it acts as a charge cover produced by the streamer zone of the leader.

One should not think that the channel does not expand through the ionization mechanism at all. This process is just much slower than the forward motion of an ionization wave towards the earth, so most of the Joule heat is released into the yet unexpanded channel having a smaller radius. The radial field leads to the channel expansion only at the beginning, as is the case with common streamers (section 2.2.2). When the radial field is somewhat reduced, the channel becomes the source of a radial streamer corona which does not require a high field. Radial streamers rapidly lose their conductivity in cold air, and their immobile charges form a cover of the type that surrounds a common leader channel. Now, though with some delay, the mechanism of radial field attenuation and hot channel stabilization comes into action. Thus, a dart leader, being essentially a streamer (i.e., an ionization wave having no streamer zone in front of the tip), must possess a charge cover, as a leader. Unlike the case with a common leader, the cover is not inherited from a streamer zone but is formed entirely behind the tip, which is the seat of the principal processes driving the dart leader. (In a classical leader, the cover formation partly continues behind the tip, as described in section 2.2.4.)

The streamer mechanism of the dart leader development due to impact ionization of the gas in the strong field of the tip can manifest itself only if the conductivity in the channel of the previous component has dropped below a critical value by the time the next lightning component is to arise. This is unambiguously supported by the following observations. The M-component is produced against the background of a continuous current of the inter-pause, whereas the dart leader arises some time after this current is cut off. As long as the medium preserves a high conductivity, the diffuse penetration of the field and current prevents the ionization wave propagation. The diffu-sion wave has practically no ionization due to a direct action of the low field. The medium pre-ionization does not stimulate but rather hampers the propa-gation of the ionization wave. The latter requires a strong field, but the high conductivity of the medium in front of the wave leads to the field dissipation. In order to focus the potential drop to a narrow region, one must stop the charge flux (electric current) by concentrating charge in a narrow region to produce a strong field. The charge flux can be 'locked in' only by creating resistance to it if one places a poor conductor in front of the well-conducting portion of the channel.

4.8.2 The non-linear diffusion wave front

At this point, we have to make a short digression to discuss the structure of the near-front region of a diffusion potential wave. One will see later that this is directly related to the ionization wave problem. The diffusion wave velocity v is determined by the propagation process along the whole wave length. Its variation along the path from the cloud to the earth is illustrated in figure 4.24. By order of magnitude, the velocity of a non-linear wave is $v \approx \chi_{av}/x_f$, where x_f is its total length from the source to the initial front point and χ_{av} is an averaged diffusion coefficient in the transformed channel behind the front, which better fits the final linear resistance of the channel than to its initial resistance. If constant potential U_i is applied to the initial channel end, the value of χ_{av} does not change much. The velocity changes appreciably over the time, during which the wave covers a distance compar-able with that between the cloud and the earth. But its change is relatively small over the time the wave covers a distance of the order of its front width where the potential $U(x)$ rises steeply. This means that we have $U(x, t) \approx U(x - vt)$ in the wave front, and the distributions of all parameters along the x-axis are quasistationary in the coordinate system related to the moving front (as in a non-linear heat wave [12]). With this circumstance in mind, we can rewrite the potential diffusion equation (4.35) as

$$-v\frac{\mathrm{d}U}{\mathrm{d}x} = \frac{\mathrm{d}}{\mathrm{d}x}\chi\frac{\mathrm{d}U}{\mathrm{d}x}, \qquad \chi = \frac{1}{R_1 C_1}. \tag{4.47}$$

Taking into account $E = -dU/dx = 0$ and $U = 0$ in front of the wave at $x \to \infty$, the integral of this equation is

$$vU = -\chi \frac{dU}{dx}, \qquad i = \frac{E}{R_1} = C_1 Uv = \tau v. \tag{4.48}$$

The familiar relation $i = \tau v$ is valid at every point of the quasistationary wave portion but not only at the site of the current cut-off in front of the streamer or leader tip.

The energy W_1 per unit length of the quasistationary channel is described as

$$\frac{dW_1}{dt} = -v\frac{dW_1}{dx} = i^2 R_1 = \frac{v^2 C_1 U^2}{\chi} \tag{4.49}$$

and is expressed directly through the local potential. Indeed, reducing the rank of the set of equations (4.48) and (4.49) by dividing them by one another, we find

$$W_1 - W_{10} = C_1 U^2 / 2 \tag{4.50}$$

where W_{10} is the initial energy in the channel far out the wave front. Thus the statement repeatedly used in evaluations that the energy dissipated in the channel is of the same order as that stored in its capacitance is valid exactly in the stationary case.†

We shall consider moderate waves, when the gas is heated at constant pressure, and the Joule heat is released at constant mass $m = \pi r^2 \rho = \pi r_0^2 \rho_0$ per unit channel length (r_0 and ρ are the initial radius and gas density in front of the wave). Then we have $W_1 = mh$, where h is the specific gas enthalpy. Assume for simplicity that thermodynamically equilibrium ionization is established at every point of the wave, so that conductivity σ and $\chi = m\sigma(\rho C_1)^{-1}$ are the functions of temperature T or $h(T)$. Then χ is unambiguously related to U through formula (4.50).‡ With (4.48)–(4.50), finding the distributions along the wave reduces to the quadrature

$$\int \frac{\chi(h)\,dh}{h - h_0} = -2vx, \qquad U = \left[\frac{2m(h - h_0)}{C_1}\right]^{1/2}, \qquad h_0 = \frac{W_{10}}{m}. \tag{4.51}$$

Let us calculate the integral by approximating the relationship $\chi \approx \sigma/\rho$ by the power function $\chi = Ah^n$ in the temperature range typical of the wave. The coordinate origin $x = 0$ is taken at an arbitrary point of the wave front

---

† This is quite natural because under the problem conditions the channel is not created anew but exists from the very beginning with its linear capacitance C_1. Then every channel portion is charged as lumped capacitance (cf. the comment on formula (2.17) in section 2.2.4).

‡ In sections 4.7 and 4.4, the quantity χ was related to electrical parameters through relation (4.38) which refers to strong waves with a high energy release. If desired, one can use this relation after substituting $\partial G/\partial t$ by $-v\,dC/dx$ and doing the above operations.

start, in front of which ($x > 0$) the channel transforms very slightly, so that χ increases slightly, followed by ($x < 0$) where it changes noticeably. The parameters of the initial front point will be marked by the subindex 1, assuming for definiteness $\chi_1 = 2\chi_0$, where χ_0 corresponds to the initial channel conductivity. Then we have $h_1 - h_0 = \delta h_0$, where $\delta = 2^{1/n} - 1$. An exponentially damping tail of the electric field and current extends forward along the wave where the diffusion is 'linear':

$$\frac{U}{U_1} = \frac{E}{E_1} = \exp\left(-\frac{x}{\Delta x}\right), \qquad \Delta x = \frac{\chi_0}{v}, \qquad E_1 = \frac{U_1}{\Delta x} = \frac{vU}{\chi_0}. \qquad (4.52)$$

Within the front, where h exceeds h_0 considerably or, asymptotically at $x \to -\infty$, we have

$$\chi \approx 2nv(-x), \qquad U \approx \left(\frac{2m}{C_1}\right)^{1/2}\left[\frac{2nv(-x)}{A}\right]^{1/2n}, \qquad E = \frac{U}{2n(-x)}. \qquad (4.53)$$

By matching the approximate solutions asymptotically valid at $x \to +\infty$ and $x \to -\infty$, the parameters of the front start and the matching point coordinate x_1 can be found as

$$U_1 \approx \left[\left(\frac{2m}{C_1}\right)\delta\left(\frac{\chi_0}{A}\right)^{1/n}\right]^{1/2}, \qquad E_1 = \frac{U_1}{2nx_1}, \qquad x_1 = \frac{\Delta x}{2n}. \qquad (4.54)$$

This point is closer to the *a priori* position of the front start $x = 0$ than Δx, which justifies the approximations.

Let us illustrate this situation numerically with reference to the conditions typical of the M-component (figure 4.25). Suppose the diffusion wave has velocity $v = 10^8$ m/s running along a channel with the initial radius $r_0 = 1$ cm, temperature $T_0 = 5900$ K ($h_0 = 14.8$ kJ/g) and $\rho_0 = 5 \times 10^{-5}$ g/cm^3 which is by a factor of 25 less than the normal; $m = 1.54 \times 10^{-4}$ g/cm; $n_e \approx 1.8 \times 10^{14}$ cm^{-3}, the initial linear resistance $R_1 = 10\,\Omega/$m, $\chi_0 = 10^{10}$ m^2/s and $C_1 = 10$ pF/m. For the temperature range $T \approx 6$–$10\,000$ K in air at 1 atm, we have $\sigma/\rho \approx 17h^3$ (where $\sigma[(\Omega \cdot \text{cm})^{-1}]$, $\rho[\text{g/cm}^3]$, and $h[\text{kJ/g}]$). Hence, $A = 2.7 \times 10^6$ (m^2/s)(kJ/g)$^{-3}$ and $\delta = 0.25$. From formulas (4.51)–(4.53), we find for the initial front point $U_1 = 3.5$ MV, $E_1 = 2.2$ kV/cm, and

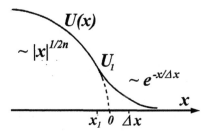

Figure 4.25. Schematic diagram of the non-linear diffusion wave front.

the effective field length before the wave front $\Delta x = 100\,\text{m}$. The point behind the wave with $U = 10\,\text{MV}$ ($h \approx 50\,\text{kJ/g}$, $T \approx 10\,000\,\text{K}$) lies at a distance $x = 500\,\text{m}$ from the front. There, $\chi \approx 3 \times 10^{11}\,\text{m}^2/\text{s}$, the resistance is by a factor of 30 lower than before the front, and the field drops to $33\,\text{V/cm}$. The field maximum lies near the initial front point. The qualitative picture presented in figure 4.25 agrees with the numerical results of figure 4.24(*a*).

4.8.3 The possibility of diffusion-to-ionization wave transformation

Let us define the conditions, under which a diffusion wave can transform to an ionization wave which is supposed to be a dart leader. Consider a simple situation. It is suggested that potential U_i is applied to the upper end of a grounded conductive channel of the previous lightning component. It begins to diffuse into the channel. It is assumed that there is no transformation and the initial conductivity corresponding to the diffusion coefficient χ_0 is preserved. The diffusion is 'linear' in this case. The potential and field vary as

$$U = U_i\left[1 - \text{erf}\left(\frac{x}{(4\chi_0 t)^{1/2}}\right)\right],$$

$$E = iR_1 = \frac{U_i}{(\pi\chi_0 t)^{1/2}} \exp\left(-\frac{x^2}{4\chi_0 t}\right). \tag{4.55}$$

At every point x, the field first rises with time but then falls after the maximum $E_{\text{max}} = 2(\pi e)^{-1/2} U_i/x$ at moment $t = x^2/2\chi_0$. The point E_{max} moves at velocity $v_g = \chi_0/x$, and the potential at this point is $U_m = 0.33 U_i$.

An ionization wave can be formed if the maximum field is sufficiently high and exceeds a certain critical value E_i. The ionization wave is assumed to propagate at velocity v_s supposed to be equal to that of a dart leader. Since $E_{\text{max}} \sim x^{-1} \sim t^{-1/2}$, the ionization wave could principally arise at earlier times when $E_{\text{max}} > E_i$, but if its velocity is $v_s < v_g$, is immediately overcome by a diffusion wave. This will not happen if v_g drops below v_s while E_{max} is still higher than E_i, i.e., if the conditions $v_s \geqslant v_g$, $E_{\text{max}} \geqslant E_i$ are fulfilled together. For this to happen, the diffusion coefficient must be smaller and the linear channel resistance larger:

$$\chi_{0\text{cr}} = \frac{2}{(\pi e)^{1/2}}\frac{v_s U_i}{E_i}, \qquad R_{1\text{cr}} = \frac{1}{C_1 \chi_{0\text{cr}}}. \tag{4.56}$$

For this, the gas temperature in the initial channel should not be high. On the other hand, for the 'waveguide' properties to manifest themselves, the temperature must be as high as possible to make the air rarefied. Because of the very sharp temperature dependence of conductivity (when it is low), these conditions are met only in a very short temperature range, $T \approx 3000\text{–}4000\,\text{K}$, where the air density is by a factor of $10\text{–}15$ lower than normal. For the estimations, we take $E_i = 3\,\text{kV/cm}$ corresponding to a field characteristic of initial air ionization, 30–$40\,\text{kV/cm}$ under normal conditions. Suppose $v_s = 10^7\,\text{m/s}$, $U_i = 5\,\text{MV}$ (such potential usually provides current $I_M \approx 10\,\text{kA}$ for the next component at $Z = 500\,\Omega$ during the return stroke), and $C_1 = 10\,\text{pF/m}$. We find $\chi_{0\text{cr}} \approx 1.1 \times 10^8\,\text{m}^2/\text{s}$ and $R_{1\text{cr}} = 880\,\Omega/\text{m}$. The resistance is two orders of magnitude higher than that supposed to precede the M-component. For this reason, a dart leader can appear only after the current cut-off during the interpause and a partial cooling of the channel.

In reality, the channel undergoes transformation due to the diffusion wave, its conductivity rises, and the field dissipates faster than what is expected from the second formula of (4.55). To provide for the critical conditions, the initial conductivity may seem to be lower than the estimated value. So we should consider the other extremal case when the diffusion wave heats a limited amount of air and an equilibrium ionization is established. The diffusion wave is now non-linear. Its maximum field is near the initial point of the wave front, and one should use, instead of (4.56), the last relation of (4.52) similar to it with $E_1 = E_i$. One should keep in mind that of interest are the temperatures 3000–$6000\,\text{K}$, at which σ varies with T much more strongly: $\sigma/\rho \approx 1.8 \times 10^{-6} h^9$ (the dimensionalities are the same as in the illustration of section 4.8.2). Now we have $n = 9$; at $v_s = 10^7\,\text{m/s}$ and channel radius $r_0 = 1\,\text{cm}$, we have $U_1 = 1.2\,\text{MV}$, $\chi_0 = 4 \times 10^7\,\text{m}^2/\text{s}$, and $R_{1\text{cr}} \approx 2500\,\Omega/\text{m}$. The field extends before the wave front only for $\Delta x = 4\,\text{m}$. The electron density in the initial channel under critical conditions is $n_e \approx 6 \times 10^{12}\,\text{cm}^{-3}$, corresponding to its temperature $4000\,\text{K}$.

4.8.4 The ionization wave in a conductive medium

The values obtained in section 4.8.3 on two extremal assumptions do not differ much and seem to be reasonable. The problem of the conditions necessary for a dart leader to arise may seem to have been solved. This optimism will, however, disappear as soon as one evaluates the parameters of an ionization wave when it propagates through a medium with critical conductivity.

Consider a wave in the front-related coordinate system, as was done in section 4.8.2. The equation for field (4.48) will be supplemented by an ionization kinetics equation written directly for χ because $\chi \sim \sigma \sim n_e$:

$$-v\frac{d\chi}{dx} = \nu_i \chi, \qquad \nu_i = N f(E/N). \tag{4.57}$$

This equation describes a new law for the channel transformation. Owing to (4.48), the ionization frequency ν_i turns to the potential function. Then, by dividing (4.48) and (4.57), the problem is reduced to the equation for $\chi(U)$ and the quadrature, as in section 4.8.2.

To advance further, one should choose the function $f(E/N)$ in a way suitable for integration.† But difficulties and doubts arise immediately here. The approximation of ν_i by the power function $\nu_i = bE^k$, as in the streamer theory when this approximation with $k = 2.5$ provided fairly good results, does not work in this case. The ionization wave propagating through a conductive medium appears absolutely diffuse, as in any other channel transformation law: (4.38) or on the assumption of equilibrium ionization (section 4.8.2). Let us impart a threshold nature to the function $\nu_i(E)$ in a simple way – $\nu_i = 0$ at $E < E^*$ and $\nu_i = \text{const}$ at $E > E^*$. This is what was done by the authors of [43] when solving a similar problem for the laboratory ionization wave in a tube. The integration of (4.48) and (4.57) by the above method yields the following result. The change in the electron density and χ in the ionization wave is defined by the ratio of potentials U_2 and U_1 at the points of ionization outset and onset, where $E = E^*$. A potential 'tongue' of effective length $\Delta x = \chi_0/v$ extends in front of the ionization wave, as in other diffusion modes. The potential at the front is $U_1 = E^* \Delta x$. The parameter ratios at the wave boundaries are

$$\frac{n_{e2}}{n_{e0}} = \frac{\chi_2}{\chi_0} = \frac{U_2}{U_1} \approx \exp\left(\frac{v^2}{\nu_i \chi_0}\right). \tag{4.58}$$

This relation can be regarded as the dependence of the wave velocity on an 'external potential' U_2 applied to its back. On the other hand, the velocity is expressed by a formula similar to (2.2) for the streamer:

$$v = \frac{\nu_i \Delta x_i}{\ln(\chi_2/\chi_0)}, \qquad \Delta x_i = \frac{v^2}{\nu_i^2 \Delta x}, \qquad \frac{v^2}{\nu_i \chi_0} > 1 \tag{4.59}$$

where Δx_i is the extension of the ionization region from the initial to the final point of the wave. For a wave to survive, its parameters must meet the last inequality of (4.59). Otherwise, the field within the wave will be unable to exceed E^*, so no ionization will occur.

The capabilities of an ionization wave are limited, and this limit increases with increasing initial conductivity of the medium. For example, if the initial parameters $n_{e0} \approx 10^{12}\,\text{cm}^{-3}$ and $\chi_0 \approx 10^7\,\text{m}^2/\text{s}$ were even lower than the critical values found in section 4.8.3 and if the threshold field was $3\,\text{kV/cm}$, it would be necessary to have the ionization frequency $\nu_i = 2.1 \times 10^6\,\text{s}^{-1}$ and potential $U_2 = 300\,\text{MV}$ in order to increase n_e and χ by three orders of magnitude ($\Delta x = 1\,\text{m}$, $U_1 = 0.3\,\text{MV}$) and $U_2 = 30\,\text{MV}$ by two orders. The wave width in this case is $\Delta x_i \approx 22\,\text{m}$, i.e., it is very extended. Only when

---

† Sometimes, it seems better to describe ν_i the function of E and, on the contrary, to remove U from (4.48) and (4.57). Instead of (4.48), we then get

$$\frac{dE}{dx} = E\left(\frac{\nu_i}{v} - \frac{v}{\chi}\right), \qquad \nu_i = \nu_i(E)$$

the initial conductivity is still an order of magnitude lower ($n_{e0} = 10^{11}$ cm^{-3}, $\chi_0 = 10^6$ m^2/s, and $R = 10^5$ Ω/m), the wave width begins to approach what would be desired for a dart leader. In the same medium at the same v and E^*, the parameters necessary for the ratio $\chi_2/\chi_0 = 10^3$ would be $v_i = 1.4 \times 10^7$ s^{-1}, $U_2 = 30$ MV, $\Delta x = 10$ cm, $\Delta x_i = 5$ m, and $U_1 = 30$ kV. A still narrower region of the potential rise would be obtained at a still lower initial conductivity. But then we approach the applicability limits of the basic concepts of the theory of perturbation propagation in a conductive medium and of the long line theory, and we are probably coming closer to the understanding of criteria for the dart leader production.

4.8.5 The dart leader as a streamer in a 'nonconductive waveguide'

The diffusion mechanism of field evolution in a channel, or in a long line, is incompatible with abrupt potential changes and, hence, with strong fields. If abrupt changes do arise, they are rapidly smeared by diffusion. We believe for this reason that neither a narrow ionization wave nor a dart leader can be formed in a well-conducting channel. To find the conditions, in which a very strong field can be induced, we should remind ourselves of the prerequisites for the long line equations.

The electrostatics equation for cylindrical geometry has the form:

$$\frac{\partial E_x}{\partial x} + \frac{1}{r}\frac{\partial}{\partial r}rE_r = \frac{\rho}{\varepsilon_0} \tag{4.60}$$

where ρ is space charge density. By integrating, in the cross section, a conductor of radius r_0 and neglecting the dependence of the longitudinal field E_x on r, we obtain

$$\pi r_0^2 \frac{\partial E_x}{\partial x} + 2\pi r_0 E_{r_0} = \frac{\tau}{\varepsilon_0}, \qquad \tau = \int_0^{r_0} 2\pi r \rho \, dr \tag{4.61}$$

where E_{r_0} is the radial field on the surface of a conductor of length $l \gg r_0$:

$$E_{r_0} \approx \frac{U}{r_0 \ln(l/r_0)}, \qquad C_1 = \frac{2\pi\varepsilon_0}{\ln(l/r_0)}. \tag{4.62}$$

If the longitudinal field varies along the channel so slowly that the axial divergence can be neglected (the characteristic length for the variation of E_x is $\Delta x \gg r_0$), we arrive at one of the basic conceptions of the long line theory, $\tau(x) = C_1 U(x)$, whose implication is the potential diffusion mechanism. It is suggested implicitly that the resistance varies very slowly along the channel, so this variation cannot be an obstacle to a charge flux, making the flux velocity decrease abruptly and create a space charge due to its local accumulation (a long line has no 'jams').

However, space charge does accumulate at a sharp boundary between a poorly- and a well-conducting channel portion. A charged tip is formed at the end of an ideal (or non-ideal) conductor, the potential in front of it drops abruptly, at distances about equal to r_0, inducing there a strong field capable of sustaining an ionization wave. This is what happens in a common streamer in a non-conductive medium. It is then clear what is necessary to support a sharp potential drop at the ionization wave front for a long time. The conductivity along the perspective trajectory must drop to a value low enough for the diffusion field tongue to be unable to smear the sharp potential drop. Therefore, the tongue length must become comparable with the channel radius $\Delta x \approx \chi_0/v \sim r_0$. Because of the strong temperature dependence of the degree of equilibrium ionization in air at low temperatures, a drop to $T \approx 3000 \, \mathrm{K}$ would be sufficient. The equilibrium electron density established for the long zero-current pause will be $n_{e0} \sim 10^{10}\!-\!10^{11} \, \mathrm{cm}^{-3}$; hence, $R_{10} \sim 10^5\!-\!10^6 \, \Omega/\mathrm{m}$, and $\chi_0 \sim 10^6\!-\!10^5 \, \mathrm{m}^2/\mathrm{s}$. But the air density in the cooled channel of the previous component at $T \approx 3000 \, \mathrm{K}$ is by an order of magnitude lower than that of cold air, so that the conductivity drop will not interfere with the 'waveguide' properties of the track.

The velocity of a dart leader as an ionization wave is defined, in order of magnitude, by the same formula (2.2) as the streamer velocity. But the 'pre-ionization' in this case $(n_{e0} \sim 10^{10}\!-\!10^{11} \, \mathrm{cm}^{-3})$ is considerable, and a much smaller number of electron generations $(\ln(n_{e2}/n_{e0}) \approx 5)$ is to be produced in the wave. With the account of the similarity law for ν_i at an order of magnitude lower gas density, the ionization frequency is $\nu_i \sim 10^{10} \, \mathrm{s}^{-1}$ and $r_0 \sim 1 \, \mathrm{cm}$; then we obtain a correct order of the velocity $v \approx \nu_i r_0 / \ln(n_{e2}/n_{e0}) \sim 10^7 \, \mathrm{m/s}$.

One cannot say that all the details of the dart leader behaviour have been clarified by the above considerations. For the dart leader channel to be well conductive, the electron density in it must be at least 5–6 orders of magnitude higher than the initial value for the track. But the capabilities of the ionization wave to produce more electrons are limited. The maximum conductivity of an ionization wave propagating through a non-conductive medium is defined, in order of magnitude, by the relation $\sigma_{max}/\varepsilon_0 \sim \nu_i$ (section 2.2.2), because the space charge of the streamer tip, providing a strong ionization field is dissipated with the Maxwellian time $\tau_M = \varepsilon_0/\sigma$.† After the wave

---

† It also determines the rate at which the linear charge $\tau = C_1 U$ is established, if it is, in the channel. Let us integrate the relation for charge conservation in the conductor cross section. Neglecting, for simplicity, the variation in σ along the channel length, we obtain

$$\frac{\partial \rho}{\partial t} + \sigma \frac{\partial E_x}{\partial x} = 0, \qquad \frac{\partial \tau}{\partial t} + \pi r_0^2 \sigma \frac{\partial E_x}{\partial x} = 0.$$

Using (4.61) and (4.62), we arrive at a refined equation for the relation between τ and U:

$$\frac{\varepsilon_0}{\sigma} \frac{\partial \tau}{\partial t} + \tau = C_1 U.$$

The postulate of the long line theory, $\tau = C_1 U$, is valid if the changes in the system, which also define $\tau(t)$, occur slower than with $\tau_M = \varepsilon_0/\sigma$. When applied to the wave front moving at velocity v in a line with conductivity σ_0, this happens at $\sigma_0 \gg v\varepsilon_0/r_0$ and $\chi_0 \gg vr_0$.

has passed, the channel still needs to be heated and ionized, but both processes are to occur in a moderate electric field, as in a classical leader channel. Besides, this must take place before a strong radial field makes the channel expand beyond the hot gas tube, or if it has already become enveloped by a stabilizing charge cover (section 4.8.1).

There are still many questions about the processes in a dart leader that remain to be answered; the development of its quantitative theory is also a task of further research.

To conclude, it is worth noting some specific features of a current impulse in the return stroke of subsequent components. Generally, the impulse duration is related to the time it takes the return stroke to run along the whole channel. For the subsequent components, this time must be longer than for the first component due to the attached intercloud leader. But the impulse duration in the subsequent components is about twice as short, although the return wave velocities are generally the same. The reason for this difference is likely to be the absence of branches in a dart leader. It is quite possible that the relatively slow process of their recharging elongates the current impulse tail of the first component. The impulses of the subsequent components do not reverse the sign, similarly to those of the first one. In the absence of branches, the action of the reflected wave can no longer be screened by the randomly reflected waves of the numerous branches (section 4.4.5). The hypothesis of 'white noise' should, probably, be discarded as being inadequate. This problem, like the others above, awaits its solution.

4.9 Experimental checkup of subsequent component theory

The theoretical treatment of processes occurring in the channel of the previous component has been reduced to the various wave propagation modes – the diffusion mode in the M-component and the ionization wave mode in the dart leader. The former has a strongly elongated front with a slowly varying potential, and the latter must possess a tip with a concentrated charge, producing an abrupt potential change. Indirect evidence for the significant difference in the field distribution is the registrations of current impulses at the earth. The impulse front durations are found to differ by 2–4 orders of magnitude between an M-component and a dart leader. There is a possibility for a direct experimental evaluation of the potential distribution in a wave approaching the earth. This can be done by measuring the electric field gain at the earth during the wave motion. If the potential slowly rises along the whole wave length, as in an M-component (figure 4.24(a)), the distributions of the potential and linear charge from the initial front point, located at height h, to the cloud can be considered to be linear, $\tau(x) = a_q(x - h)$ (x is counted from the earth and $x \geqslant h$). For the field at distance r from a vertical channel, we find

$$\Delta E(r) = \frac{a_q}{2\pi\varepsilon_0}\left[\ln\frac{H+(H^2+r^2)^{1/2}}{h+(h^2+r^2)^{1/2}} - \frac{H-h}{(H^2+r^2)^{1/2}}\right] \qquad (4.63)$$

where H is the height of the grounded channel. If H is, at least, several times larger than r, the dependence of field ΔE on the distance between the registration point and the channel line will be only logarithmic. The same is true of the front duration of a field pulse.

The situation must be quite different for a dart leader with the abrupt potential drop at the wave front, since the first approximation in the field calculation may assume a uniform potential along the channel and $\tau(x) = $ const at $x > h$. This gives formulae (3.6) and (3.7), which yield the maximum value $\Delta E_{max}(r) \sim r^{-1}$. Such a large difference in the field variation is easily detectable experimentally, especially if we remember that it concerns not only the field pulse amplitude but also its front rise time. To see that this is so, it is sufficient to introduce into (4.63) and (3.6) the h-coordinate for the wave front, expressed through the respective velocities: $h = H - vt$.

Triggered lightning is a perfect source for such measurements. A triggered lightning is initiated by launching a small rocket raising a very thin wire which evaporates during the development of the first component. The point of contact of the lightning with the earth is strictly defined, so it is easy to position current detectors at the necessary distances. Besides, the channel at the earth follows the wire track and is strictly vertical, as is implied in the numerical formulae. Such measurements have been partly made [44–45]. In section 3.5, we discussed the measurements of field ΔE at distances $r_1 = 30$ m and $r_2 = 500$ m from the channel during the dart leader development. These measurements were not synchronized. However, the ratio $\Delta E(30)/\Delta E(500) = 17.4$ for approximately equal currents is nearly the same as $r_2/r_1 = 16.7$.

The field measurements for M-components have been reported only for $r = 30$ m [42]. The oscillogram of $\Delta E(t)$ is accompanied by a simultaneous registration of a current impulse with the amplitude of 800 A and the front rise time ~ 100 µs. The duration of the impulse front ΔE is approximately the same, but the field reaches its maximum of 1350 V/m earlier, when the current has reached half of its maximum amplitude (until the potential wave arrives, the current at the earth is zero, whereas the field begins to rise since its start). Figure 4.26 shows the calculated functions $i(t)$ and $\Delta E(t)$ at the observation points with $r = 30$ m and 500 m. The long line model described in section 4.7.1 was used with the same $C_1 = 10$ pF/m, $L_1 = 2.7$ µH/m, and $R_1(0) = 10\,\Omega$/m. The length of the grounded channel was 4000 m and that of the intercloud leader contacting it was 2000 m. The experimentally observed current of 800 A was reproduced in the calculation at the leader potential $U_i = 9.7$ MV. Under these conditions, the field

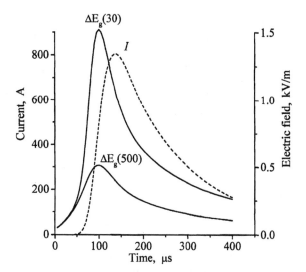

Figure 4.26. Calculated variations of the electric field at the earth's surface due to the M-component under the conditions of figure 4.24. The dashed curve shows the current impulse I.

amplitude of 1500 V/m at the point $r = 30$ m is close to the measured value. It follows from figure 4.26 that the temporal parameters of the current impulse are also consistent with the measurements. At the point $r = 500$ m, the calculated field amplitude is a factor of three smaller and the time for the maximum amplitude is nearly the same as for $r = 30$ m. Both parameters would differ by an order of magnitude in a dart leader with this increase in r. Therefore, the diffusion model of the M-component reproduces fairly well the available observations. It would, certainly, be most desirable to make simultaneous field registrations at different distances from a grounded lightning channel.

References

[1] Berger K, Anderson R B and Kroninger H 1975 *Electra* **41** 23
[2] Idone V P and Orville R E 1985 *J. Geophys. Res.* **90** 6159
[3] Antsurov K V, Vereschagin I P, Makalsky L M *et al* 1992 *Proc. 9th Intern. Conf. on Atmosph. Electricity* **1** (St Peterburg: A I Voeikov Main Geophys. Observ.) 360
[4] Vereschagin I P, Koshelev M A, Makalsky L M and Sysoev V S 1989 *Izvestiya. Akad. Nauk SSSR, Energetika i transport* **4** 100
[5] Simpson G C and Robinson G D 1941 *Proc. R. Soc. London A* **117** 281
[6] Kasemir H W 1960 *J. Geophys. Res.* **65** 1873

[7] Gorin B N and Shkilev A V 1976 *Elektrichestvo* **6** 31
[8] Proctor D A 1971 *J. Geophys. Res.* **76** 1078
[9] Mazur V, Gerlach J C and Rust W D 1984 *Geophys. Res. Lett.* **11** 61
[10] Mazur V, Rust W D and Gerlach J C 1986 *J. Geophys. Res.* **91** 8690
[11] Raizer Yu P 1991 *Gas Discharge Physics* (Berlin: Springer) p 449
[12] Zel'dovich Ya B and Raizer Yu P 1968 *Physics of Shock Waves and High-Temperature Hydrodynamic Phenomena* (New York: Academic Press) p 916
[13] Schonland B 1956 *The Lightning Discharge. Handbuch der Physik* **22** (Berlin: Springer) 576
[14] Orvill R E 1999 *J. Geophys. Res.* **104**
[15] Gorin B N and Shkilev A V 1974 *Elektrichestvo* **2** 29
[16] Bazelyan E M and Raizer Yu P 1997 *Spark Discharge* (Boca Raton: CRC Press) p 294
[17] Abramson I S, Gegechkori N M, Drabkina S I and Mandel'shtam S L 1947 *Zh. Eksper. i Teor. Fiz.* **17** 862
[18] Drabkina S I 1951 *Zh. Eksper. i Teor. Fiz.* **21** 473
[19] Dolgov G G and Mandel'shtam S L 1953 *Zh. Eksper. i Teor. Fiz.* **24** 691
[20] Braginsky S N 1958 *Soviet Phys. JETP* **7** (Eng. Trans.) 1068
[21] Zhivyuk Yu N and Mandel'shtam S L 1961 *Soviet Phys. JETP* **13** (Eng. Trans.) 338
[22] Plooster M N 1971 *Phys. Fluids* **14** 2111 and 2124
[23] Paxton A N, Gardner R L and Baker L 1986 *Phys. Fluids* **29** 2736
[24] Sneider M N 1997 *Unpublished report*
[25] Gorin B N and Markin V I 1975 in *Research of Lightning and High-Voltage Discharge* (Moscow: Krzhizhanovsky Power Engineering Inst.) p 114 (in Russian)
[26] Bazelyan E M, Gorin B N and Levitov V I 1978 *Physical and Engineering Fundamentals of Lightning Protection* (Leningrad: Gidrometeoizdat) p 223 (in Russian)
[27] Gorin B N 1985 *Elektrichestvo* **4** 10
[28] Gorin B N 1992 *Proc. 9th Intern. Conf. on Atmosph. Electricity* **1** (St Peterburg: A I Voeikov Main Geophys. Observ.) 206
[29] Jordan D M and Uman M A 1983 *J. Geophys. Res.* **88** 6555
[30] Rakov V A and Uman M A 1998 *IEEE Trans. on EM Compatibility* **40** 403
[31] Berger K 1977 in *Lightning, vol. 1, Physics Lightning* (R Golde (ed) New York: Academic Press) p 119
[32] Berger K and Vogrlsanger E 1966 *Bull. SEV* **57**(13) 1
[33] Schonland B, Malan D and Collens H 1935 *Proc. Roy. Soc. London Ser A* **152** 595
[34] Schonland B, Malan D and Collens H 1938 *Proc. Roy. Soc. London Ser A* **168** 455
[35] Orvill R E 1968 *J. Geophys. Res.* **73** 6999
[36] Orvill R E and Idone V P 1982 *J. Geophys. Res.* **87** 11177
[37] Krider E P 1974 *J. Geophys. Res.* **79** 4542
[38] Uman M A 1969 *Lightning* (New York: McGraw Book Company) p 300
[39] Fisher R G, Rakov V A, Uman M A *et al* 1993 *J. Geophys. Res.* **98** 22887
[40] Fisher R G, Rakov V A, Uman M A *et al* 1992 *Proc. 9th Intern. Conf. on Atmosph. Electricity* **3** (St Petersburg: A I Voeikov Main Geophys. Observ.) p 873

[41] Bazelyan E M 1995 *Fiz. Plazmy* **21** 497 (Engl. transl.: 1995 *Plasma Phys. Rep.* **21** 470)

[42] Rakov V A, Thottappillil R and Uman M A 1995 *J. Geophys. Res.* **100** 25701

[43] Sinkevich O A and Gerasimov D N 1999 *Fiz. Plazmy* **25** 376 (Engl. transl.: 1999 *Plasma Phys. Rep.* **25** 339)

[44] Rubinstein M, Rachidi F, Uman M A *et al* 1995 *J. Geophys. Res.* **100** 8863

[45] Rakov V A, Uman M A, Rambo K J *et al* 1998 *J. Geophys. Res.* **103** 14117

Chapter 5

Lightning attraction by objects

In this chapter, we shall describe the way a lightning channel chooses a point to strike (a terrestrial or a flying body). This is the principal issue for lightning protection technology. In any case, a direct stroke is more hazardous than a remote lightning effect via the electromagnetic field or shock wave in the air. Historically, direct lightning strokes were observed earlier than indirect ones, and the first research into lightning protection problems was associated with direct strokes.

Everyday experience and scientific observations, including those made as far back as the 18th century, indicate that lightning most often strikes individual structures elevated above the earth. These may be towers, churches, houses on high open hills, and just high trees. Today, this list is much longer and includes power transmission lines, transmitting and receiving antennas, and the like. The experience in maintaining such structures indicates that the frequency of strokes increases with the object's height. This observation was used as a basis for the most common lightning protection techniques. A grounded rod higher than the object to be protected – a lightning rod – put up in the vicinity of the object is supposed to attract most strokes, thus protecting the object. The underlying principle of this approach has not changed since the first lightning rod was constructed two and a half centuries ago. What has changed is the requirement for the protection reliability, which have become extremely stringent. For this reason, the specialists have to deal with exceptions rather than the rules, focusing on the rare cases of lightning breakthroughs to the object being protected, because they lead to emergencies and sometimes to catastrophes.

The study of lightning attraction mechanisms is extremely time-consuming and expensive. Even a simple measurement of the number of lightning strokes at objects of various heights is very hard to arrange. Most apartment houses and industrial premises in Europe are less than

50 m high. On the average, a lightning strokes a 50 m building once in five years. Every kilometre of a power transmission line 30 m high attracts approximately one lightning discharge per year. Long-term observations of a large number of buildings and multi-kilometre transmission lines are necessary to accumulate a representative statistics. The difficulties increase many-fold when one needs to extract information on the protection reliability from the observational statistics. To illustrate, 10–20 years of continuous observations of a 50 m building would be required to obtain information on the lightning discharge frequency, and at least 1000 years would be necessary to check whether its lightning rod can really provide a '99% protection' promised by the rod producers.

In a situation like this, one has to resort to theoretical evaluations, and this is one reason why lightning attraction theory has been the focal point of research for many lightning specialists. Here, as in many other lightning problems, there is an acute lack of factual data. The available evidence obtained from laboratory investigations on long sparks does not always provide an unambiguous interpretation, and this makes one treat with caution many, even generally accepted, concepts. We shall focus on the most advanced approaches, discussing, where necessary, alternative hypotheses.

5.1 The equidistance principle

This approach is oldest and clearly correct in its theoretical formulation. Suppose that an object of small area and height h is located on a flat earth's surface (it is a rod electrode in laboratory simulations). Let us assume further that a lightning channel is shifted from it horizontally at a distance r, and the channel tip is at an altitude H_0 (figure 5.1). In order to predict whether the lightning will strike the object or the earth, we shall take into account the breakdown voltage measurements of long air gaps with a sharply non-uniform electric field. They show that the longer the gap, the higher the average voltage required for its breakdown and the longer the time necessary for the discharge formation. This means that the shortest gap has the best chance to experience a breakdown, provided that the same voltage is applied simultaneously to several gaps. Let us keep in mind that the distance from the lightning tip to the object, $[(H_0 - h)^2 + r^2]^{1/2}$, is shorter than that to the earth's surface H_0 at

$$r \leqslant R_{eq} = h \left(\frac{2H_0}{h} - 1 \right)^{1/2}. \tag{5.1}$$

The distance R_{eq} is known as the equivalent attraction radius for an object of height h. It indicates the surface area, from which lightning discharges that

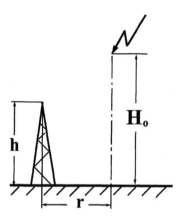

Figure 5.1. Estimating the equivalent attraction radius.

have descended to the altitude H_0 are attracted by the object. For a compact object of small cross section, this is a circle of area $S_{eq} \approx \pi R_{eq}^2$; for an extended object of length $L \gg h$ and width $b \ll h$ (e.g., a power transmission line), this is a stripe of area $S_{eq} \approx 2R_{eq}L$. The average number of strokes per storm season is evaluated from S_{eq} as

$$N_1 = n_1 S_{eq} \tag{5.2}$$

where n_1 is the year density of lightning discharges into the earth at the object's site. Global and regional maps of storm intensity are made from meteorological survey data [1, 2]. The n_1 data are usually given per $1\,km^2$ per year. Quite often, the maps indicate the number of storm days or hours, together with empirical formulae to relate this parameter to n_1.

The equidistance principle, simple and clear as it may seem, is of little use, because one can employ formulas (5.1) and (5.2) to advantage only if one knows the altitude H_0 (the attraction altitude), at which a descending lightning leader begins to show its preference and selects the point to strike. The condition of the earth's surface and the objects located on it cannot influence the lightning behaviour high up in the clouds. A lightning develops by changing its path randomly. As it approaches the earth, the field perturbation by charges induced by terrestrial objects become increasingly comparable with random field fluctuations. Eventually, the perturbation begins to play the dominant role, determining the channel path more or less rigorously. The average altitude H_0 at which this happens is known as the attraction altitude.

It is unlikely that the altitude H_0 should be determined only by the terrestrial object's height h. It must also depend on the leader field varying statistically with the lightning due to the variations in the storm cloud charge, the starting point of the descending leader, its path, number of branches, etc. This diversity of lightning conditions is uncontrollable. The only parameter that can, to some extent, depend on observations is the attraction altitude averaged over all descending discharges. It deserves attention because the averaging will require only the statistics of descending lightnings which have struck objects of various heights. These statistics cannot be said to be reliable but it provides some factual information important for lightning protection practice.

Before we use the stroke statistics in a theoretical treatment, we think it worthwhile defining the range of object heights. Unfortunately, one has to discard the stroke data concerning high constructions. Ascending discharges become dominant at heights $h > 150$ m. Data on such strokes cannot be included in the statistics without reservations, even though they were obtained from well-arranged observations, in which every discharge was identified unambiguously. The point is that ascending lightnings partly discharge the clouds, reducing the number of descending discharges. This interference into the storm cloud activity is so appreciable that a further increase of h above 200 m does not practically change the stroke frequency of an object by descending lightnings. Of little use are the data on low structures (10–15 m). The number of strokes in this case is greatly affected by the nearest neighbours and the local topography. Account should be taken of the statistics for low buildings, but such observations are scarce. The overall data have a too large spread.

The authors of [3] selected the most reliable data and, by averaging many observations, derived the relationship between the number of descending strokes and the terrestrial object height. Figure 5.2 shows individual representative values to demonstrate the data spread. All of the results are normalized to the intensity of the storm cloud activity, which is 25 storm days per year. In the range $h \leqslant 150$ m considered, we can admit, with some reservations, the existence of a quadratic height dependence of the number of lightning strokes for concentrated objects and a linear dependence for extended ones. Both dependencies mean $R_{eq}/h \approx$ const.

Figure 5.2 shows that the expression $R_{eq} = 3h$, sometimes used for rough estimations of the expected number of strokes, agrees fairly well with the averaged values of R_{eq} derived from observations. The substitution of $R_{eq} = 3h$ into (5.1) yields for the average attraction altitude for descending leaders

$$H_0 = 5h. \tag{5.3}$$

This does not seem to be a large height. A lightning is insensitive to the earth's surface along most of its path and it is only its last 50–500 m which are predetermined. Below, we shall discuss the mechanism of the more or less rigid determination of the leader behaviour by a particular site on the earth (section 5.6).

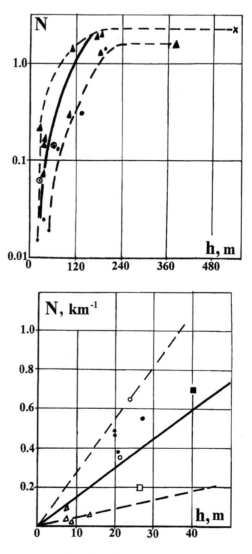

Figure 5.2. The average number of strokes per year for compact (top) and extended (bottom) objects of height *h*. The dashed curves bound spread zones in observation data. Solid curves are plotted using the equivalent attraction radius.

5.2   The electrogeometric method

Popular among some lightning specialists, this method of calculating the number of lightning discharges into a grounded structure [4–8] should be considered as a modification of the equidistance principle. The main

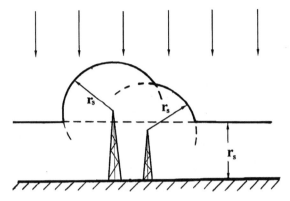

Figure 5.3. Lightning capture regions.

calculation parameter in this method is the striking distance r_s. Surfaces located at a distance r_s from the upper points of a structure (roof), the adjacent buildings, and from the earth's surface define by means of their interception lines the lightning capture regions (figure 5.3).

The further path of a lightning channel which has reached the capture region is considered unambiguously predetermined. The leader will move to the object (or to the earth), whose capture surface it has intercepted. With these initial assumptions, the calculation of the number of strokes N_1 reduces to geometrical constructions, since the lightning density n_1 at an altitude $z > h + r_s$ is considered to be uniform, and the value of N_1 can be calculated if one knows the area S_s of the capture surface projection on to the earth's plane, $N_1 = n_1 S_s$.

The long history of the electrogeometric method has witnessed only one improvement – that of the selection principles concerning the striking distance r_s [1, 3]. Discarding inessential details, the quantity r_s is found from an average electric field E_a between the object's top (or the earth's surface) and the lightning leader tip which has reached the capture region. Usually, the values of E_a were taken to be equal to the average breakdown strengths of the longest laboratory gaps. As the laboratory study of increasingly longer sparks progressed, the values of E_a introduced into the calculation method decreased from 6 to 2 kV/cm, entailing larger striking distances. In this approach, the parameter r_s is independent of the grounded object's height but is sensitive to the leader tip potential U_t ($r_s \approx U_t/E_a$). However for applications, attempts are made to find the relation to the current amplitude I_M of the return stroke, rather than U_t, using simulation models of the kind discussed in section 4.4. If the function $r_s = f(I_M)$ and, hence, $S_s(I_M)$ are known, the number of lightning strokes at an object with current $I_M > I_{Mo}$ is found as

$$N_1(I_{Mo}) = n_1 \int_{I_{Mo}}^{\infty} S_s(I)\varphi(I)\,\mathrm{d}I \qquad (5.4)$$

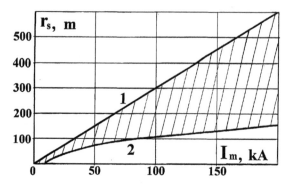

Figure 5.4. The dependence of striking distance on lightning currents. The lower curve is plotted using [4] data, the upper using [5] data. The spread region is hatched.

where $\varphi(I)$ is the probability density of current of amplitude I found from natural measurements. To find the total number of strokes, the lower limit of the integral in (5.4) should be taken to be zero.

The generally correct idea of differentiating distances r_s in current amplitude actually fails to refine the calculation of N_1. There is no factual information to determine the function $r_s = f(I_M)$ experimentally, while theoretical evaluations suffer from an unacceptably large spread, so that the values obtained by different authors differ several times (figure 5.4). The stroke statistics for objects of various heights could, to some extent, be used for fitting the calculated total number of strokes N_1, but it proves unsuitable for finding the function $r_s = f(I_M)$.

The calculation procedures in the electrogeometric approach do not involve a strong dependence of stroke frequency on an object's height. Indeed, for a single construction, like a tower, the capture region projection on to the earth's plane is a circle of radius

$$R = (2r_s h - h^2)^{1/2} \quad \text{at} \quad r_s \geqslant h$$
$$R = r_s \qquad\qquad \text{at} \quad r_s < h \qquad\qquad (5.5)$$

and for an extended object, like a transmission line, it is a stripe of width $2R$. Therefore, the number of strokes of low power lightnings with a small stroke distance ($r_s < h$) will be entirely independent of the object's height, while the frequency of powerful discharges with $r_s > h$ must increase with height slower than h for compact objects and as $h^{1/2}$ for extended ones. Actually both of these dependencies are steeper.

5.3   The probability approach to finding the stroke point

A predetermined choice of the discharge path through an air gap contradicts the experience gained from long spark investigations. Neither a spark nor a lightning travel along the shortest path. When voltage is simultaneously applied in parallel to several air gaps of various lengths, it is the longest gap that is sometimes closed by a spark. This is supported by the large spread of breakdown voltages: the standard deviation σ for multi-metre gaps with a sharply non-uniform field is 5–10% of the average breakdown voltage.

If two gaps, tested individually, possess the distributions of breakdown voltage of the probability densities $\varphi_1(U)$ and $\varphi_2(U)$, then, provided that the voltage of a common source is applied simultaneously, the breakdown probability for one of the gaps, say, the first one, is described as

$$P_1 = \int_0^\infty \varphi_1(U)[1 - \Phi_2(U)]\,dU, \qquad \Phi_2(U) = \int_0^U \varphi_2(U)\,dU \qquad (5.6)$$

where Φ_2 is the integral distribution defining the probability of the gap breakdown at a voltage less than U. If the distributions φ_1 and φ_2 are described by the normal law with the standard deviations σ_1 and σ_2 and by the average values of U_{av1} and U_{av2}, the breakdown voltage difference $\Delta U = U_1 - U_2$ also obeys this law, with $\Delta U_{av} = U_{av1} - U_{av2}$ and $\sigma = (\sigma_1^2 + \sigma_2^2)^{1/2}$. This allows us to rewrite (5.6) using the tabulated probability integral:

$$P_1 = \frac{1}{2}\left[1 - \sqrt{\frac{2}{\pi}}\int_0^A \exp\left(-x^2/2\right)dx\right], \qquad A = \frac{U_{av1} - U_{av2}}{(\sigma_1^2 + \sigma_2^2)^{1/2}}. \qquad (5.7)$$

The expressions of (5.7) are valid for $U_{av1} \geqslant U_{av2}$. Otherwise, one should find the breakdown probability P_2 for the second gap, writing for the first one $P_1 = 1 - P_2$.

The formal relations of the probability theory (5.6) and (5.7) are valid if the discharge processes in the gaps do not affect one another and if every individual breakdown can be considered as an independent event. Multi-electrode systems of this kind can be termed uncoupled. A classical example of an uncoupled multi-electrode system is an insulator string of a power transmission line. The distance between the adjacent towers is so large that there is no electrical or electromagnetic effect of discharges occurring in one string on those of its neighbours. The earth's surface and an object located on it can also be regarded as an uncoupled system, with a descending lightning leader acting as a common high voltage electrode. Such systems have been studied in laboratory conditions [9], in which the distribution of breakdown voltage was used as the indicator of an uncoupled nature of the system. If the individual gaps comprising a system are tested individually and have the integral distributions $\Phi_1(U)$ and $\Phi_2(U)$ with the probability densities $\varphi_1(U)$ and $\varphi_2(U)$, the system will have the following distribution of the breakdown voltages:

$$\Phi_{sys}(U) = \int_0^U \{\varphi_1(U)[1 - \Phi_2(U)] + \varphi_2(U)[1 - \Phi_1(U)]\}\,dU. \qquad (5.8)$$

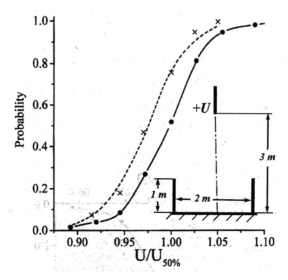

Figure 5.5. The breakdown voltage probability for the uncoupled multielectrode system involving the high-voltage and two grounded electrodes. ×: measured $\Phi_{sys}(U)$, •: measured $\Phi(U)$ for the single gap. The dashed curve is evaluated for the system using $\Phi(U)$.

In the particular case of equal gap lengths with $\Phi_1(U) = \Phi_2(U) = \Phi(U)$, we have

$$\Phi_{sys}(U) = 1 - [1 - \Phi(U)]^{1/2}. \tag{5.9}$$

Therefore, the uncoupled character of a system can be tested experimentally by comparing the measured distribution of its breakdown voltages with those calculated from formulas (5.8) and (5.9) and the distributions in the individual gaps. Experiments show that if the distance between grounded electrodes is comparable with their height, the leader processes in each gap develop independently, so that the system they comprise can, indeed, be regarded as an uncoupled one (figure 5.5).

Suppose now that the attraction of a descending leader begins when its tip reaches the altitude H_0. The problem reduces to finding the breakdown path in an uncoupled system with a common high voltage electrode – the lightning leader – and two grounded electrodes, namely, the earth's surface and an object of height h located on it. The probability of lightning attraction towards the object from the point with the x- and y-coordinates in the attraction plane is equal to that of the gap bridging between the leader tip and the object's top, $P_a(x,y)$. This probability is defined by the integral of (5.7). When the relative standard deviations are identical, $\sigma_1/U_{av1} = \sigma_2/U_{av2} = \sigma_a$, at identical average breakdown voltages, the upper probability limit is expressed through the shortest distance from the leader tip to the earth, $d_e(x,y)$, and to the object, $d_0(x, y)$:

$$A_a = \frac{d_0 - d_e}{\sigma_a (d_0^2 + d_e^2)^{1/2}}. \qquad (5.10)$$

The expected total number of lightning strokes at the object, N_1, is found by integrating P_a over the attraction plane. If the earth's surface is flat and the lightning discharge density n_1 is constant, then we have for a compact object of height h and for an extended object of average height h and length L, respectively:

$$N_1 = 2\pi n_1 \int_0^\infty P_a(r) r \, dr, \qquad N_1 = 2\pi n_1 L \int_0^\infty P_a(y) \, dy,$$

$$A_a = \frac{[z^2 + (H_0 - h)^2]^{1/2} - H_0}{\sigma_a [z^2 + (H_0 - h)^2 + H_0^2]^{1/2}}, \qquad z = r, y. \qquad (5.11)$$

The relations obtained from the equidistance principle are identical to (5.11) at $\sigma_a = 0$.

In virtue of the approximate symmetry of the function $P_a(r)$ relative to the point with $P_a = 0.5$ ($r/h \approx 3$; see figure 5.6), the calculations of N_1 slightly depend on the standard attraction deviation σ_a. When σ_a varies from zero to 10% (there are practically no greater deviations in pure air), the value of N_1 increases only by 15% for a compact object and by less than 5% for an extended one. It would be unreasonable to discard the simple and clear equidistance principle for the sake of this small correction, but for the greatly inclined (almost horizontal) paths of lightnings attracted by objects.

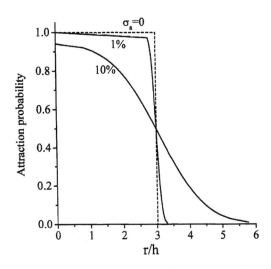

Figure 5.6. Evaluated attraction probability for the attraction altitude $H_0 = 5h$ (h is the object height, r is the object–lightning stroke distance).

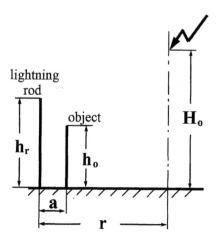

Figure 5.7. Why a lightning rod is less effective when an inclined lightning approaches from a side of the protected object.

It is clear from the foregoing that the larger the distance between the lightning and the object (compared with that from the lightning to the rod) the greater the protective effectiveness of a lightning rod. For a lightning travelling in the attraction plane strictly above the lightning rod, the difference between the two paths (figure 5.7) is largest:

$$\Delta d = [(H_0 - h_0)^2 + a^2]^{1/2} - H_0 + h_r$$

$$\Delta d \approx h_r - h_0 \quad \text{at} \quad a \ll [2(h_r - h_0)(H_0 - h_0)]^{1/2}.$$

With increasing side shift of the lightning in the attraction plane r, the value of Δd decreases, and this decrease is especially noticeable when the lightning approaches from the side the object being protected, as is shown in figure 5.7:

$$\Delta d = [(H_0 - h_0)^2 + (r - a)^2]^{1/2} - [(H_0 - h_r)^2 + r^2]^{1/2}.$$

In the limit $r \to \infty$, the distance to the object is smaller than to the lightning rod ($\Delta d \approx -a$), and it is quite ineffective for lateral strongly inclined lightnings coming from the object side. In the equidistance approach, there should be no events like this, since lightnings are not to strike an object at a distance $r > R_{eq}$. In reality, the proportion of lateral strokes is found to be fairly large. The fact that the probability method considers this circumstance correctly (figure 5.6) is very important for the evaluation of the lightning rod effectiveness.

5.4 Laboratory study of lightning attraction

Laboratory investigations of lightning attraction were initiated in the 1940s by simulating a descending lightning by a long spark and placing small model rods and objects to be protected on the grounded floor [10, 11]. At that time, experimental researchers expected to derive information necessary for a numerical evaluation of lightning rod effectiveness. The naive optimism has long vanished. The measurements showed that the attraction process did not obey similarity laws. Essentially different results were obtained from gaps of different lengths and different time characteristics of the voltage pulses applied [12–14]. But the interest in laboratory investigations of lightning has survived, and they are currently performed in an attempt to understand the attraction mechanism of long leaders.

The primary question is when the attraction begins. Clearly, the condition of the earth's surface does not affect the leader propagation while its tip is far from the earth. Here, the spark paths become distributed randomly. If one projects a multiplicity of paths on a sheet of paper and finds the mean deviation Δx from the normal passing through a high-voltage electrode with the account of the sign (e.g., plus on the right and minus on the left), one obtains $\Delta x = 0$ for altitudes $z > H_0$. The mean path in a gap perfectly symmetrical relative to the normal proves strictly vertical down to the altitude H_0. The attraction onset is indicated by the mean path deviation towards the electrode simulating a terrestrial object (figure 5.8). Data treatment for determining the attraction altitude was made in [14] for spark discharges of up to 12 m in length. The path statistics involved different time characteristics of the voltage pulse. In the case of a steep pulse front (6 μs), a leader was attracted from the moment of its origin; for a smooth front (250 μs), it had enough time to cover an appreciable gap length before the deviation towards a grounded electrode became noticeable (figure 5.9).

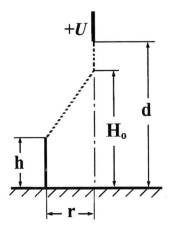

Figure 5.8. Determination of the attraction altitude H_0 by the bend point of a mean path deviation onset.

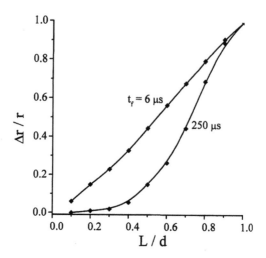

Figure 5.9. The average leader deviation from the high-voltage electrode axis toward the grounded electrode, Δr, depending on the leader length L. The results are given for configuration of figure 5.8 and two voltage fronts, t_f; $d = 3\,\text{m}$. In fact each curve presents a leader trajectory in cylindrical $z - r$ coordinates, averaged over many experiments.

The reason for this is as follows. When the voltage rises rapidly, the streamers of the initial corona flash reach the grounded cathode, and the leader develops in the jump mode from the very beginning. But when the voltage rises slowly, the leader channel covers about one third of a 3 m gap before the transition to the jump mode. This suggests that the streamer zone imposes a definite direction on the leader as soon as it touches the grounded electrode. If this is the case, the attraction of a laboratory leader must begin later in a long gap than in a short one. The experimental data presented in figure 5.10 show that the attraction delay time does increase with the length of the discharge gap d: $H_0 \approx d$ at $d = 0.5\,\text{m}$ but $H_0 \approx (0.4\text{--}0.5)d$ at $d = 10\,\text{m}$. The ratio of the streamer zone length to d decreases nearly as much by the moment of the final jump.

Another independent method for the study of spark attraction is to use a blocking electrode. Suppose we are able to set up instantaneously a metallic electrode on a grounded plane in the right place at the right moment of time. Let us do this many times for different lengths of a developing leader channel L and plot the probability of the electrode striking, Φ_0, as a function of L. The possible curves are presented in figure 5.11. The first version corresponds to the 'instantaneous' choice of the striking point at an altitude $H_0 < d$ (at the critical leader length $L_{\text{cr}} = d - H_0$). A probability of the electrode striking falls sharply if the electrode is set up with delay (at $L > L_{\text{cr}}$) since the leader has already chosen some other point for stroke at the moment of the leader start ($H_0 = d$, $L_{\text{cr}} = 0$) while in the third version the leader chooses a striking point gradually also but beginning from the altitude $H_0 < d$ when $L_{\text{cr}} > 0$.

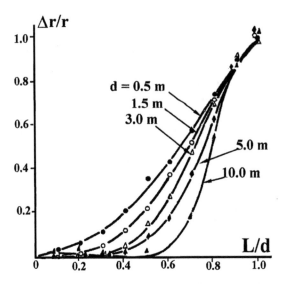

Figure 5.10. The deviation Δr at various gap length d for $t_f = 250\,\mu s$ under the conditions of figure 5.9.

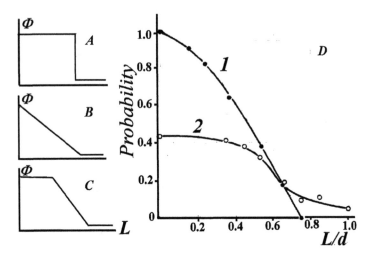

Figure 5.11. 'Blocking electrode' experiment. A supposed qualitative probability Φ of a stroke to the electrode when: (A) a leader is instantly chooses the striking point when its length reaches the critical length L_{cr}, (B) a leader is gradually attracted from the very beginning, (C) a leader is gradually attracted reaching L_{cr}. (D) measurements for $d = 3\,m$, $t_f = 6\,\mu s$ (curve 1) and $250\,\mu s$ (curve 2).

This can be done experimentally if the electrode displacement is replaced by its screening by an electric field. The electrode should be insulated from the earth and a high voltage of the same sign as that of the leader should be applied to it. This will create a counter-propagating field which will block the electrode from the leader. The electrode will become accessible only after the blocking voltage is cut off. The electric circuit provides a precise control of the voltage cut-off [13]. The experimental relationships in figure 5.11 show again that the attraction begins since the leader origin, if it develops in the final jump mode from the very beginning. In long gaps with a smooth voltage pulse, when the initial leader phase is well defined, the attraction is delayed as much as the transition to the final jump (curve 2 in figure 5.11).

Experiments on negative leaders have yielded similar qualitative results [15], but the attracting effect of a grounded electrode on a negative leader proves to be stronger.

5.5   Extrapolation to lightning

The scale of laboratory experiments, 1 : 100 or 1 : 1000, is too small to resolve the details or to make long-term predictions. Laboratory studies have so far failed to clarify an important point: Does the attraction onset really coincide with the moment of the leader transition to the final jump, or is this process controlled by a counterleader starting from the grounded electrode? The interest in the counterleader and its relation to lightning attraction arose long ago [4]. The counterleader seems to increase the altitude of the grounded electrode. The difference between the tip potential and the external potential, ΔU, increases, so the counterleader goes up with acceleration (section 4.1.2) to meet the descending leader.

One of the difficulties is that the moments of the descending leader transition to the final jump and of the counterleader origin in a laboratory are hardly discernible. The experiment accuracy is insufficient to separate them reliably in conventional laboratory gaps of about 10 m long. For lightning, these moments may be considerably separated, but a direct measurement is practically unfeasible. So, one has, as usual, to rely on numerical evaluations.

Let us first evaluate the field perturbation in the atmosphere by the charge of the grounded electrode of height h and radius r_0 before a counterleader starts from it. The external threshold field necessary for a counterleader to arise and develop is defined by formula (4.11), in which d must be equalized to h. The field E_0 for an industrial building of height $h = 50$ m was found to be 350 V/cm at the parameters used in section 4.1.1. Note that this field results not so much from cloud charges as from the charge of the descending leader approaching the earth. The field induces a charge on the grounded rod, whose density per unit length can be considered to depend linearly on height: $\tau(z) = a_q z$ (section 3.6.2). The value of a_q is defined by (3.11) where $d = h$ and $r = r_0$. For $E_0 = 350 \, \text{V/cm}$, $h = 50 \, \text{m}$ and $r_0 = 0.1 \, \text{m}$, we have $a_q \approx 3 \times 10^{-7} \, \text{C/m}^2$.

The field gain ΔE_0 at the altitude z_0 above the rod, associated with the rod charge and its reflection by the earth, is

$$\Delta E_0 = \frac{a_q}{4\pi\varepsilon_0} \int_{-h}^{h} \frac{z \, dz}{(z_0 - z)^2} = \frac{a_q}{4\pi\varepsilon_0} \left(\frac{2z_0 h}{z_0^2 - h^2} - \ln\frac{z_0 + h}{z_0 - h} \right). \tag{5.12}$$

At the attraction altitude $z_0 = H_0 \approx 5h = 250 \, \text{m}$ and the found value of a_q, we get $\Delta E_0 \approx 30 \, \text{V/m}$. This is about 10^{-3} of the unperturbed atmospheric field at the altitude H_0 and 3×10^4 times lower than the field in the streamer zone of a negative leader. It is hard to imagine a lightning leader which would respond to such weak perturbations. In any case, laboratory experiments have failed to reveal changes in the breakdown probability of a gap for such a small relative increase in the voltage. Consequently, the attraction process cannot begin before the counterleader is excited.

Let us follow the excitation of a counterleader by the field of a descending leader, relating the tip altitude of the latter to the grounded rod height. For this, expression (4.11) should be supplemented by the dependence of the average near-earth field on the descending leader charge. Consider a simple situation. Suppose a descending leader starts at altitude H_1 and moves together with its partner, a positive ascending leader, vertically without branching right above the grounded rod of height h. At the moment the descending channel acquires the length L, with its tip having descended to the altitude $H_0 = H_1 - L$, the potential of the leader charge at the rod top, together with the charges reflected by the earth, is

$$\varphi_q = \frac{a_q}{4\pi\varepsilon_0} \left[(H_1 - h) \ln\frac{2H_1 - H_0 - h}{H_0 - h} - (H_1 + h) \ln\frac{2H_1 - H_0 + h}{H_0 + h} \right]. \tag{5.13}$$

The field average in the rod height is $E_{av} = \varphi_q/h + E_0$ (with the account of the cloud field E_0). By equating E_{av} to the threshold field necessary for the excitation of a viable counterleader (formula (4.11)), we find the attraction altitude H_0 from (5.13), assuming the attraction to begin at the moment of the counterleader start.

We shall not be interested in the quantity H_0 linearly dependent on the poorly known parameter a_q to be averaged over all descending lightnings. Rather, we shall focus on the tendency in the variation of the H_0/h ratio with varying h in the range 10–150 m. Buildings lower than 10 m are rarely affected by lightning, while the picture for high structures is greatly distorted by ascending lightnings, as pointed out above. Suppose that the attraction altitude for an object of average height, say, $h = 50 \, \text{m}$, is found from (4.11) and (5.13) to be really close to the experimental value $H_0 = 5h$. This yields the estimate for a_q (which is $a_q/4\pi\varepsilon_0 \approx 1.5 \, \text{kV/m}$ at $E_0 = 100 \, \text{V/cm}$ and

$H_1 = 3 \times 10^3 \, \text{m}$), permitting the calculation of H_0/h for constructions of other heights. The calculations are

h, m	10	20	30	50	100	150
H_0/h	9.3	7.0	6.0	5.0	4.0	3.6

Of course, this is not the linear dependence $H_0 \approx h$ obtained from a preliminary treatment of observational data. The attraction altitude definitely rises with the object's height, as $H_0 \approx h^{0.65}$ according to the calculation. A better agreement could hardly be expected since the observational data are limited and have a very large spread.

Another result would be obtained if one related the attraction onset to the moment of the leader transition to the final jump. The attraction altitude would then be determined by the streamer zone length L_s, $H_0 = h + L_s$. The length L_s only slightly depends on the grounded rod height. At its zero height, the streamer zone is totally created by anode-directed streamers of the negative descending leader, which require an average field of $10 \, \text{kV/cm}$ (there is no counter-discharge). If the rod has a large height, the active voltage is shared equally between the streamer zones of the descending leader and the positive counterleader. Cathode-directed streamers of the latter can develop in a $5 \, \text{kV/cm}$ field, thereby decreasing the average field in the common streamer zone, at most, by a factor of 1.5–2. This would set a limit to the possible variation in the attraction altitude. It may seem that the result obtained is quite promising. The attraction altitude at the leader potential $U \approx 100 \, \text{MV}$ is also found to be close to 100 m for low objects and about 300 m for structures 100–150 m high. But one should keep in mind that only unique unbranched leaders are capable of delivering to the earth such a large cloud potential (section 4.3.2). Such leaders occur rarely in nature; the potential of a normally branched lightning is several times lower. As smaller will be the streamer zone length proportional to U. The attraction altitude would then become equal to the object's height, provided it is not too low.

In other words, the attempt to relate the attraction process to the final jump unambiguously relates the quantity H_0 to the potential of a descending leader, making it strongly dependent on the factors discussed in section 4.3.2, which change this potential (e.g., branching). If one relates the attraction to the excitation and development of a counterleader, the dominant factor will be the total charge delivered by all the components of a descending leader to the earth.

The idea that the attraction onset is associated with the excitation of a counterleader leaves little hope for an unambiguous relationship between H_0 and the return stroke current I_M, as was implied, for example, by the electrogeometric method. Indeed, the current I_M is determined by the potential U_i delivered by the leader to the earth (section 4.4.1), whereas the field at the grounded rod is due to the total charge of the descending leader. The branching and path bending typical of a descending lightning greatly affect the value of U_i and, hence, I_M (section 4.3.3), but they do not much change the total leader charge.

5.6 On the attraction mechanism of external field

There is no doubt that lightning attraction is due to the electric field which is related to the object. It is difficult to imagine another remote way to affect a leader. As for the field source, the evaluations made in section 5.5 show that the field created by the charge induced in the object itself proves very weak. When the distance between the descending leader tip and the object is sufficient for an attraction effect to reveal itself, the object charge field at the leader tip is by a factor of $10^2 - 10^3$ lower than the cloud charge field. There are no reasons why such a slight perturbation should make the leader change its path, which is subject to various random bendings even without the influence of any terrestrial objects. No doubt, a counterleader excited by the object serves as a mediator between the object and the descending lightning. It looks as if it elongates the object, thereby increasing the charge acting on the descending leader. The counterleader travelling towards its tip attracts it to itself, and this eventually results in the lightning stroke at the object. The mutual attraction of the two leaders becomes especially pronounced when the fields they excite at the tip are comparable with or, better, exceed the differently directed cloud field. It is only then that the descending leader changes its path to go to the object, and the counterleader is attracted by the descending leader rather than by the cloud charge centre, as is usually the case. It is the excess of the perturbation field over the cloud charge field which imparts a quasi-threshold character to the attraction process.

    This unquestionable and fairly trivial reasoning is certainly useful for lightning protection practice. Physically, however, it remains quite meaningless until the mechanism of the external field effect on the leader is known. This is equally true of the cloud field which is also involved in the attraction of the leader, generally directing it to the earth. It is not clear at first sight what exactly is affected by the external field, which may be very weak. The fact is that the leader moves along the field even at $E_0 \approx 100\,\text{V/cm}$. Fields of this scale cannot affect directly the leader development – we have emphasized this several times above. The leader propagation, which occurs via turning the air into the streamer and leader channel plasmas, requires much stronger fields. These are present in the leader tip, in the tips of numerous streamers, as well as in the streamer zone where the strength (the lowest of the three) exceeds $10\,\text{kV/cm}$ in a negative leader and $5\,\text{kV/cm}$ in a positive one. High driving fields are created by the charges of the tips, streamer zones and, partly, by the nearest portions of the channel and leader cover. They cannot be created by the cloud or any other remote objects. The instantaneous leader velocity is entirely independent of the low external strength E_0 but is determined by the potential difference between the leader tip U_t and the external field U_0 at the tip site. This great difference, $\Delta U = |U_t - U_0| \approx 10\text{–}100\,\text{MV}$, along the relatively short length of the streamer zone creates in it the field $E_s \approx 5\text{–}10\,\text{kV/cm} \gg E_0$ necessary for the streamer and, eventually, leader development. What is then the instantaneous effect of the negligible field E_0 and its weak perturbations produced by the remote counterleader on the motion of the descending leader?

Apparently, this effect is that the external field accelerates the leader. We mentioned this at the end of section 4.1.3 and shall now discuss it at length. The underlying mechanism is as follows. Voltage determines the leader velocity, while the voltage gradient determines its acceleration. Velocity is a function of the absolute potential change at the tip, $V_L = f(\Delta U)$, with U_0 in the expression for ΔU being a function of the space coordinates or of the tip vector radius r. A particular form of the function $f(\Delta U)$ in this case does not matter; what is important is that V_L grows with ΔU. So, retaining the generality, we can use the empirical approximation of (4.3), $V_L \sim |\Delta U|^\gamma$ ($\gamma = \frac{1}{2}$). The algebraic value of the leader acceleration is

$$\frac{dV_L}{dt} = \pm\frac{dV_L}{d\Delta U}\left(-\frac{dU_t}{dt} + \frac{dU_0}{dr}\frac{dr}{dt}\right) = \pm\gamma\frac{V_L}{\Delta U}\left(-\frac{dU_t}{dt} - E_0 V_L\right) \quad (5.14)$$

where plus refers to a negative leader and minus to a positive one. The first term in the sum of (5.14) does not depend on the direction of the external field. One of the reasons for the variation of U_t with time was discussed in section 4.3.2. Another reason is the increasing voltage drop across the channel with its elongation. Normally, a variation in U_t has a retarding effect on descending leaders of both signs.

The second term in (5.14) leads to acceleration if the negative leader moves in the direction opposite to the field vector, with the positive leader moving along the field. The accelerating effect of the external field increases as the field becomes higher and the angle between the field and velocity vectors becomes smaller. Both terms have been estimated to have the same order of magnitude ($10^7\,\text{m/s}^2$); the second term may sometimes be even larger. For this reason, the attractive action of the external field proves essential.

We can now make clear the attraction mechanism. The actual mechanism, by which a leader chooses its propagation direction, has a statistical nature. This is indicated by numerous random path bendings and branching. Clearly, there is a high probability that the leader moves towards a site where it can acquire the greatest acceleration or the least retardation. It will be able to develop a maximum velocity in this direction, bypassing other competitors on its way. Large-scale leader photographs taken with a very short exposition nearly always show several leader tips on short, variously oriented branches

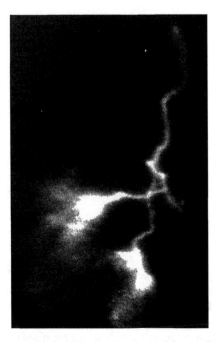

Figure 5.12. A still photograph of the leader channel front with exposure of 0.3 μs.

(figure 5.12). Among these, only one tip has a real chance of survival – for a positive leader, it is the one which belongs to the branch oriented along the external field; for a negative leader, the respective branch must be oriented against the field vector. The other tips usually die.

The mutual attraction of the descending leader and the counterleader, mediated by the electric fields created by their charges, is a self-accelerating process. This is due to a positive feedback arising between them. An enhanced field of one leader accelerates the other leader towards the first one. Because the distance between the leaders becomes shorter, the field of each leader rises at the site of the other leader tip, and the mutual acceleration proceeds at an increasing rate. This goes on until the streamer zones of the leaders come in contact and their channels unite. As a result, the common channel appears to be tied up to the object, from which the counterleader started.

5.7 How lightning chooses the point of stroke

Suppose the descending lightning leader has deviated from the vertical line to go to some high terrestrial structures. The highest structure is a lightning rod, or several lightning rods. If the objects to be protected are much lower, the

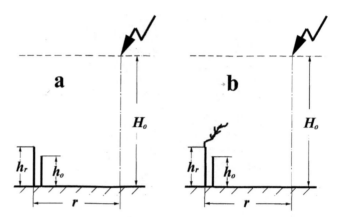

Figure 5.13. Increasing the grounded electrode effective height by a counterleader.

lightning usually bypasses them to strike one of the rods. This can be predicted from the equidistance principle. But when designing lightning protection devices, one usually focuses on exceptions rather than the rules. So the question arises of how large is the probability that the leader will miss the rod and strike the object, having taken a longer path. It seems justifiable to apply the concepts of a multi-electrode system to this problem. The lightning which has become oriented towards a group of grounded 'electrodes' has to choose among them. Let us make an estimation from formulas (5.10) and (5.11), substituting the distance from the leader tip to the earth, d_e, by the distance to the rod top, d_r (d_0 is, as before, the distance to the object's top). Lightning protection experience shows that there is no need to make a lightning rod much higher than typical terrestrial constructions ($h_0 < 50$ m). Arranging them close to each other, one can provide a reliable protection of the 0.99 level (of 100 lightnings, 99 are attracted by a protection rod) if the rod height h_r is only 15–20% larger than h_0. For an 'average' lightning, displaced at a distance equal to the attraction radius $R_{eq} \approx 3h_0$ relative to the grounded system, we have $\Delta d = d_0 - d_r \approx (0.12-0.15)h_r$ at $H_0 = 5h_r$ (figure 5.13(a)). The substitution of these values into (5.10) with $\sigma_a \approx 10\%$ gives $A_a \approx 0.2$. After taking the integral of (5.7,) one gets the probability of the lightning stroke at the object $P_0 \approx 0.4$ instead of the experimental value 0.01.

The complete failure of the theory was predictable. A system with a close arrangement of grounded electrodes cannot be considered to be disconnected. Its counterleaders affect one another. The first leader that has started from one of the electrodes decreases the electric field behind it, via its cover space charge, preventing the upward development of counterleaders from the other electrodes. Appearing with a delay, if they do, these counterleaders cannot retard their faster competitor, because the field is enhanced in the direction of the first leader propagation (figure 5.13(b)). This makes all of

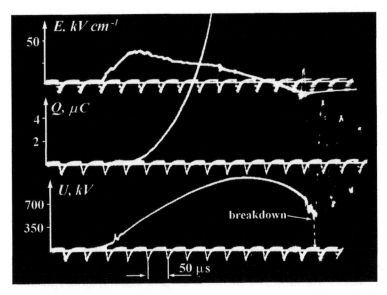

Figure 5.14. The oscillogram shows how the counterleader started from the 'active' grounded electrode of 1.1 m height screens an electric field on similar 'passive' electrode located at a distance of 10 cm. The gap length is 3 m. E – the field at the passive electrode tip, Q – the counterleader charge, U – the gap voltage.

the counterleaders interconnected; therefore, one now deals with a connected multielectrode system.

Turn to laboratory experiments [9]. The oscillograms in figure 5.14 illustrate the field variation on the grounded, 'passive' electrode when a counterleader develops from the nearby 'active' electrode. To simulate this process for a sufficiently long time, a plane–plane gap 3 m long was used with two rod electrodes on the grounded plane. A high negative voltage pulse was applied to the other plane. A possible discharge from the passive electrode was excluded by placing a thin dielectric screen totally covering the rod top. Before discharge processes came into action, the passive electrode field rose in a way similar to the voltage pulse. After a leader had started from the active electrode, the field rise on the passive electrode became slower, and the shorter the distance between the electrodes, the greater the rate of slowdown. At a very short inter-electrode distance, the passive electrode field stopped rising with voltage and even decreased somewhat. This obvious result indicates that the degree of mutual effects of discharge processes and grounded electrodes becomes greater with decreasing distance between them. Eventually, the role of passive electrodes becomes negligible – the grounded electrode system behaves as if it is replaced by one active electrode which attracts nearly all descending leaders.

Owing to the feedback mechanism considered, the choice of the stroke point made by a lightning become more definite. Even the slightest

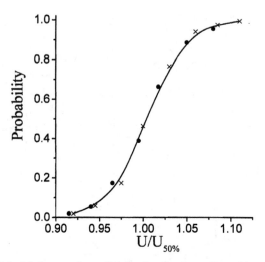

Figure 5.15. The breakdown voltage distribution for the system of figure 5.5, but with a small distance 10 cm between the grounded electrodes, which makes the system coupled.

advantages in the conditions in which a counterleader arises acquire an additional significance, being enhanced by the weakening electric field in the vicinity of the passive electrode, below the leader channel. It seems as if the passive electrode entirely disappears from the system. The breakdown voltage distribution in it nearly exactly coincides with that characteristic of a solitary active electrode (cf. figures 5.15 and 5.5). Formally, this can be accounted for by introducing a smaller relative standard deviation for the distribution of the breakdown voltage difference σ_c in expressions (5.10) and (5.11). We shall term it a choice standard. The upper limit of the probability integral

$$A_c = \frac{d_0 - d_r}{\sigma_c (d_0^2 + d_r^2)^{1/2}} \tag{5.15}$$

defines, as in (5.7), the probability of choosing the stroke point on grounded electrodes:

$$P_c = \frac{1}{2}\left(1 - \sqrt{\frac{2}{\pi}} \int_0^{A_c} \exp\left(-x^2/2\right) dx\right). \tag{5.16}$$

Formula (5.16) describes the probability of a lightning striking a body more remote from its leader, and expression (5.15) is valid as long as $d_0 > d_r$. Otherwise, instead of finding the probability of a lightning stroke at an object (P_{co}), one should find this probability for a lightning rod (P_{cr}), then defining P_{co} as $1 - P_{cr}$.† If the height of a lightning rod is h_r, that of an

† At $A_c \gg 1$ and, hence, $P_c \ll 1$, one can use the approximate expression $P_c \approx (2\pi)^{-1/2} A_c^{-1} \exp(-A_c^2/2)$, which is valid and more convenient for estimations.

object is h_0 and the distance between their top projections on to the earth's surface is Δr, we have

$$A_c = \frac{[(H_0 - h_0)^2 + (r - \Delta r)^2]^{1/2} - [(H_0 - h_r)^2 + r^2]^{1/2}}{\sigma_c[(H_0 - h_0)^2 + (r - \Delta r)^2 + (H_0 - h_r)^2 + r^2]^{1/2}}. \tag{5.17}$$

Here, as before, σ_c is given in relative units, r is the horizontal distance from the descending leader tip to the lightning rod axis, and H_0 is the attraction altitude. With increasing r, A_c and the probability integral values become smaller. As a consequence, the probability of a lightning stroke at the object increases. Therefore, remote lightnings make protection measures complicated, especially when their paths are greater deflected from a vertical line (section 5.3).

It would be useless today to try to define the choice standard from theoretical considerations. One should also bear in mind that the final result of the integration of (5.16) in area for finding the number of lightning strokes at an object strongly depends on σ_c, in contrast to (5.11). The quantity σ_c can no longer be taken to be constant, since it must decrease as the distance between the rod and object tops is made shorter. It is only the practical experience gained with various lightning protection systems which can give some hope. The choice of objects to be observed is strictly limited. Bulk registrations of stroke locations are made only for power transmission lines of high and ultrahigh voltages. Sometimes, registration equipment is mounted on unique constructions such as skyscrapers or very high television towers [3, 16]. In order to derive the values of choice standard from observations, it is necessary to calculate the expected number of lightning strokes at the object of interest at various σ_c values, trying to get the best possible agreement with the observations. As a first approximation, one may consider the lightning attraction by a system of grounded electrodes and the choice of the stroke point within the system to be independent events described by the probabilities P_a and P_c. Then, by analogy with (5.11), the expected number of breakthroughs to a compact and an extended object (of length L) will be described as

$$N_b = 2\pi n_1 \int_0^\infty P_a(r) P_c(r) r \, dr, \qquad N_b = 2\pi n_1 \int_0^\infty P_a(y) P_c(y) \, dy. \tag{5.18}$$

The observational data processing made in [3, 17] revealed the dependence of the choice standard σ_c on the distance between the object and the lightning-rod tops D. For a relative choice standard, the following formula is recommended:

$$\sigma_c = 7 \times 10^{-3} + 8 \times 10^{-5} D, \qquad D \text{ [m]}. \tag{5.19}$$

Its use in the calculations of (5.16)–(5.18) provides reasonable agreement with observations of 0.9–0.999% reliability rods. There are no data on rods with a higher reliability.

Note the following important circumstance concerning the protective action of a lightning rod. Even common sense indicates that the rod height must be increased with increasing distance between the object and the rod. Let us discuss the opposite situation when a rod is mounted directly on an object of small area. How large must be the excess $\Delta h = h_r - h_0$ to provide a given protection reliability? Essentially, we deal here with the frequency of lightning strokes below the lightning rod top. This question is justified by observations of such a high construction as the Ostankino Television Tower in Moscow (540 m). During the 18 years of observations, descending lightnings have struck it at various distances below the top, down to 200 m (figure 1.10). The rod has been unable to protect itself. This sounds ridiculous, but this is the reality.

The results of a numerical integration of the first expression in (5.18), using (5.10), (5.11), (5.16), and (5.17) at $\Delta r = 0$, are presented in figure 5.16. The stroke probability $\Phi_b = N_b/N_1$ shows the fraction of lightnings, attracted by the whole system of grounded electrodes, N_1, which have missed the lightning rod to strike the object. The calculations were made with the attraction standard $\sigma_a = 0.1$ for objects of height $h_0 = 30$–150 m. The actual protective effect is achieved only if the height of the lightning rod considerably exceeds that of the object. For short constructions with, say, $h_0 = 30$ m, a 99% protection reliability ($\Phi_b = 10^{-2}$) requires the lightning rod height excess of $\Delta h \approx 0.2h_0$, which is quite feasible technically because it is equal only to 6 m above the object. An object 150 m high will require a lightning rod 50 m higher than the object ($\Delta h \approx 0.3h_0$), which

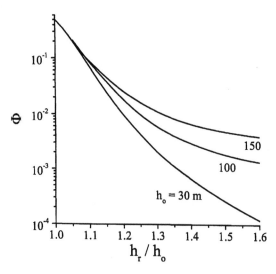

Figure 5.16. The evaluated probability of a lightning breakthrough to an object of height h_0, protected by an adjacent lightning rod of height $h_r > h_0$.

will be more expensive and complicated. In technical applications, the tendency of the $\Phi_b(\Delta h)$ curves to saturation is very important. This tendency becomes greater with increasing construction height, which means that a single rod will be ineffective for a high protection reliability. It is hard to protect an object with $h_0 > 100$ m with a reliability above 0.999% ($\Phi_b = 10^{-3}$), an object with $h_0 > 150$ m above 0.99%, etc. The higher the construction, the more complicated is the problem, and this is the reason why the Ostankino Tower is unable to protect itself. Nine lightning strokes were registered photographically along its length of 200 m from the top [18].

The protection efficiency decreases as the distance between the top of a high lightning rod and that of an object of similar height increases, reducing the mutual effect of counterleaders. Formally, this manifests itself as a larger choice standard σ_c, in accordance with (5.19). Sooner or later, its effect begins to dominate over that of lengths in formula (5.17), so the upper limit of the probability integral A_c stops rising.

5.8 Why are several lightning rods more effective than one?

The answer to this question can be found geometrically. Let us consider two lightnings which travel in the same vertical plane going through an object and its lightning rod in opposite directions. Suppose both leader tips are at an attraction altitude H_0 at the same distance from the rod. They have, therefore, an equal chance to be attracted by the object–rod system. The only difference is that one leader will approach it on the lightning rod side (version 1) and the other on the side of the object to be protected (version 2). Assume, for definiteness, that the displacement of the lightnings relative to the rod axis is equal to the attraction radius $R_{eq} = 3h_r$ (i.e., an average displacement), $H_0 = 5h_r$, and the horizontal distance between the rod and the object is $\Delta r = h_r - h_0 \ll h_r$. From (5.17), the upper limit of the probability integral for version 2 is nearly seven times less than for version 1:

$$A_{c1} \approx \frac{7\Delta r}{\sigma_c 25\sqrt{2}h_r}, \qquad A_{c2} \approx \frac{\Delta r}{\sigma_c 25\sqrt{2}h_r}.$$

Consequently, the probability integral from (5.16) for version 2 also decreases, increasing sharply the probability of striking the object. To illustrate, for $\Delta r = 0.2h_r$ and $\sigma_c = 0.01$, the parameter A_c takes the values of 4 and 0.57, respectively. When the lightning approaches on the lightning rod side, the probability of striking the object is, according to (5.16), nearly zero, but on the object side it is 0.28. Therefore, a single lightning rod can protect an object reliably only from the 'back', while its protection efficiency from the 'front' is much lower. This situation can be rectified if the object to be protected is placed half-way between two rods; it is still better if there are three rods and so on – this becomes only a matter of cost. No rod palisades are known from the protection practice; nevertheless, it is tempting to surround the object of interest with a protecting wire, especially if it is not very high but occupies a large area.

As an illustration, let us consider a case simple for the calculations. This will allow us to get numerical results and demonstrate the calculation procedures. Suppose a circle of radius $R_0 = 100\,\text{m}$ is densely filled by constructions of height $h_0 = 10\,\text{m}$. All of them must be protected with a 0.99% reliability, i.e., the probability of a lightning stroke should not exceed $\Phi_{b_{\max}} = 10^{-2}$. Let us now place a circular grounded wire at a distance of $10\,\text{m}$ from the external perimeter of the premises. This distance is necessary for technical considerations. For example, we must prevent a sparkover between the grounded wire and the communications systems and other structures, whether it occurs across the earth's surface or through the air due to high current pulses of the lightning discharge. Therefore, the circular grounded wire will have a radius $R_r = 110\,\text{m}$. Let us find the wire height h_r, which will provide the necessary value of $\Phi_{b_{\max}}$. For the radial symmetry, the probability of the lightning breakthrough is found from formulae (5.11) and (5.18) as

$$\Phi_b = \frac{N_b}{N_1} = \int_0^\infty P_a(r)r\,dr \Big/ \int_0^\infty P_a(r)P_c(r)r\,dr. \qquad (5.20)$$

The probabilities of attraction $P_a(r)$ and point choice $P_c(r)$ for a lightning, whose tip (in the horizontal plane at the attraction altitude H_0) is at the instantaneous distance r from the area being protected, are defined by similar expressions (5.7) and (5.16). These differ only in the values of the upper limit of the probability integral. For the attraction probability, the limit A_a, according to (5.10), is described by the difference between the minimal distances from the leader tip at the attraction altitude H_0 to the system of grounded electrodes and to the earth, $\Delta d_a = d_a - d_e$. In the case being considered, A_a is defined by the smaller of the values (at $r \leqslant R_0$):

$$\Delta d_{a1} = [(R_r - r)^2 + (H_0 - h_r)^2]^{1/2} - H_0, \qquad \Delta d_{a2} = h_0.$$

At $r > R_0$, we have $\Delta d_a \equiv \Delta d_{a1}$. In the calculation of the choice probability, the upper integral limit is given by formula (5.15). When calculating the difference between the minimal distances to the object and the protector, $\Delta d = d_0 - d_r$, one has to keep in mind that we have $d_{0\min} = H - h_0$ at $r < R$ and $d_{0\min} = [(r - R_0)^2 + (H_0 - h_0)^2]^{1/2}$ at $r > R_0$.

The calculation procedure reduces to finding, for every value of r, the upper limits A_a and A_c† in the integrals of (5.7) and (5.16) to calculate (extract from tables) these integrals, which give $P_a(r)$ and $P_c(r)$, and to calculate the integrals of (5.20). Practically, it is sufficient to make the

---

† The value of the choice standard σ_c necessary for the calculation of A_c is found from formula (5.19), taking into account the distance D between the protector top and the point on the object's surface nearest to the lightning with the instantaneous coordinate r; $\sigma_a \approx 0.1$.

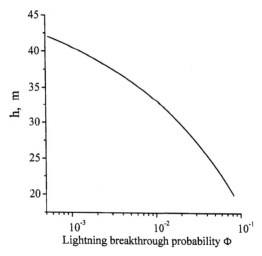

Figure 5.17. The object of 10 m height and of 100 m radius is protected by a bounding circular wire. In the graph is presented the evaluated wire height *h* necessary to decrease the probability of a lightning breakthrough to the object up to the value of Φ shown on the abscissa axis.

calculations with the step $\Delta r \approx (0.1 - 0.2)h_r$ and finish them when $P_a(r)$ drops to 10^{-6}–10^{-7} with growing r. If the probability integral is given reasonably (by an empirical formula or by borrowing it from a table, e.g., using a spline), the volume of calculations proves so small that they can be made with a programmed calculator. With a modern computer, the time necessary for numerical computations is only that for the data input.

The calculations made for the above example are shown in figure 5.17. The probability of a lightning breakthrough to the object decreases to the given value of 10^{-2} when the protective wire is suspended at a reasonable height $h_r \approx 34$ m. Note, for comparison, that a single lightning rod placed at the centre of a similar area provides the same protection reliability only if its height is $h_r > 150$ m. Even if one builds such a rod, the result may prove disappointing. Quite often, it is impossible to provide a safe delivery to the earth of a high lightning current impulse, when conductors with current pass close to structures being protected. Electromagnetic induction, sparking capable of setting a fire, etc. may also be dangerous.

5.9   Some technical parameters of lightning protection

5.9.1   The protection zone

It follows from the foregoing that a lightning-rod has a better chance of intercepting descending lightnings if it has a greater height above the object and is closer to it. Practically, it is important to identify a certain area around a protector, which would be reliably protected. This is the protection zone. Any object located within this zone must be considered to be protected with a reliability equal to or higher than that used for the calculation of the zone boundary. There is no doubt that this idea is technically constructive. When the configuration of the protection zone is known, the determination of the grounded rod or wire height reduces to a simple calculation or geometrical construction – this was an important factor in the recent age of 'manual' protection designing. At that time, the general tendency was to simplify the zone configuration as much as possible. In Russia, for instance, a single lightning rod zone was usually a circular cone, whose vertex coincided with the rod top [10]. When lightning protection engineers realized that the height of the rod was to exceed that of the object to be protected (section 5.7), the cone vertex was placed on the rod axis under its top [19]. The greater the protection reliability required, the more pointed and lower was the zone cone. For a grounded wire, the protection zone had a double pitch symmetry; when intersected transversally by a plane, it produced an isosceles triangle with nearly the same dimensions as those of a vertical cross section made through the rod cone half. Lightning protection manuals give a set of empirical formulas to design protection zones for simple types of lightning protector [2, 4].

The long-term practice has somewhat screened the principal ambiguity of the notion of protection zone. Indeed, having only one parameter – the admissible probability of a lightning stroke $\Phi_{b_{max}}$ – one is unable to determine exactly the zone boundary. So one has to resort to some additional considerations of one's own choice. In particular, there is nothing behind the concept of a conic zone except for the consideration of an axial symmetry and the desire to make the geometry simple. The value of $\Phi_{b_{max}}$ corresponds to a wide range of zone configurations, so the chosen configuration may appear to be far short of optimum. A protection zone is rarely filled up. When an object occupies a small fraction of this area, which is frequently the case in practice, the lightning rod height proves excessive. For high objects and still higher lightning rods, this results in unjustifiably large costs, which increase when high reliability is required. When the engineer places an object within a protection zone, he has no idea about its actual protection. But by decreasing the distance from the zone boundary inward, the probability of a lightning stroke may decrease by several orders of magnitude. To specify its value, one has to make numerical calculations similar to those illustrated in section 5.8.

Finally, the most important thing is that protection zones can be built with sufficient validity only for two types of lightning-rods – rods and wires. Even an attempt to combine them causes much difficulty. The same is true of multirod protectors, non-parallel two-wire protectors, and sets of rods of different height. All of them find application, especially when natural 'protectors' are used, such as neighbouring well-grounded metallic structures or high trees. The analysis of protection practice shows that preference is often given to easily-calculated designs rather than to effective designs. However, the statistical techniques used for the calculation have no limitations on the protector type, their number, or the geometry of the objects to be protected. Some problems may arise only in finding the shortest distance from the lightning leader tip to the lightning-rod and to the object. But they are surmountable with the use of modern computers. One should also bear in mind that the calculation provides the engineer not only with the breakthrough probability but with the number of expected breakthroughs over the time a particular object is in use. The latter parameter is more definite and cost-significant.

5.9.2   The protection angle of a grounded wire

The concept of protection angle α is used in designing wire protectors for power transmission lines (figure 5.18). The protection angle is considered positive when the power wires are suspended farther from the axis than the grounded wires, so they are open, to some extent, to descending lightnings. The value of $|\alpha|$ decreases with the grounded wire suspension height and with decreasing horizontal displacement of the power wire relative to the grounded one. The protection reliability is lower when the positive angle is larger. The angle was introduced as a parameter necessary for the generalization of observations of lightning strokes at transmission lines of various designs. It turned out that the angle α could not serve as an unambiguous characteristic of the protective quality of a grounded wire. A transmission line must also be described in terms of the grounded wire height above power line wires, Δh, and of the grounded wire height above the earth, h_r.

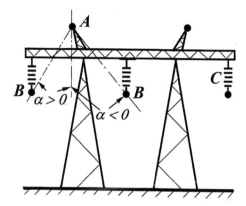

Figure 5.18. Positive and negative protection angles. *A*: grounding wire; *B*: power wire; *C*: insulator string.

This determines the distance between the grounded wire and the power wire at fixed α, which defines, through the choice standard σ_c, the degree of the system connectivity. Of lines with an identical protection angle, the best protected line is the one with the largest value of Δh and the lowest value of h_r.

Empirical formulas, which relate the lightning breakthrough probability to wires with α and h_r, have found wide practical application. Their accuracy, however, is not very high because they do not include Δh. For example, there are expressions identical in composition [21, 22]:

$$\lg \Phi_b = \frac{\alpha h_r^{1/2}}{90} - 4, \quad \lg \Phi_b = \frac{\alpha h_r^{1/2}}{75} - 3.95, \quad h_r \text{ [m]}, \quad \alpha \text{ [degree]}. \quad (5.21)$$

They give the probability of a lightning stroke with a 300% error related to a value supported by practical observations. These formulae should be treated with caution when the line supports are higher than 50 m at small positive and, especially, at negative protection angles. This is because most maintenance data refer to lines of up to 40 m high with positive protection angles of 20–30°. Besides, very few of the data used for deriving empirical formulas represent direct measurements. Usually, the data are derived from registrations of storm cut-offs minus the calculated return sparkovers (section 1.6.1). The latter calculations often give a large error. Still, expressions (5.21) demonstrate that negative protection angles are quite attractive. The action of protection wires placed farther from the tower axis than line wires ($\alpha < 0$) is similar to that of a closed grounded wire surrounding a region being protected (section 5.8). This type of protector could provide an exceptionally low probability of a lightning stoke at line wires, but the implementation of negative angle protection requires larger towers and, hence, a higher cost. This approach is, for this reason, unpopular.

5.10 Protection efficiency versus the object function

No doubt, there is a close relationship between the protector efficiency and a particular function (purpose) of a protected construction, especially when it is under high potential relative to the earth (e.g., ultrahigh voltage wires) or ejects a highly heated gas into the atmosphere. By raising the object potential to values comparable with the absolute potential of a descending leader, one can either increase the field at its tip, making the leader move towards the object, or lower it, suppressing the lightning attraction. It is a matter of the quantitative effect produced rather than its principal feasibility. High object potential U_{ob} may affect both the process of lightning attraction and the choice of the stroke point. The latter is more sensitive to external effects owing to the positive feedback in the connected system. Expressions (5.10) and (5.15) are suitable for estimations. They include the standards of attraction, σ_a, and of choice, σ_c. It is easier to control the process when their values are lower. For objects of regular height (~30 m), we have $\sigma_a \approx 10^{-1}$ and $\sigma_c \approx 10^{-2}$. This enables us to focus on the choice only. The effects of the high potential action of the object, U_{ob}, will noticeable at $U_{ob} \approx \sigma_c U_t$, where U_t is the tip potential at the moment the leader has descended to the attraction altitude. An 'average' lightning has $U_t \approx 50\,MV$. Therefore, in order to get an effect on the process of choice, one must apply $U_{ob} \approx 500\,kV$ to the object (or to the protecting wire). In order to affect the attraction process, the applied voltage must be ~5 MV. The latter value is, certainly, not feasible for the present power industry, but available operation voltage of an ultrahigh voltage (UHV) line is high enough to affect the lightning preference to a protecting wire or to a line wire [22, 23].

Most UHV lines operate at alternative voltage of frequency $f = 50\,Hz$ (60 Hz in the USA). Over the time $H_0/V_L \approx 10^{-3}\,s$ along the flight path H_0, during which the lightning chooses a point to strike, the wire potential changes but little, and its values $U_{ob}(t) = U_{f_{max}} \sin \omega t$ $(\omega = 2\pi f)$ can be taken to be equally probable. By the initial moment of attraction, $U_{ob}(t)$, may have the same or opposite sign relative to the lightning. If the sign is the same, the development of a counterleader from the wire will be delayed, so the probability of the lightning striking the wire will be reduced. In the other situation, the effect will be opposite. To get a total result over a long-term observation of the line operation (or a short-term observation of a very long line), one should average the operating voltage effects over an oscillation period. For this, expression (5.15) for the parameter A_c must be extended to the case in question. Expression (5.15) was based on the difference in the average fields along the lengths from the leader tip at the attraction altitude to the protector and to the object. Now, this difference can be calculated with the potential U_{ob} to get, instead of (5.15),

$$A_c \approx \frac{d_0 - d_r}{\sigma_c(d_0^2 + d_r^2)^{1/2}} + \frac{U_{ob}(t)}{\sigma_c\sqrt{2}U_t}, \qquad U_{ob}(t) = U_{f_{max}} \sin \omega t \qquad (5.22)$$

where U_t is the descending leader tip potential at altitude H_0.

The qualitative result of the calculations to be given below is predictable. We are interested in the effect of alternative voltage on the preferential choice of the stroke point between a protection wire and a power wire, since the operating line voltage is too low to affect appreciably the lightning attraction. In the half period when $U_{ob}(t)$ and U_t have the same sign (suppose it is a negative descending leader and negative voltage), the lightning is 'repelled' by the power wire; in the positive half period, it is attracted by it. Owing to the protecting wire, the probability of a lightning stroke at the wire in the off-voltage mode is low, 10^{-2}–10^{-3}. Therefore, the favourable effect of all negative half periods is small. Even if no lightning strikes the power wire during this time, the number of strokes at it will, for a long time, be reduced only by half relative to the no-load mode, because negative half periods take

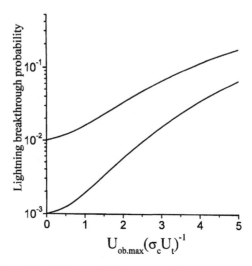

Figure 5.19. Effect of the AC transmission line operation voltage on the lightning breakthrough probability. The lower and upper curves correspond to probabilities of 10^{-3} and 10^{-2}, respectively, without voltage.

only half of the on-voltage time. The unfavourable effect of positive half periods may be much stronger. In principle, the potential difference $U_{ob} > 0$ and $U_t < 0$ at a high alternative voltage amplitude may produce such a strong 'attracting' field that all lightnings going to the power line will strike its wires. The probability of a strike at the line wire during the positive half periods may rise by 2–3 orders of magnitude (even as much as unity) against its two-fold reduction during the negative half periods. As a result, the stroke probability averaged over a long time for the line wire grows. The operating voltage effect on power lines reduces the reliability of lightning protection.

The numerical calculations of this effect are illustrated in figure 5.19. The probability of lightning breakthrough to AC lines increases by an order relative to the probability $\Phi_b \approx 10^{-2}$ for the off-voltage mode at $\gamma = U_{ob\,max}/\sigma_c U_t \approx 3.75$; at $\Phi_b \approx 10^{-3}$, this effect is produced at 1.5 times lower voltage. For the typical size of modern power line towers with $\sigma_c \approx 0.008$, the stroke probability for the power wire at $U_t \approx 30\,MV$ rises from 10^{-3} to 10^{-2} at phase voltage amplitude $U_{ob\,max} \approx 625\,kV$. Such are the line voltages (750 kV) in some countries. Only the next generation of power lines with 1150 kV can be expected to produce as strong effect on lightning at $U_t \approx 50\,MV$. An experimental line of this kind has been in use in Russia for a short time.

Direct current line has a more pronounced effect on lightning. Lightning separation is possible in DC lines: a positive line wire more strongly attracts negative lightnings and a negative wire more strongly attracts positive ones. Since the frequency of positive descending lightnings is an order of magnitude smaller, a positive UHV DC line wire will attract a larger fraction of strokes. This effect may become well pronounced at the wire potential of ±500 kV and higher.

The treatment of a hot air flow from an object to be protected is generally similar to the above analysis. The density and electrical strength of hot air are lower, and the strength is proportional to the density in the first approximation [25–26]. Formally, this is equivalent to the reduction of distance d_0 from the lightning to the object in expression (5.15), as if the object height were increased. As a result, the lightning protector has a lower efficiency. Consider, as an illustration, a chimney 100 m high with a 10 m lightning rod fixed on its top. With the practically zero horizontal distance between the rod and the object and in the absence of hot smoke gases, the rod will intercept about 90% of all lightnings attracted by the chimney (figure 5.16). But if the chimney ejects a hot gas flow with the temperature of 100°C along the length of 30 m, the probability that the lightning will miss the rod to strike the chimney will rise from 10 to 50%. Actually, the lightning rod becomes ineffective. The question is whether it is worth constructing this purely decorative device on the chimney top.

5.11 Lightning attraction by aircraft

Protection of aircraft and spacecraft has always been a complex and demanding problem – poor protection may have serious repercussions. It has been mentioned that an aircraft can be damaged by an ascending lightning starting from its surface or by an attracted descending discharge in the atmosphere, as happens with a terrestrial construction. Naturally, the concept of attraction refers only to descending lightnings. There are no observational data on the interaction between aircraft and descending lightnings, and one has to resort again to laboratory experiments. Figure 5.20 gives a set of static photographs taken from the screen of an electron optical converter. Of many pictures, we have selected the most typical ones. The electronic shutter was shut at different moments of time, so the result is not exactly a movie film but something close to it. One can see that a vertical rod insulated from the earth has attracted one of the leader branches together with its streamer zone, having first excited a streamer flash and then a counterleader. Its contact with the descending leader has produced a short luminosity enhancement of their, now common, channel, like a step of the negative leader with its miniature return stroke (sections 2.7 and 4.6). As a result, the channel and the rod have become the extension of a high voltage electrode. The leader has started off towards the earth from the lower end of the rod which now seems to be part of the leader channel.

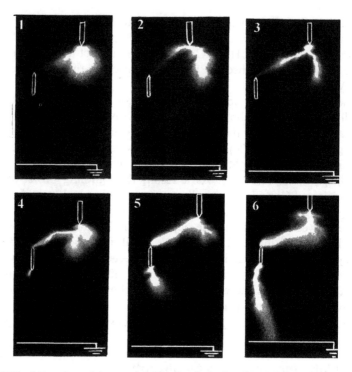

Figure 5.20. Attraction of the spark leader by the isolated metal rod suspended in the gap middle.

It appears that the attraction of a descending lightning by an insulated conductor, as well as by a grounded one, is stimulated by the excitation of a counterleader. The similarity in their mechanisms accounts for the similarity in the basic parameters of attraction. Below, we present some laboratory measurements of equivalent radii R_{eq} for spark attraction by a vertical metallic rod of length $l = 0.5\,\mathrm{m}$, suspended at height H above a grounded plane. The spark was produced by a positive voltage pulse with a $100\,\mu s$ front in a rod–plane gap of $3\,\mathrm{m}$. The front provides a more or less reliable field rise time for a real object during the development of a descending lightning leader. The measured values of $R_{eq}(H)$ are normalized to the value of $R_{eq}(0)$ for a rod that has descended to a plane to become grounded:

$$H/l \qquad 0 \quad 1 \quad 2.8 \quad 3.4$$
$$R_{eq}(H)/R_{eq}(0) \quad 1.0 \quad 0.9 \quad 0.9 \quad 0.8$$

The response to the conductor rise above the earth is fairly weak. A 10–20% decrease in R_{eq} seems to be regular, although it lies within the experimental error range. To extrapolate this result to lightning, one should assume that the number of descending lightning strokes for aircraft with the maximum size l is not larger than that for a grounded object of the height $h = l$. This limit for the number of strokes does not follow only from the experimental fact of a certain decrease in R_{eq} with H. Of greater importance are the possible variations in the aircraft position relative to the external field vector, $\mathbf{E}_0$, during the flight. The field enhancement at the ends of its fuselage of length l is defined by field projection on to the aircraft axis, rather than by the value of E_0. High terrestrial constructions are always aligned with the field since it is vertical at the earth.

Let us now estimate the possible number of descending lightning strokes at an aircraft of length $l = 70$ m, using the concept of attraction radius $R_{eq} \approx 3l$. We shall have $N_d \approx n_1 \pi R_{eq}^2 k_h$, where n_1 is an average annual frequency of lightning strokes at the earth and k_h is the ratio of the total flight hours per year to the total number of hours in a year. For $k_h = \frac{1}{3}$ and $n_1 \approx 3$ km^{-2} per year, we get $N_d \approx 0.1$ per year. This is at least an order of magnitude less than what follows from official statistics. One should not think that the discrepancy is due to the neglect of intercloud discharges, whose number is 2–3 times larger than that of lightnings striking the earth. In order to be attacked by intercloud lightnings, aircraft must penetrate through the storm front, but this is absolutely forbidden and may happen only as an accident. Rather, the result was overestimated because any pilot tries to stay as far away from a storm as possible.

Therefore, descending lightnings are responsible for fewer than 10% of strokes at aircraft. The other 90% or more are due to ascending lightnings excited by aircraft and spacecraft themselves (section 4.2). However, the interest in descending lightnings remains active because of the poor predictability of the stroke points on the aircraft surface. A similar situation but for high terrestrial constructions was discussed in section 5.7. The probability of a lightning striking much below the top is rather high. This situation can be readily simulated in the laboratory for a long positive spark excited by a voltage pulse with a smooth front, $t_f \approx 100$ μs and higher. The photograph in figure 5.21 illustrates a spark stroke almost at the rod centre, together with the integral distribution of the stroke points along its length. The wide, if not random, spread of stroke points over the aircraft surface creates additional problems. The aircraft has many vulnerable areas. In addition to the cockpit and fuel tanks, these are hundreds of antennas and external detectors providing a safe flight. It would be desirable to hide them from descending lightnings but the chances for this are quite limited. One consolation is that most lightnings affecting aircraft are of the ascending type starting mostly from the ends of the fuselage and wings, where the external electric field is greatly enhanced.

The excitation of ascending lightnings by aircraft was considered in section 4.2. Formula (4.11) allows estimation of the hazardous field E_0 for an aircraft of length $l = 2d$. The field E_0 decreases with growing d, some slower than $d^{-3/5}$. Note that the parameter $2d$ is not necessarily the fuselage

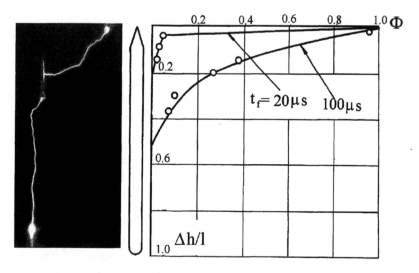

Figure 5.21. The stroke probability at various points of an isolated rod for two voltage front durations. The photograph shows how the spark has struck at the rod centre.

length; this may be the wing length if it is larger. In general, the experience indicates a direct relationship between the aircraft size and the frequency of lightning strokes. There are exceptions, of course. The statistics of flight accidents shows that aircraft of identical size may differ considerably in the capacity to excite lightnings. In one design, the engines are mounted on the wing pylons, and the ejected hot gas jet passes near the metallic fuselage, where the low fields cannot excite a leader. In another design characteristic of rockets also, the engine nozzle is placed in the tail, so that the hot jet serves as the fuselage extension. This is a perfect site for a counterleader to be excited since the leader development needs a lower field in a low density gas.

In the estimation, we shall assume the jet length to be half the fuselage length, $l_j = d$, and its average temperature to be twice as high as the ambient air temperature. Suppose that the jet radius is large enough for the streamer zone to be entirely within it and that the leader develops in a gas of relative density $\delta = 0.5$. When the gas density becomes lower due to the heating, the field providing the streamer propagation decreases at a rate δ [25, 26]. The rate of the electric strength decrease in long gaps is approximately the same for mountainous regions, although the density variation range in these experiments was narrower, $\delta \approx 0.7$ [25]. We shall assume from these data that a leader developing within a hot jet requires a potential drop δ^{-1} times smaller than that given by formula (2.49), i.e., $\Delta U = \frac{5}{3}\delta A^{3/5}(3bd/2)^{2/5}$ (here, the leader length L has been replaced by the jet length d). The total length of a conductor consisting of a fuselage of $2d$ long and a leader of length d is equal to $3d$. Hence, we have $\Delta U = 3E_0 d/2$, and the estimated external field providing the leader formation in the jet is

$$E_0(\delta) \approx 0.9\delta \left(\frac{A}{d}\right)^{3/5} \left(\frac{3b}{2}\right)^{2/5}.$$

The field is found to be $E_0(\delta) \approx 165\,\mathrm{V/cm}$ at the values of $A = (2\pi\varepsilon_0 a)^{-2/3}$, $a = 1.5 \times 10^3\,\mathrm{V^{-1/2}\,cm/s}$, $b = 300\,\mathrm{V\,A/cm}$, used in chapters 2 and 4, and $d = 35\,\mathrm{m}$. At this field, there will be a breakdown of the jet, increasing the aircraft size by the jet length value. This will create favourable conditions for an ascending lightning to develop from the fuselage in a low field. To estimate the field, d should be replaced by $L_{eq} = (2d + l_j)/2$ in formula (4.11); for the present example, it should be $1.5d$. The 25% decrease of the threshold field which will follow may greatly change the total number of lightning strokes at the aircraft.

5.12   Are attraction processes controllable?

We gave an affirmative answer to this question, when discussing the effects of operating voltage in ultrahigh voltage lines and hot gas flows. The further consideration of this problem should be concerned with quantitative aspects and particular methods of lightning control. Lightning control has two aims: to raise the reliability of lightning protection of nearby objects and to expand the area being protected by using conventional techniques. These may only seem to be two sides of the same effect. For example, increasing the lightning rod height increases both the protection reliability for a particular object and the maximum radius of the protected area. This, in principle, is the case, but quantitatively the two results differ considerably.

Turn to the estimations above. It follows from the calculations in figure 5.15 that the increase in the lightning rod height h_r by only $4\,\mathrm{m}$ (from 36 to $40\,\mathrm{m}$) reduces the probability of a stroke at an adjacent $30\,\mathrm{m}$ object from 10^{-2} to 10^{-3}, or by an order of magnitude. The effect is significant. As for the expansion of the protected area on the earth, its radius Δr does not grow faster than h_r. This can be demonstrated by putting $h_0 = 0$, $r = 0$, and $H_0 = 5h_r$ in formula (5.17). Then we shall have $\Delta r \sim h_r$ for a given probability of choosing a stroke point, i.e., at fixed A_c and $\sigma_c = \mathrm{const}$. In actual reality, the standard σ_c grows with h_r, due to which Δr rises still slower. In our example, Δr increases by less than 10%, and this insignificant effect is of no interest to us.

Lightning control eventually reduces to a change either in the electrical strength of the discharge gaps between the descending leader tip and the protector and the earth or in the gap voltage. For this reason, the particular conclusion that follows from the above example can be extended to any control measures – their effectiveness falls with distance between the object top and the lightning-rod, since the mutual effects of the components in a multi-electrode system become weaker. Formally, this weaker effect manifests itself in increasing standard σ_c. It appears that the lightning control is easier for objects of low height and area, when conventional protectors are sufficiently effective. It is much more difficult to deviate a lightning from an object without mounting a metallic rod on top of it. The application of destructive technologies to storm clouds and their charge neutralization are not discussed in this book, because this is a special problem having no direct relation to lightning processes.

The physics of the effect of a voltage pulse rise on a descending lightning leader is clearer than that of other effects. The effect can be expected to be favourable when the potential applied to the lightning rod is of opposite sign to that of the lightning, or the potential applied to the object is of the same sign. In the former case, the conditions for a counterleader to start from the rod are quite favourable. To initiate a preventive start of a counterleader from a lightning rod is to deviate the stroke point from the object. But in order to produce a noticeable effect, the counterleader must have a channel length comparable with the length difference between the object and the rod, or between their tops (the latter quantities are comparable). Only then does the effective rod height really grow and the charge space of the counterleader considerably limits the field at the object top. Therefore, one deals with channels of metre lengths, sometimes of tens or even hundreds of metres, especially if one takes into account the multi-fold increase in the radius of the area to be protected. This is a fairly complicated task.

A short-term 'elongation' of the rod by exciting a plasma channel from its top is very similar to the counterleader behaviour. A laser spark or a short-term long plasma jet would be sufficient for this. Laboratory studies have shown that a man-made plasma conductor affects a long spark path as a metallic conductor. The problem is the technological complexity and considerable cost of the project rather than the principal possibility of control.

Imagine an ideal pulse generator, whose effectiveness is so high that it blocks a lightning breakthrough to the object with 100% probability. The protection reliability will then be determined by the reliability of the generator itself, primarily by its synchronizing unit. It is a difficult task to design a reliable synchronizing unit capable of responding to a nearby descending lightning leader. A leader always chooses a complicated, poorly predictable path and has many branches. It is necessary either to distinguish a branch from the main channel or to trigger the control unit repeatedly. The latter is undesirable not only because this is resource-consuming. A control pulse can stimulate a branch to become the main channel, which is the first to reach the grounded electrode, producing a powerful return stroke pulse. The close vicinity of a strong current may be as hazardous to the object being protected as a direct stroke. Finally, we should not discard multi-component lightnings – 50% of subsequent components do not follow the channel of the first component [27]. So it is necessary to design a control unit capable of generating a series of pulses with millisecond intervals. Such a project would be very costly.

High costs have been the main reason for the decreasing interest in lightning control among specialists. They think of using nonmetallic rods or other unconventional measures only in exceptional situations when the common approaches are incompatible with the technological functions of the object being protected. Designers have suggested some exotic ways of lightning protection. Specialized firms advertise lightning rods with radioactive, piezoelectric and other wonder tops. The performance of radioactive sources has been tested in a laboratory, and no noticeable effect has been registered even on the leader start, let alone its propagation along the discharge gap. This should have been expected, because a leader arises from a pulse corona flash resulting from a long application of an electric field (as happens during a storm). Every pulse flash represents a streamer branch with a channel electron density of 10^{12}–10^{14} cm^{-3} [28]. A radioactive top can hardly add anything to this density, unless its power is so high that it kills everything alive around it.

It appears that leader suppression may be more promising than its excitation. Laboratory experiments have long been known [29], in which an ultracorona was successfully used to suppress the leader start. The corona arises as a thin uniform cover on the anode or the cathode made of a thin wire ($\sim$0.1–1 mm). A slow voltage rise does not change the corona structure or the ionization region thickness. The electric field strength on the electrode is stabilized by the space charge of ions drifting slowly on the corona periphery. The field stabilization prevents the formation of an ionization wave, or a streamer flash, which would otherwise produce a leader. With no consequences, the average field in a gap of several dozen centimetres long could be raised to 20–22 kV/cm, whereas 5 kV/cm was commonly sufficient to produce a breakdown in the absence of an ultracorona.

It would be tempting to extend the laboratory effect to lightning protection practice to suppress the counterleader start from the object being protected. An obstacle here is the rate of external field variation at the top of a grounded electrode of height h, rather than the much greater gap length. The electrode possesses zero potential, $U = 0$. The potential of the external field E_0 at the top is $U_0 = E_0 h$, so that the air at the electrode top is affected by the potential difference equal to $U - U_0 = -U_0$. The linear charge of a leader descending directly on to the object creates field ΔE_0 at the earth, given by formula (3.5). As the leader approaches the earth, potential U_0 rises at the rate

$$A_u = \frac{dU_0}{dt} = h\frac{d(\Delta E_0)}{dt} = \frac{\tau_L V_L h}{2\pi\varepsilon_0 z^2} \tag{5.23}$$

where z is the altitude of the descending leader tip and $V_L = -dz/dt$ is its velocity.

Let us find the maximum rate of the field rise at which the ultracorona can still survive. Assume, for simplicity, that a corona (positive, for definiteness) arises at a sphere of radius r_0, attached to the electrode top. To prevent the corona transformation to an ionization wave capable of initiating a streamer flash and then a leader, the field on the sphere should not rise in time with ΔE_0. The surface with maximum field should not detach from the sphere to move into the gap interior. In an ultracorona, the field on the sphere is stabilized by space charge on the level of E_c depending on radius r_0. The sphere concentrates a constant charge $Q_c = 4\pi\varepsilon_0 r_0^2 E_c$. A short time Δt after the corona ignition, the voltage increases by the value $\Delta U = A_u \Delta t$, which is supposed to increase the positive sphere charge by $\Delta Q_1 = C\Delta U = 4\pi\varepsilon_0 r_0 A_u \Delta t$. To avoid this, the sphere charge ΔQ_1 must be compensated. The compensation occurs owing to the gas ionization in the thin surface layer. Positive ions transport the charge ΔQ for the distance $\Delta r = \mu_i E_c \Delta t$ (where μ_i is the ion mobility), so that the negative charge induced in the sphere $\Delta Q_i = -\Delta Q r_0/(r_0 + \Delta r)$ is able to neutralize ΔQ_1. The charge actually induced in the sphere is transported into it by electrons produced in the near-surface layer, whose number is excessively large since $|\Delta Q_i| = \Delta Q_1 < \Delta Q$. 'Excessive' electrons leave for the external circuit and then to the 'opposite' electrode – the earth. The field on the radius $r = r_0 + \Delta r$ now becomes equal to $E(r) = (Q_c + \Delta Q)/[4\pi\varepsilon_0(r_0 + \Delta r)^2]$ and should not exceed E_c. To the small value of about $\Delta r/r_0$, this requirement is met at $A_u \leqslant 2\mu_i E_0^2 \approx 3.6\,\text{kV}/\mu\text{s}$ ($E_c \approx 30\,\text{kV/cm}$, $\mu_i \approx 2\,\text{cm}^2/\text{V s}$).

We have analysed the other extremal situation when the corona exists so long that charge $Q \gg Q_c$ is incorporated into space and the ion cloud radius becomes $r_1 \gg r_0$. A well-developed corona can exist at the sphere for a long time if the voltage U_0 does not grow in time faster than $U_0 = A_u t$. The maximum admissible growth rate A_u coincides, in order of magnitude, with the above estimate but is slightly lower. At a fast voltage growth, say, $U \approx t^n$ with $n > 1$, there necessarily comes the moment when the ion cloud field becomes higher than E_c, stimulating the transition to a streamer flash. For the typical values of $h = 50\,\text{m}$, $\tau_L \approx 5 \times 10^{-4}\,\text{C/m}$, and $V_L \approx 3 \times 10^5\,\text{m/s}$, the voltage growth rate reaches the estimated critical value when the leader descends to the altitude $z \approx 200\,\text{m}$, at which the attraction process begins. A little later, $A_u \sim z^{-2}$ becomes even more critical, and the ultracorona dies giving way to a counterleader.

To conclude, lightning can be controlled but this task is costly and very complicated technologically. So it would be unreasonable to discard conventional protection technologies where they can solve the problem successfully. One should not expect miracles in lightning protection. If particular circumstances make one turn to unconventional measures, one must be ready to create complex devices, whose protection reliability will be determined by their operation, rather than by the interaction with a lightning.

5.13   If the lightning misses the object

This is likely to happen more often than direct strokes at an object. Some-times, the object attracts a lightning branch which could hit the object if it had enough time before the return stroke develops from the main channel. Such a situation is illustrated in figure 5.22. The counterleader, which has started from the television tower top, has no time to transform to an ascend-ing lightning or intercept the descending leader, because the latter has struck a metallic tower below the tower top. As a result, the counterleader remains uncompleted. The counterleader channel has, however, become several dozens of metres longer. This is now a mature channel, whose temperature is at least 5000–6000 K. If it had touched a hot gas jet, it would inevitably ignite the gas. Practically a leader of any length is suitable for ignition of inflammable exhausts into the atmosphere. To excite and develop a leader in air under normal conditions, a voltage of 300–400 kV would be sufficient. Such a potential difference $\Delta U = E_0 h$ can be produced in objects of height $h > 30$ m even in the absence of lightning because this would require a storm cloud field of $E_0 \approx 100$ V/cm. If the object is lower, uncompleted counter-leaders can be excited even by remote lightnings. From formula (3.7), a descending leader that has started at an altitude of $H = 3$ km and has touched the earth creates a field $E_0 = 100$ V/cm at a distance $R = 1$ km from the stroke point if it carries the linear charge $\tau_L \approx 8 \times 10^{-4}$ C/m. This charge is characteristic of a descending leader with average parameters. This is one of the long-range mechanisms of lightning, which should be

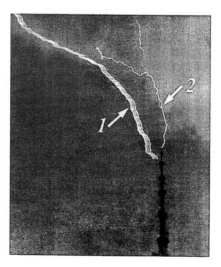

Figure 5.22. The long incomplete counterleader (2) started from the top of the Ostankino Tower while the descending lightning struck lower than the top (1).

taken into account when treating possible emergencies for objects containing large amounts of inflammable fuels.

References

[1] Uman M A 1987 *The Lightning Discharge* (New York: Academic Press) p 377
[2] *Operating Instruction for Lightning Protection of Buildings and Works RD 34.21.122-87* 1989 (Moscow: Energoatomizdat) p 56 (in Russian)
[3] Bazelyan E M, Gorin B N and Levitov V I 1978 *Physical and Engineering Fundamentals of Lightning Protection* (Leningrad: Gidrometeoizdat) p 223 (in Russian)
[4] Golde R H 1967 *J. Franklin Inst.* **286** 6 451
[5] Linck H and Sargent M 1974 *CIGRE, Sec. N 33/09* (Paris) 11
[6] Wagner C F 1963 *AIII Trans.* **83** (Pt 3) 606
[7] Wagner C F 1967 *J. Franklin Inst.* **283** (Pt 3) 558
[8] Darveniza M, Popolansky F and Whitehead E R 1975 *Electra* **41** 39
[9] Bazelyan E M, Levitov V I and Pulavskya I G 1974 *Elektrichestvo* **5** 44
[10] Stekolnikov I S 1943 *Lightning Physics and Lightning Protection* (Moscow, Leningrad: Izdatelstvo Akademii Nauk SSSR) p 229 (in Russian).
[11] Akopyan A A 1940 *Res. All-Union. Electr. Inst (Moscow)* **36** 94
[12] Bazelyan E M, Sadychova E A and Filippova E B 1968 *Elektrichesrvo* **1** 30
[13] Bazelyan E M and Sadichova E A 1970 *Elektrichesrvo* **10** 63
[14] Aleksandrov G N, Bazelyan E M, Ivanov V L *et al* 1973 *Elektrichesrvo* **3** 63
[15] Bazelyan E M, Burmistrov M V, Volkova O V and Levitov V I 1973 *Elektrichesrvo* **7** 72
[16] Cann G 1944 *Trans. AIEE* **63** 1157
[17] Gorin B N and Berlina N S 1972 *Elektrichesrvo* **6** 36
[18] Gorin B N, Levitob V I and Shkilev A V 1977 *Elektrichesrvo* **8** 19
[19] Bazelyan E M 1967 *Elektrichesrvo* **7** 64
[20] *International Standard: Protection Structures against Lightning* 1990 **IEC** 1021 p 48
[21] Burgsdorf V V 1969 *Elektrichesrvo* **8** 31
[22] Kostenko M V, Polovoy I F and Rosenfeld A N 1961 *Elektrichesrvo* **4** 20
[23] Bazelyan E M 1981 *Elektrichesrvo* **5** 24
[24] Larionov V P, Kolechitsky E S and Shulgin V N 1981 *Elektrichesrvo* **5** 19
[25] Bazelyan E N, Valamat-Zade T G and Shkilev A V 1975 *Izvestiya. Akad. Nauk SSSR, Energetika i transport* **6** 149
[26] Aleksandrov N L and Bazelyan E M 1996 *J. Phys. D: Appl. Phys.* **29** 2873
[27] Rakov V A, Uman M A and Thottappillil R 1994 *J. Franklin Inst.* **99** 10745
[28] Bazelyan E M and Raizer Yu P 1997 *Spark Discharge* (Boca Raton: CRC Press) p 294
[29] Uhlig C A 1956 *Proc. High Voltage Symp. Nat. Res. Council of Canada*

Chapter 6

Dangerous lightning effects on modern structures

This chapter is concerned with the mechanisms of hazardous lightning effects on various objects in the atmosphere, having no contact with the earth, on terrestrial constructions and underground communications lines. The discussion will be restricted to those effects which are, in this way or other, produced by the electrical and magnetic fields of lightning. No doubt, a hot lightning channel can ignite inflammable material but their direct contact is a rare phenomenon, whereas a remote excitation of sparks in such material due to electrostatic or magnetic induction is a regular thing. Lightning can destroy constructions by a purely mechanical action but this does not happen often. The burn-offs and holes at the site of contact of a hot lightning channel with metal are hazardous only to thin (one millimetre thick) metallic coatings. On the other hand, the range of electromagnetic effects is very wide. They can damage both microelectronic devices and ultrahigh voltage lines. The test maintenance of a 1150 kV transmission line in Russia has shown that it is not resistant to powerful lightning discharges. Most of the material presented in this chapter concerns the physical mechanisms of electrical, magnetic and current effects of lightning. We shall discuss simple and clear qualitative models illustrating the physics of these processes. We believe that this is the key issue to lightning protection theory. The process of equation solution, so important two decades ago, is not so essential today. If a physical model describes the reality adequately and the respective equations are available, modern computers are able to overcome almost any mathematical complexity.

When a lightning strikes a grounded metallic construction, a high return stroke current I_M passes through it. Because of an imperfect grounding having a resistance R_g, the construction potential rises by the value $U = I_M R_g$, for example, by 1 MV at $I_M = 50$ kA and $R_g = 20\,\Omega$. This is one of the reasons for the overvoltage due to a direct lightning stroke. Another reason is the emf of magnetic induction (the intrinsic induction due to an abrupt current change in the construction and the mutual induction produced by the current wave running through the lightning channel). But lightning overvoltages may result not only from a direct stroke but from remote lightning discharges as well. Their effect is associated with electrostatic and electromagnetic inductions. In the former case, an overvoltage results from the time variation of the electric field strength at the object, created by the lightning channel charges during the leader and return stroke stages (sometimes, by the slowly changing charge of the storm cloud). Another reason for a remote excitation of overvoltage is the varying magnetic field of the rapidly changing lightning current. Overvoltages became a very serious hazard at the beginning of the twentieth century when the first power transmission lines were built, and the engineer still associates an overvoltage with a powerful effect of tens and hundreds of kilovolts. This is true of transmission lines of high and ultrahigh voltages (UHV lines). However, an overvoltage as small as several hundreds or dozens of volts may become hazardous to electric circuits with a low operating voltage. Especially vulnerable in this respect are the circuits of microelectronic devices.

Historically, the theory of overvoltages has developed with reference to power transmission lines. Naturally, the mechanisms of ultrahigh voltage excitation were the first to attract the researchers' attention. So this theory is now very detailed [1–4] and the numerical procedures suggested are capable of solving engineering problems with a desired accuracy. We shall not describe these approaches here but rather focus on the physical aspects of the overvoltage problem, because in many practical applications they are not as self-evident as in a lightning stroke at a power line.

The calculation of overvoltage includes the solution of two equally important problems. One is to find the electromagnetic field of a lightning discharge at the site where the object to be protected is located. These calculations may prove very cumbersome and time-consuming, especially when one tries to take into consideration such parameters as the real path and length of a leader channel, the non-uniform charge distribution along the channel length, and the lightning current spread over the metallic parts of a particular object and underground service lines. The physical aspects of this problem, however, are quite clear and the numerical methods are well known. The other problem is to determine the response of an object and its electrical circuits to the electromagnetic field of lightning. The physical aspects of this problem are much more diverse, and the basic mechanisms of overvoltage excitation are not always obvious. So the latter are the subject of special interest in this chapter.

An induced overvoltage is normally smaller than an overvoltage produced by a direct stroke, especially by remote strokes, but it affects the object more frequently. When one calculates the frequency of emergencies for a high-voltage circuit with an insulation designed for hundreds of kilovolts, one usually deals with direct strokes, because induced overvoltages cannot damage the insulation. Objects with metallic shells which can screen well the internal electric circuits (including low-voltage ones) are designed in a similar way. However, unscreened low-voltage circuits suffer equally from overvoltages due to direct strokes and from induced overvoltages. Since the latter are more numerous, they should not be discarded when choosing the protective measures.

6.1 Induced overvoltage

6.1.1 'Electrostatic' effects of cloud and lightning charges

The atmospheric electric field varies in time during a storm. The slowest changes, lasting for several seconds or tens of seconds, are due to the accumulated charges of the storm cloud cells and their transport by the wind. Field variations associated with the leader propagation last for several milliseconds. Changes of microsecond duration arise from the charge redistribution during the return stroke. In any field variation, the electrostatic potential of a perfectly grounded object would remain equal to zero. In reality, however, the grounding resistance R_g is always finite. If the change in the charge induced on the object surface creates current $i_g = dq_i/dt$ through the grounding rod, the object acquires potential $U = -i_g R_g$ relative to the earth.

A grounded body of capacitance C possesses a potential difference $\Delta U = U - U_0$ relative to the adjacent space (here, U_0 is the average potential of the external field E_0 at the object's site). The charge induced on the body is $q_i = C\Delta U$; hence, current i_g is defined by the equations

$$\frac{di_g}{dt} + \frac{i_g}{R_g C} = -\frac{A_u}{R_g}, \qquad A_u = \frac{dU_0}{dt}.$$

$$i_g = -\frac{\exp(-t/R_g C)}{R_g} \int_0^t A_u(t') \exp(t'/R_g C)\, dt' \tag{6.1}$$

where we assume $i_g(0) = 0$. In a simple case with $A_u = \mathrm{const}$, we have

$$i_g = -A_u C[1 - \exp(-t/R_g C)], \qquad U = A_u R_g C[1 - \exp(-t/R_g C)]. \tag{6.2}$$

For the estimation, we put $C = 100\,\mathrm{pF}$, corresponding to a sphere of 1 m radius, and set the overvoltage amplitude below $1\,\mathrm{kV}$. During a storm without lightning discharges (the field variation $A_u \sim 10^4\,\mathrm{V/s}$), the desired grounding resistance should be $R_g < 1000\,\mathrm{M\Omega}$. But in the presence of a close descending lightning leader with the field variation $A_u \sim 10^9\,\mathrm{V/s}$, the grounding resistance must be reduced to $10\,\mathrm{k\Omega}$. With the account of the return stroke at $A_u \sim 10^{11}\,\mathrm{V/s}$, this value must be decreased further to $100\,\Omega$. Therefore, a good grounding of an object seems to be an effective tool for its protection against overvoltages excited by electrostatic induction. No doubt, faster variations in the external field impose more stringent requirements on the grounding rod. Resistances exceeding $1000\,M\Omega$ are hardly realistic because of the leakage across the unclean surface of even a perfect insulation. For this reason, overvoltages due to a slow variation in the storm cloud charge present a problem only in exceptional situations (for example, in providing protection to the explosives industries or to storages of explosives). It is not difficult to provide a $100\,\Omega$ resistance but special designs are necessary.

We should like to mention an exotic but fairly realistic situation when the object capacitance is subject to a change. This happens, for example, in apparatus with remote wire control. When the apparatus goes away horizontally from the operator and the cable elongates with a constant velocity $v = const$, the capacitance grows linearly in time, $C(t) = C_1 v t$. At a constant external field, the grounding electrode current and the object voltage relative to the earth do not change in time and are

$$i_g = U_0 \frac{C_1 v}{1 + R_g C_1 v}, \qquad U = U_0 \frac{R_g C_1 v}{1 + R_g C_1 v}. \qquad (6.3)$$

During the object motion up to a cloud, the overvoltage will be larger because of the higher average potential of the conductor, $U_0 \approx E_0 v t / 2$.

Let us calculate the overvoltage due to the return stroke current. Its specificity results from a high velocity of the recharging wave through the channel, v_r, which is comparable with light velocity c. Strictly, this requires account to be taken of the delay time of an electromagnetic signal in the calculation of charges induced on the object. When faced with this complex task, engineers sometimes feel a mystic horror. In actual fact, the delay changes little in many situations, especially in the case of a compact object. To illustrate this, consider the limiting case when a terrestrial compact object is located right under a vertical, descending leader, more exactly, when the horizontal distance to the stroke point is $r \ll z$, where z is the height of the return wave front. At the moment of time t, the leader charge is neutralized, and the channel is recharged along its portion from the earth to the altitude $z = v_r t$. But the object 'is aware' of the charge change along a shorter portion only, $z_e = c v_r t / (c + v_r)$. The effect of the delay is equivalent to a decrease in the return wave velocity by a factor of $(1 + v_r/c)$. The equivalent velocity is $v_{re} = z_e/t > v_r/2$, because $v_r < c$. For a lightning of medium power with $v_r \approx 0.25c$, the velocity is $v_{re} \approx 0.8 v_r$. A 20% correction is of little importance, particularly as the neglect of the delay leads to an overestimated overvoltage, thus providing a certain reserve for the engineering solution. The effect of the delay will be smaller at comparable values of r and z. Indeed, the distance between the charge neutralization front and the object, $(r^2 + z^2)^{1/2}$, increases more slowly than z. It remains nearly unchanged at $r \gg z$. Therefore, the time evolution of the field, $E_0(t)$, at the object's site will not differ from that calculated neglecting the delay. The phase delay which acts for the time $\Delta t \approx r/c$ does not affect the overvoltage.

Let us make a direct evaluation of the 'electrostatic' component of overvoltage during the return stroke, assuming that a rectangular charge neutralization wave (section 4.4) is moving along a vertical, perfectly conducting channel towards a cloud. At any point of the channel behind the wave front $z = v_r t$, the charge changes by the same value τ. The electric field follows the charge variation. Without the account of the delay, its change ΔE_r at the distance r from the channel is described by an expression similar to (3.5) (with $h = 0$, $H = z$ and $R = z$):

$$\Delta E_r = \frac{\tau}{2\pi\varepsilon_0}\left[\frac{1}{r} - \frac{1}{(z^2 + r^2)^{1/2}}\right],$$

$$A_E = \frac{d\Delta E_r}{dt} = \frac{\tau}{2\pi\varepsilon_0}\frac{v_r^2 t}{(v_r^2 t^2 + r^2)^{3/2}}. \tag{6.4}$$

The time constant for real electric circuits, $R_g C < 0.1\,\mu s$, is several orders of magnitude smaller than the time of the return stroke flight from the earth to the cloud. Then, according to (6.2), the electric component of the overvoltage (relative to the earth) for a compact object is defined as

$$U_e \approx R_g Ch\frac{d\Delta E_r}{dt} = \frac{\tau R_g Ch}{2\pi\varepsilon_0}\frac{v_r^2 t}{(v_r^2 t^2 + r^2)^{3/2}} \tag{6.5}$$

where h is an average object height. The short-term action of this overvoltage load must be endured by all the insulation gaps separating the object from the adjacent constructions and service lines, whose potentials were not changed by the lightning or, if they were, to a different extent.

At the moment of time t_{max}, the pulse $U_e(t)$ reaches its maximum

$$t_{max} = \frac{r}{\sqrt{2}v_r}, \qquad U_{e_{max}} = \frac{R_g Ch I_M}{3\sqrt{3}\pi\varepsilon_0 r^2}. \tag{6.6}$$

In the second formula of (6.6), we have substituted $I_M = \tau v_r$. A lightning of medium current $I_M = 30\,kA$, which has contacted the earth at the distance $r = 100\,m$ from the object of medium height $h = 10\,m$ and capacitance $C = 1000\,pF$ (a wire 100 m long), is capable of exciting an overvoltage pulse with an amplitude $U_{e_{max}} \approx 2\,kV$ at $R_g = 10\,\Omega$ because of the channel recharging during the return stroke.

Most of the parameters in (6.6), are beyond the engineer's capacity when he requires a high protection reliability. It is hardly possible to change the capacitance or average height of the object being protected. It seems more feasible to reduce the overvoltage to a safe level by decreasing the grounding rod resistance R_g. This is an effective way of overvoltage protection against electrostatic induction. However, this measure, like any other technological tool, has its limitations. It is difficult to provide $R_g < 1\,\Omega$ in a impulse mode. The obstacles are the relatively low conductivity of the earth and the inductance of the grounding conductors, which are fairly long when the grounding mat occupies a large area. After the potentialities of R_g reduction have been exhausted, there remains only one way – increasing the distance r to the nearest lightning discharge. To do this, one has to protect from direct strokes not only the object itself but the area around it together with the other constructions located on it, some of which are higher than the object to be protected. In that case, all lightning rods must necessarily be mounted outside this area; otherwise, the protectors will be able to attract lightnings, bringing their charges close to the object.

In contrast to the amplitude, the duration of the overvoltage pulse front is practically independent of the object's parameters, being primarily determined by the distance to the stroke point, r. From the first formula of (6.6), we have $t_{max} \approx 0.7\,\mu s$ at $r = 100\,m$ and $v_r \approx 0.3c$. Overvoltages of microsecond duration are typical of the lightning return stroke. Pulses with the front duration of 1–1.2 μs are still used as standards in insulation tests for resistance to lightning overvoltages, although they do not always reflect the reality.

6.1.2 Overvoltage due to lightning magnetic field

The problem of overvoltage induced by the magnetic field of a lightning discharge is the most common one among overvoltage problems. The lightning current varying in time and space induces the emf in any circuit. If a circuit is formed by conductors, the emf excites electric current. If the circuit is disconnected, the voltage equal to the induced emf is applied to the break. Let us estimate the maximum effect produced by an infinitely long straight conductor with current i. At the distance r from the conductor, the magnetic field is $H = \mu_0 i / 2\pi r$. Consider a rectangular frame in a plane intercepting the conductor (figure 6.1). Suppose the side parallel to the conductor has a length h and the side normal to it has $r_2 - r_1 = d$; the shortest distance between the frame and the conductor is r_1. The magnetic flux through the frame is defined as

$$\Phi = \frac{\mu_0 i h}{2\pi} \int_{r_1}^{r_1 + d} \frac{dr}{r} = \frac{\mu_0 i h}{2\pi} \ln \frac{r_1 + d}{r_1}. \tag{6.7}$$

The emf induced in the circuit, $U_M = -d\Phi/dt$, is

$$U_M(t) = \frac{\mu_0 A_i(t) h}{2\pi} \ln \frac{r_1 + d}{r_1}, \qquad A_i = \frac{di}{dt}. \tag{6.8}$$

At the maximum rate of the current change, $A_i \approx 10^{11}\,A/s$, characteristic of the return stroke of subsequent lightning components, the emf induced in a circuit

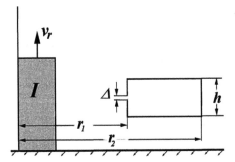

Figure 6.1. Estimating the overvoltage magnetic component.

with the sides $h = d = 10$ m at the distance $r_1 = 100$ m from the conductor with current is $U_M \approx 19$ kV. The emf for a smoother current impulse of the first component of a moderate lightning with $A_i \approx 5 \times 10^9$ A/s is $U_M \approx 1$ kV.

Overvoltages excited electrostatically and electromagnetically are generally comparable. The former can be coped with using an effective grounding of the object, but overvoltages due to the electromagnetic mechanism do not respond to the grounding efficiency. Imagine metallic columns buried deep in the ground, which support rails for a mobile overhead-track crane mounted high up at the ceiling of industrial premises. The whole construction has a perfect grounding owing to the metallic columns which provide a complete absence of electrostatic overvoltages from close lightning strokes. However, a pair of columns with a rail and the conducting earth forms a closed circuit with an area of several hundreds of square metres, in which the time-variable lightning current excites an emf. The same thing occurs in a circuit formed by columns, fixed at the opposite sides of the premises, and an overhead crane. A possible disconnection at any site of the metallic construction cannot be ignored either. A disconnection may arise due to metal erosion, poor welding or inadequate contact between the crane wheel and the rail. In that case, practically all emf of the circuit will appear to be applied to the site of defect, provoking a spark discharge through the air or a creeping discharge across the surface to bypass the defective site. A spark-induced emergency is inevitable if there is an explosive gas mixture in the premise.

The fact that any construction may serve as a circuit capable of inducing an emf increases the hazard – this may be a metallic ladder on a conductive floor, a metallic pipe leaning against a wall, etc. Such casual circuits present an even more serious hazard, because their parts may have only a slight contact between them, so that the probability of a spark gap is extremely high. An explosion would, no doubt, destroy the casual circuit, creating a mystery to the fire brigade in the spirit of Agatha Christie's stories.

The sequence of procedures for the calculation of overvoltages due to lightning current is similar to that for lightning charge calculation. One should first find the magnetic flux through the circuit in question and calculate the induced emf. The magnetic flux is often replaced by the vector potential $\mathbf{A}(t)$ to simplify the calculations. For current i in a thin conductor such as a lightning channel, the vector-potential is

$$\mathbf{A} = \frac{\mu_0}{4\pi} \int_1 \frac{i\,\mathrm{d}l}{r} \tag{6.9}$$

where the integral is taken in the conductor length and r is the distance from the current element $i\,\mathrm{d}l$ to the point, at which $\mathbf{A}$ is determined. The emf induced in the circuit of interest is defined as

$$U_M = \oint \mathbf{E}_M\,\mathrm{d}l, \qquad \mathbf{E}_M = -\frac{\partial \mathbf{A}}{\partial t} \tag{6.10}$$

where $\mathbf{E}_M$ is the strength of a vortex electric field excited by the time-variable magnetic field of the lightning. For a straight conductor with current, the vector E_M is parallel to the current. If the lightning channel is vertical, the vector E_M is also vertical.

Let us represent a lightning return stroke as a rectangular wave of current I_M propagating at velocity v_r along a vertical channel from the earth up to the cloud. Without accounting for the delay, leading to a certain overestimation of the result, we have

$$A(t) = \frac{\mu_0 I_M}{2\pi} \ln \frac{v_\mathrm{r} t + (v_\mathrm{r}^2 t^2 + r^2)^{1/2}}{r}, \qquad E_M = -\frac{\mu_0 I_M}{2\pi} \frac{v_\mathrm{r}}{(v_\mathrm{r}^2 t^2 + r^2)^{1/2}}. \tag{6.11}$$

Factor 2, instead of 4, in the denominator results from the allowance for the current spread in the earth. The field E_M is vertical, so the horizontal sections of the circuit do not contribute to U_M. In the vertical sections, the values of E_M are summed algebraically. For a metallic frame with an air gap, like the one shown in figure 6.1, the magnetic component of overvoltage in a small gap of $\Delta \ll h$ is

$$U_M = h[E_M(r_1) - E_M(r_2)] \tag{6.12}$$

where E_M are taken to be values averaged over the conductor heights h.

All the results obtained within the model of a rectangular current wave of the lightning return stroke overestimate the overvoltage; the smoother the front of the real current wave, the greater is the overestimation.

6.2   Lightning stroke at a screened object

6.2.1   A stroke at the metallic shell of a body

Overvoltages due to a lightning stroke at the metallic shell of a body, such as a plane or other objects, occur very frequently. To get an idea of what happens in this case, let us look at the schematic diagram in figure 6.2.

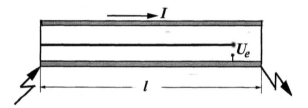

Figure 6.2. Lightning current flows along a pipe with a conductor inside.

Suppose lightning current runs along a closed metallic shell of an object, inside which there is a conductor connected to the shell at one of its ends, say, at the lightning current input. The potential at this contact will be taken to be zero. If R_1 is the linear resistance of a shell of length l and L_1 is its linear inductance, the voltage applied to the shell will be $U_f = -(L_1 \, di/dt + R_1 i)l$. The lightning current does not branch into the inner conductor disconnected at the other end (the capacitance is neglected). The conductor potential changes only due to the mutual inductance, $U_c = M_1 l \, di/dt$. Since the magnetic flux of the shell current is entirely attributed to the inner conductor, the linear mutual inductance M_1 is equal to the linear inductance of the frame, L_1. Then the potential difference between the shell and the inner conductor at the far end of the latter is described as

$$U_e = U_c - U_f = iR_1 l. \tag{6.13}$$

The remarkable property of a cylindrical system with an inner wire to compensate completely the induction emf is well known to impulse measurement technology. This property is the basis for making shunts for measuring current impulses with very short fronts (to a few fractions of a nanosecond). The respective theory, useful for the understanding of the overvoltage mechanism, is discussed in detail in [5]. We shall turn to it when evaluating the skin-effect in a shell. Here, it should be noted that the shape of an overvoltage pulse, $U_e(t)$, in the absence of a skin-effect is similar to that of a current impulse, $i(t)$. This is valid as long as the time of the electromagnetic wave propagation along the frame is much shorter than the impulse duration.

No principal changes will occur when the conductor ends are connected to the shell via resistances R_{k1} and R_{k2}. The voltage U_e will then appear to be operative in the inner closed circuit consisting of the shell, conductor, and resistors. When the resistances of the conductor and the shell are small, the current $i = U_e/(R_{k1} + R_{k2})$ arising in the circuit will distribute the overvoltage U_e between the turned-on resistors in reverse proportion to their values. The same will happen when the conductor is connected to the shell via spurious capacitances. Of course, if a massive aluminium shell has a cross section of $100 \, \text{cm}^2$ and the linear resistance is $R_1 \approx 3 \times 10^{-6} \, \Omega/\text{m}$, the overvoltages may be small, $U_e \approx 30\,\text{V}$, even at the maximum lightning current $I_M = 200\,\text{kA}$ and a long $l = 50\,\text{m}$. But one should bear in mind that modern microelectronic units stuffed into an aircraft or spacecraft can be damaged easily even by lower voltages.

6.2.2 How lightning finds its way to an underground cable

The problem of lightning access to an underground cable deserves special attention, because the spark propagation under these conditions has a peculiar physical mechanism. A direct stroke of a descending lightning at an underground cable is a rare event, since the leader cannot 'sense' its presence. The current flows to the cable along a spark channel creeping along the earth's surface. Its path can sometimes be identified easily because of the bulging loosened soil. Normally, the spark path is as long as several dozens of metres, or even hundreds of metres in low conductivity soils. It seems unlikely that a creeping spark should move towards a cable purposefully; rather, this is a matter of chance. But the local topography can stimulate the spark access to the cable. Suppose a cable is laid along a forest path, and the current of the lightning that has struck a nearby tree flows down to its roots, giving rise to a spark channel, which propagates across the path until it hits the cable.

A high current spreading through a poor conductor such as soil induces a fairly high electric field which initiates ionization. This fact has long been known, so the calculations of grounding resistance take into account the increasing radius of the metallic conductor owing to the larger ionization zone around the metal. It has also been suggested that a strong electric field induced by high current may cause a breakdown of some gaps by a spark filament in the soil [6]. The soil ionization creates a natural 'grounding electrode', when a lightning channel contacts the earth's surface. The mechanism of current spread through the soil can become clear from analysis of the simple case of a spherically symmetrical distribution of current I_M. At the distance r from the point of lightning stroke, the current is sustained by the electric field $E = \rho I_M/(2\pi r^2)$, where ρ is the soil resistivity. The soil represents a porous medium with the pores filled by air. Experiments show that the soil air is ionized more readily than the atmospheric air at $E > E_{ig} \approx 10\,\text{kV/cm}$ [6, 7]. This is due to the local field enhancement around sand grains, etc. (cf. section 4.3.1). Therefore, the medium in a hemisphere of radius $r_i = (I_M\rho/2\pi E_{ig})^{1/2}$ is ionized to become a natural well-conducting grounding electrode. The grounding electrode resistance, i.e., the resistance to the current spread through a non-ionized soil, is defined as

$$R_g = \frac{1}{I_M}\int_{r_i}^{\infty} E\,\mathrm{d}r = \frac{\rho}{2\pi r_i}. \qquad (6.14)$$

For example, the grounding resistance is found to be $R_g = 72\,\Omega$ at $I_M = 30\,\text{kA}$, $\rho = 10^3\,\Omega\cdot\text{m}$ (sandy soil), and $r_i = 2.2\,\text{m}$.

This situation is very unlikely because the process is unstable. Even a slight asymmetry, which is always present in nature, say, the asymmetry created by tree roots at the site of the lightning strike, may produce a creeping discharge. A plasma channel similar to a leader channel originates at the strike site. It acts as a long grounding electrode, from which the lightning current spreads through the soil. The leakage current per unit channel length, I_1, is proportional to the channel potential U at this site, $I_1 = G_1 U$. The linear conductivity G_1 of the leakage through the channel surface contacting the soil is defined by an expression similar to (6.14) but with allowance for the cylindrical (or, rather, semi-cylindrical) geometry. The radial field at radial distances r smaller than the conductor length l is $E \approx I_1 \rho (\pi r)^{-1}$, where $I_1 = I_M/l$ is the leakage current per unit channel length. When integrating the field over the radius to find the channel potential U, one should take the upper limit $l_1 \approx l$, because at $r > l$ the field decreases as $1/r^2$ and the integral converges quickly. Hence, we have

$$G_1 = \frac{I_1}{U} \approx I_1 \bigg/ \int_{r_i}^{l_1} E \, dr \approx \frac{\pi}{\rho \ln (l/r_i)}. \tag{6.15}$$

Here, r_i is the radius of a well-conducting channel. Because of the logarithmic dependence of G_1 on r_i and l_1, these values do not affect G_1 much.

Laboratory experiments [8] have shown that the principal difference between a classical leader in air and a spark running along a conducting surface is the mechanism of current production providing the energy for the channel heating. In the former case, the current is produced by the streamer zone in front of the leader tip (section 2.4.3) and in the latter, owing to the transverse current leakage from the surface of the channel contact with a conducting medium. A streamer-free leader process was clearly observed under these conditions in laboratory experiments [5, 8]. The streak picture in figure 6.3 does not show even a trace of the streamer zone, whereas the air gap of the same length is filled by streamers nearly from the very beginning of the leader process, in the absence of a conducting surface. The spark process occurring along a conducting surface is very effective. A creeping leader requires an order of magnitude lower voltage for its development than an ordinary leader – 135 kV instead of 1300 kV – for bridging a gap of 5 m long. Of primary importance here is the medium conductivity and the current supplied to the channel. To make a streamer-free leader move on, the field at its tip must be $E > E_{ig}$ to be able to initiate the ionization, to supply the initial channel with a current as high as the ordinary leader current, $i_t > i_{t_{min}} \sim 1\,A$, to heat the gas rapidly, and to maintain the channel conductivity (section 2.4.3).

A small portion of the lightning current, $i_t \ll I_M$, delivered to the leader tip is sufficient for a creeping leader to develop successfully. The tip current is, at first, very high. But as the channel grows, more and more lightning current leaks down to the earth because of the increasing contact area of

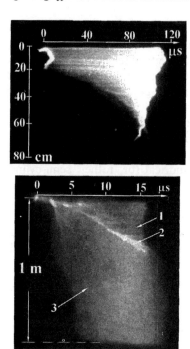

Figure 6.3. Streak photographs of a leader creeping along the soil (top) and an air leader (bottom). 1: channel, 2: tip, 3: streamer zone.

the plasma column with the conducting soil. So, the leader eventually stops. Let us evaluate the maximum length l of the leader channel. Suppose current I_M is delivered to the channel through its base at the stroke point. The current value is determined by the recharging of the lightning leader channel at the return stroke stage and is independent of the creeping spark length. For simplicity, we take the delivered current I_M and the longitudinal field E_c, supporting the creeping leader current, to be constant. At high currents $(i > 1\,\mathrm{A})$, the dependence $E_c(i)$ is, indeed, not particularly strong. By the moment the leader has stopped, the tip potential U_t and current i_t are low relative to $U(x)$ and $i(x)$ at distances x from the tip, comparable with the channel length. We then have

$$U(x) \approx E_c x, \qquad \frac{\mathrm{d}i}{\mathrm{d}x} = I_1 = G_1 U(x), \qquad i(x) = \frac{G_1 E_c x^2}{2}. \qquad (6.16)$$

With $i(l) = I_M$ at the channel base, the maximum channel length is defined, with the account of (6.15), as

$$l \approx \left(\frac{2I_M}{G_1 E_c} \right)^{1/2} \approx \left[\frac{2I_M \rho \ln (l/r_i)}{\pi E_c} \right]^{1/2}. \qquad (6.17)$$

If the longitudinal field E_c is $100 \, \text{V/cm}$, as in the case with a common leader channel which is usually close to the arc state, the channel length l will be $40 \, \text{m}$ ($r_i \approx 1 \, \text{cm}$) for an average lightning with $I_M \approx 30 \, \text{kA}$ and a well conducting soil with $\rho \approx 100 \, \Omega \cdot \text{m}$. The channel length will grow with rising lightning current and decreasing soil conductivity: its value is $l \approx 220 \, \text{m}$ at the maximum current $I_M \approx 200 \, \text{kA}$ and $\rho \approx 1000 \, \Omega/\text{m}$. These estimates are consistent with observations. If a creeping leader encounters a cable, the still available current in it will penetrate to the cable sheath.

6.2.3 Overvoltage on underground cable insulation

If one digs the soil to expose the site of the lightning current input into a cable, one can observe the cable cores with damaged insulation, which are in contact with the metallic sheath. The damage may be stimulated by the presence of a gas-generating dielectric in the cable. The dielectric is decomposed, because of the heating by high current, to produce an electrical hydraulic effect, so that the cable appears literally compressed by the shock wave. A similar effect can be produced by an explosive evaporation of soil water. The elimination of the damage at the current input may not remove the emergency, because there may be several others along a distance of several hundred metres, on both sides of the strike point. These damages result from overvoltages arising between the core and the sheath during the lightning current flow along the cable. The overvoltage mechanism is similar to that described in section 6.2.1, except that the conductor with a sheath has a longer length, sometimes of many kilometres. When the lightning has incorporated its current into the sheath, the cable in a soil of infinite volume should be regarded as a long line with distributed parameters, or, more exactly, as two lines. One is the sheath in a conducting soil. The lightning current flowing along the sheath gradually leaks into the soil and goes to 'infinity', thus raising the sheath potential $U_s(x, t)$ relative to an infinitely far point on the earth. The other line is the core with the sheath. It is affected by the magnetic field of the sheath current, giving rise to an induction emf and voltage drop in the conductive sheath due to its finite linear resistance R_{1s}. As a result, the cable core acquires potential $U_c(x, t)$ relative to infinity, which is generally different from $U_s(x, t)$. The difference $U_e = U_c - U_s$ represents the overvoltage on the cable insulation capable of damaging it.

A rigorous solution to the problem of $U_e(x, t)$ follows from a combined solution of the set of equations describing the lightning current flow along a cable sheath and the voltage wave propagation (between the core and the sheath) along the cable core. This would be a correct approach, provided that the waves in the sheath and inside the cable had approximately the same velocities. But we shall show that these velocities differ by several orders of magnitude, which necessitates the subdivision of this problem into two problems. One will describe the lightning current flow along the sheath and the other the propagation of waves, excited by this current, inside the cable.

Let us first follow the fate of lightning current $i(x, t)$ in the cable sheath. Its variation along the length due to the displacement current associated with the charging of the sheath linear capacitance C_{1s} to the voltage $U_s(x, t)$ can be assumed to be negligible, as compared with the large current leakage into the soil through the linear conduction G_1 of the sheath grounding. One can also neglect the mutual induction emf in the sheath, produced by the core current i_c, because it is small compared to the self-induction emf. Since the total magnetic flux of the sheath current i involves both the sheath and the core, the mutual inductance M_1 per unit length of the sheath–core system is equal to the linear sheath inductance L_1. However, the current in the core is $i_c \ll i$. Indeed, the current in the core screened from the earth by the sheath is only due to the charging of the cable capacitance C_{1c} to the voltage U_e acting between the core and the sheath. The value of U_e does not exceed the electrical strength of the cable insulation, $U_t \approx 2\,\mathrm{kV}$. Even if the current wave velocity in the core were close to light velocity, the core current would be of the order $i_c \approx C_{1c} U_e c \approx 10\text{–}30\,\mathrm{A}$ (for a cable of a small cross section, $C_{1c} \approx 20\text{–}50\,\mathrm{pF/m}$), which is much lower than the lightning current $i \approx 10\,\mathrm{kA}$. So, one can ignore the current deviation into the core even when its insulation is damaged at the lightning current input into the cable so that the core appears to be connected to the sheath.

Therefore, on the above assumptions, the current flow along the sheath is defined by the equations

$$-\frac{\partial U_e}{\partial x} = L_1 \frac{\partial i}{\partial t} + R_1 i, \qquad -\frac{\partial i}{\partial x} = G_1 U_s \qquad (6.18)$$

where L_1 is given by formula (4.25) and R_1 is its linear resistance. If the cable were on the earth's surface, with the lower half of the sheath touching the earth, formula (6.15) would be valid for G_1. When a cable is buried at a large depth, the current spreads radially from it in all directions uniformly, so the value of G_1 is doubled. In intermediate situations, one can use the empirical formula

$$G_1 = \frac{2\pi}{\rho \ln (l^2/2hr)}, \qquad r \leqslant h \ll l/4 \qquad (6.19)$$

where h is the cable depth. The boundary condition for (6.18) is expressed by the equality $i(0, t) = I_0$, where I_0 is one half of the current delivered by the lightning to the cable at the input $x = 0$ (the current flows in both directions from this point).

The cable sheath possesses a low active resistance and a fairly high inductance because it is a solitary conductor. The self-induction emf has a greater effect on the distribution of the rapidly varying lightning current in the sheath than the active voltage drop. If $R_1 i$ is neglected in the first

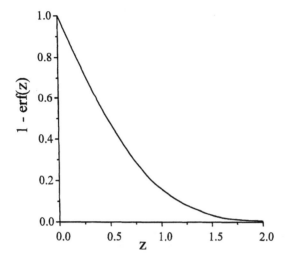

Figure 6.4. The function $1 - \mathrm{erf}(z)$.

approximation, the set of equations (6.18) changes to the familiar diffusion equation, with the only difference that the diffusion coefficient χ_0 is now defined by $L_1 G_1$ rather than by $R_1 C_1$. The current varies along the sheath length in both directions from the input as

$$i(x, t) = I_0 \left\{ 1 - \mathrm{erf}\left[\frac{x}{(4\chi_0 t)^{1/2}} \right] \right\}, \qquad I_0 = \frac{I_M}{2}, \qquad \chi_0 = \frac{1}{L_1 G_1} \approx \frac{2\rho}{\mu_0}. \quad (6.20)$$

The latter expression for χ_0 corresponds to a surface cable; at a large depth, this would be $\chi_0 = \rho/\mu_0$: $\chi_0 \approx 160$–$600\,\mathrm{m}^2/\mu\mathrm{s}$ at $\rho \approx 10^2$–$10^3\,\Omega/\mathrm{m}$ ($\rho \approx 500\,\Omega/\mathrm{m}$ for a common sandy soil). The point with a fixed value of i/I_0 is shifted with a decreasing velocity $v \approx \chi_0/x \approx (\chi_0/t)^{1/2}$, in agreement with the diffusion law $x \approx (4\chi_0 t)^{1/2}$. The current covers a $1\,\mathrm{km}$ cable length for $t \approx 2000$–$200\,\mu\mathrm{s}$, decreasing rapidly at the wave front (figure 6.4). The sheath potential from (6.18) and (6.20) is

$$U_s(x) = U_s(0) \exp(-x^2/4\chi_0 t), \qquad U_s(0) = I_0 \left(\frac{L_1}{\pi G_1 t} \right)^{1/2}. \quad (6.21)$$

The potential at the current input drops with time, from an 'infinite' value at $t = 0$, which results from the neglect of C_{1s}.† The equivalent sheath resistance

$$R_{\mathrm{eq}} = \frac{U_s(0)}{I_0} = \left(\frac{L_1}{\pi G_1 t} \right)^{1/2} = \frac{2}{\sqrt{\pi G_1} x_1}, \qquad x_1 = (4\chi_0 t)^{1/2}. \quad (6.22)$$

† If C_{1o} is taken into account, there is a weak precursor which propagates with the velocity of an electromagnetic signal $(L_1 C_{1s})^{-1/2}$, overtaking the diffusion wave described by (6.20) and (6.21) (cf. section 4.4.2).

also decreases with time. It is defined by the resistance of the soil around the elongating cylindrical surface, through which the current leaks.

Now turn to the wave process inside the cable. The type of overvoltage under consideration is dangerous only to communications lines, whose linear inductance L_{1c} is small because of the narrow gap between the core and the sheath. The self-induction term is usually small relative to the voltage drop on the active core resistance R_{1c}. The current leakage through a high quality insulation can also be neglected, since it is small relative to the displacement current charging the linear core capacitance C_{1c} (relative to the sheath). With these assumptions, the core potential U_c relative to infinity and the core current i_c are described by the equations

$$-\frac{\partial U_c}{\partial x} = R_{1c}i_c + L_1\frac{\partial i}{\partial t}, \qquad -\frac{\partial i_c}{\partial x} = C_{1c}\frac{\partial(U_c - U_s)}{\partial t} \qquad (6.23)$$

which account for $M_1 = L_1$. The electrical signal induced in the core by the lightning stroke has a much higher propagation rate than the process of filling the sheath with current. Indeed, we have $U_s = 0$ and $\partial i/\partial t = 0$ far ahead of the filled part of the sheath. Equations (6.23) transform to the diffusion equation for U_c and i_c with the coefficient $\chi_c = (R_{1c}C_{1c})^{-1} \approx 2.5 \times 10^6 - 2 \times 10^5\,\mathrm{m^2/\mu s}$ ($R_{1c} \approx 0.01 - 0.1\,\Omega\cdot\mathrm{m}$) exceeding χ_o by several orders of magnitude. This means that the charging of the cable capacitance occurs very quickly, and a quasi-stationary mode is established in the cable, in which U_c and i_c follow a relatively slow variation of the sheath current.

By subtracting the first equalities of (6.18) and (6.23) from one another and keeping in mind $R_{1c} \sim R_1$ and $i_c \ll i$, we obtain the equations for over-voltages in the cable:

$$\frac{\partial U_e}{\partial x} = R_1 i, \qquad U_e(x,t) \approx \int_0^x i(x,t)R_1\,dx + U_e(0,t). \qquad (6.24)$$

If the cable insulation at the lightning current input is intact, $U_c(\infty,t) = 0$ and $U_e(\infty,t) = 0$, the overvoltage value is maximal at the input point and is defined as

$$U_e(0,t) = -\int_0^\infty i(x,t)R_1\,dx \approx -I_0 R_1 x_1, \qquad x_1 = (4\chi_o t)^{1/2} \qquad (6.25)$$

where x_1 is the equivalent sheath length with the lightning current at the moment of time t. Overvoltages rise in time as long as the lightning current is high; more exactly, as long as its decrease is compensated by the elongation x_1 (the situation for a realistic current impulse will be discussed below).

If the cable insulation is damaged at the current input 'instantaneously' and the core contacts the sheath, then we have $U_e(0,t) = 0$ and the over-voltage grows with distance from the stroke point up to the maximum value of (6.25) at $x > x_1 = (4\chi_o t)^{1/2}$, provided of course that the lightning current is still high at the moment t.† For example, at $I_0 = I_M/2 = 10\,\text{kA}$ and $R_1 = 3.5 \times 10^{-4}\,\Omega/\text{m}$ (the aluminium sheath is $1\,\text{mm}$ thick and $30\,\text{mm}$ in diameter), we have $U_e \approx I_0 R_1 x_1 \approx 2\,\text{kV}$ at a distance $x_1 \approx 600\,\text{m}$ from the current input. This happens at the moment of time $t \approx x_1^2/4\chi_0 \approx 60\,\mu\text{s}$ (at $\chi_0 \approx 1000\,\text{m}^2/\mu\text{s}$, if the lightning current is still high relative to its amplitude. This is the duration of comparatively short current impulses of negative lightnings. For anomalously long impulses ($\sim 1000\,\mu\text{s}$) of positive lightnings, the length of the 'active' cable portion where the overvoltage arises can increase to 1–$10\,\text{km}$, with the overvoltage amplitude becoming appreciably larger. It is clear now why the repair of the damaged insulation at the lightning input is insufficient and other damaged sites must be found and removed along several kilometres of the cable length. In regions with poorly conducting soils (rocks, permafrost), a damaged line may extend to dozens of kilometres.

So far, we have evaluated the overvoltage for a rectangular current impulse. To calculate it for a real lightning pulse, we should first find a more rigorous solution for the current input into the cable with an intact insulation. This will provide the maximum value of U_e. We shall apply the operator approach to equation (6.18), omitting the term $R_1 i$, as before. As a result, we get the expression

$$\frac{\mathrm{d}^2 i(x,p)}{\mathrm{d}x^2} - \frac{p}{\chi_0} i(x,p) = 0, \qquad i(x,p) = A\exp(-\lambda x),$$

$$\lambda = (p/\chi_0)^{1/2} = (p\mu_0/2\rho)^{1/2} \tag{6.26}$$

in which the last term corresponds to a cable on the earth's surface. If unit current $i(0,t) = I_0 = 1$ flows into the sheath, the integration constant is $A = 1$. The operator form of the overvoltage is

$$U_e(0,p) = R_1 \int_0^\infty \exp(-\lambda x)\,\mathrm{d}x = \frac{R_1}{(p\mu_0/2\rho)^{1/2}} = \frac{R_1}{(\mu_0/2\rho)^{1/2}}\frac{\sqrt{p}}{p}. \tag{6.27}$$

The inverse transform of (6.27) is the function

$$U_e(0,t) = \frac{2R_1}{(\pi\mu_0/2\rho)^{1/2}}\sqrt{t} \tag{6.28}$$

which coincides, within the accuracy of the numerical coefficient of the order of unity, with (6.25) at $I_0 = 1$. Expression (6.28) for unit current $I_0(t) = 1$ represents the unit step function of $y(t)$ providing the solution for arbitrary lightning current $i(t)$ by taking the Duhamel–Carson integral. In particular, we get the following expression for a bi-exponential current impulse

---

† The equivalent core resistance $R_{1c}x_2$, with $x_2 \approx (4\chi_c t)^{1/2}$, grows in time, in contrast to the decreasing input resistance of the sheath. This is another argument in favour of the current entering primarily the sheath rather than the core, even if they come in contact at the input.

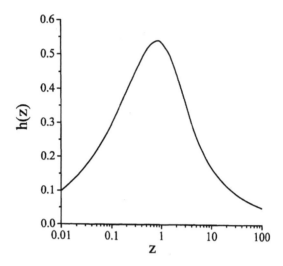

Figure 6.5. The function $h(z)$.

$i = 2I_0 = I_M[\exp(-\alpha t) - \exp(-\beta t)]$, allowing for its spread in both directions from the stroke point:

$$U_e(0, t) = \frac{I_M R_1}{(\pi\mu_0/2\rho)^{1/2}}\left[\frac{h(\alpha t)}{\alpha^{1/2}} - \frac{h(\beta t)}{\beta^{1/2}}\right]$$

$$h(z) = \exp(-z)\int_0^{z^{1/2}} \exp(y^2)\,\mathrm{d}y.$$

(6.29)

The function $h(z)$ has a maximum $h_{max} \approx 0.54$ at $z \approx 1$ and goes down to $0.5h_{max}$ at $z \approx 5$ (figure 6.5). Therefore, the overvoltage pulse front $U_e(0, t)$ is close to the duration of a lightning current impulse and U_e decrease several times more slowly than the current. The duration of the current pulse front affects the overvoltage value only slightly. The estimation from (6.29) for a current impulse of duration $t_p = 100\,\mu s$ ($\alpha = 0.007\,\mu s^{-1}$, $\beta = 0.6\,\mu s^{-1}$) gives $U_{e_{max}}/I_M R_1 \approx 145\,m$ at resistivity $\rho = 1000\,\Omega\cdot m$. For $I_M = 30\,kA$ and $R_1 = 3.5 \times 10^{-4}\,\Omega/m$ (an aluminium sheath of 1 mm thickness and 30 mm diameter), the maximum overvoltage is 1.5 kV.

The diffusion equation for sheath current can be solved numerically for any shape of the current impulse in a cable of finite length, when the core contacts the sheath at the lightning stroke point and when the insulation is intact. Figure 6.6 illustrates the results for the former situation. At $\chi_0 = 2\rho/\mu_0 = 1.6 \times 10^9\,m^2/s$ ($\rho = 1000\,\Omega\cdot m$), the bi-exponential current impulse $i(t) = I_M[\exp(-\alpha t) - \exp(-\beta t)]$ with a 5 μs front and duration of $t_p = 100\,\mu s$ (on the 0.5 level) excites an overvoltage pulse with the reduced amplitude $U_{e_{max}}/R_1 I_0 = 110\,m$ along the cable length of 500 m. The overvoltage maximum occurs at the moment of time $t_m = 60\,\mu s$; the overvoltage

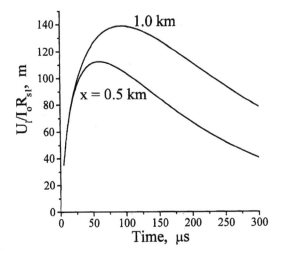

Figure 6.6. Evaluated overvoltage pulses in the cable of a length x with a core contacting a sheath at the stroke point. The bi-exponential lightning impulse current with $\alpha = 0.007\,\mu s$, $\beta = 0.6\,\mu s$; $\rho = 1000\,\Omega/m$.

drops by half in $230\,\mu s$. At a distance of $1000\,m$ from the lightning current input, the overvoltage pulse is somewhat higher, smoother and longer. For current $I_M = 30\,kA$, its amplitude rises to $1.1\,kV$ in an aluminium sheath with $R_1 = 3.5 \times 10^{-4}\,\Omega/m$ and to $7.5\,kV$ for a cable with a lead sheath of the same cross section. All of the calculated parameters are quite comparable with those estimated from (6.29).

6.2.4 The action of the skin-effect

One of the manifestations of the skin-effect is that the current turned on at a certain moment takes some time to penetrate into the conductor bulk. The characteristic time for a conductor of thickness d and conductivity σ to be filled by current is $T_s = \mu_0 \sigma d^2$; for example, $T_s \approx 6\,\mu s$ at $d \approx 1\,mm$ and $\sigma \approx 10^7\,(\Omega \cdot m)^{-1}$. One can neglect the skin-effect when treating overvoltages in underground cables with about the same sheath thickness but with an order of magnitude longer time of lightning current flow along the cable. One can assume the current to flow through the whole sheath thickness, as was suggested above in the treatment of linear sheath resistance R_1. But when one considers short sheaths, especially those of terrestrial objects, in which current runs very rapidly (at light velocity), it is often impossible to ignore the finite time of current penetration in the transverse direction, i.e., through the sheath thickness.

The electric field and current diffuse from the conductor surface into its bulk with the diffusion coefficient $\chi_s = (\mu_0 \sigma)^{-1}$ (hence, $d^2 \sim \chi_s T_s$). Due to this fact, the effective resistance of the conductor is higher than in the case of direct current. The formal use of this fact in (6.13) would result in an increase in the overvoltage which is proportional to R_1. But an opposite effect is observed in reality. Owing to the skin effect, the overvoltage pulse front becomes smoother than the current pulse front, reducing the overvoltage at a finite pulse duration.

The reason for this paradox is that the last equality of (6.13), which is strictly valid only for an infinitely thin sheath or for direct current, should not be used in any situation. If the current varies in time and the sheath has a finite thickness, its voltage can also be represented as a sum of the resistance $U_R(t) = \sigma^{-1}jl$ (j is the current density) and the induction $U_i(t) \approx d\Phi/dt$ components (Φ is the magnetic flux). But with the same sum $U_s = U_R + U_i$, the value of each component varies with the point r of the sheath cross section, for which the calculation is being made, since the proportion between the current density $j(r)$ and the magnetic flux $\Phi(r)$ varies when the total current over the cross section changes. Calculations of overvoltages between the conductor and the sheath, $U_e = U_c - U_s$, are generally indifferent to which r the value of U_s is being found, because the potential does not vary with the thickness. For simplicity, however, it is reasonable to make calculations for the inner sheath surface: this surface and the internal conductor are the only elements of the system affected by equal magnetic fluxes, mutually excluding the induction components of overvoltage on the conductor and the sheath. Consequently, formula (6.13) can be replaced, without any restrictions, by the expression

$$U_e(t) = j_{in}(t)\sigma^{-1}l = E_{in}(t)l \qquad (6.30)$$

where j_{in} and E_{in} are the current density and longitudinal electric field, respectively, on the inner surface of the object's sheath.

The current penetration into a thin sheath is described by the equation for one-dimensional plane diffusion. It has been mentioned that the diffusion coefficient is expressed by the quantity $\chi_s = (\mu_0\sigma)^{-1}$. For a rectangular current impulse of infinite duration, the longitudinal field strength on the inner surface of a sheath of thickness d can be written as

$$E_{in}(t) = R_1 I_0 \left\{ 1 + 2\sum_{n=1}^{\infty}[(-1)^n \exp(-n^2\gamma t)] \right\}, \qquad \gamma = \frac{\pi^2}{\mu_0\sigma d^2}. \qquad (6.31)$$

The exponential series at $t > \gamma^{-1}$ converges very rapidly, so one can restrict oneself to the first term only. Therefore, the field E_{in} rises with the time constant $T'_s = \gamma^{-1} = \mu_0\sigma d^2/\pi^2$; its value is $6\,\mu s$ at $\sigma \approx 5 \times 10^7\,(\Omega \cdot m)^{-1}$ and $d \approx 1\,mm$. This permits the neglect of the skin-effect action on overvoltages in long underground cables, in which the current diffusion along the sheath and, hence, the time of the overvoltage rise to the maximum take 1 or 2 orders of magnitude longer than T'_s. However, the skin-effect in objects located on the earth's surface and having relatively short sheaths,

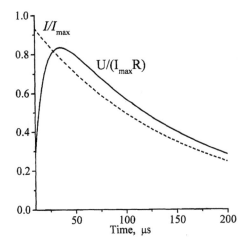

Figure 6.7. Overvoltage pulse deformation in a cable sheath due to skin-effect with the time constant $T_s = 10\,\mu s$. An exponential current impulse is duration of $100\,\mu s$ (dashed curve).

in which lightning current propagates over the time $t \ll T'_s$, decreases the overvoltage with a greater efficiency in the case of a shorter current impulse. For an exponential current impulse $i(t) = I_M \exp(-\alpha t)$, we have from formula (6.31) with the first series term only and the Duhamel integral

$$E_{in} \approx I_0 R_1 \left[\frac{\gamma + \alpha}{\gamma - \alpha} \exp(-\alpha t) - \frac{2\gamma}{\gamma - \alpha} \exp(-\gamma t) \right], \qquad t > \gamma^{-1}. \qquad (6.32)$$

The results of the calculations presented in figure 6.7 show that the skin effect elongates the overvoltage pulse front to T'_s and the amplitude decreases by several dozens of percent.

6.2.5   The effect of cross section geometry

We have assumed so far that the sheath has the shape of a circular cylinder. In that case, the current is distributed uniformly along the cross section perimeter and there is no magnetic field inside. But this model is inapplicable to many real objects, for example, the fuselage or wing of an aircraft, having very complex cross section profiles with different curvatures. The current flowing through a non-circular sheath is distributed non-uniformly along its perimeter and the magnetic field is present inside. These factors affect the mechanism of overvoltage excitation by lightning current.

Let us consider a two-dimensional sheath shaped as a cylinder of a non-circular cross section and a considerable length l, when the end effects are weak and the current and field distributions are plane-parallel. The sheath is considered to have a uniform resistivity and thickness. Let us subdivide the sheath into a set of N parallel conductors of a short length Δr_k along the cross section perimeter, such that the current J_k per perimeter unit length in the kth conductor could be regarded as varying only with time (the total current in the kth conductor is $i_k = J_k\Delta r_k$). In a steady-state mode when the current becomes direct, all J_k values are the same, since they are determined by equal ohmic voltage drops in all of the conductors. A mere summing of the magnetic fields of the conductors will indicate that a magnetic field may be present inside a sheath of an arbitrary cross section geometry.

When lightning current is introduced into the sheath very quickly, the magnetic induction emf in the conductors greatly exceeds the ohmic voltage drop. But in this approximation, all of the conductors will indeed form an integral 'perfectly conducting' sheath, namely, they will be connected in parallel. This means that all of them will have equal potentials. Hence, the magnetic flux coupling Φ for each conductor is the same. This provides a set of equations for finding the currents i_k at the initial stage of the process:

$$L_k i_k(0) + \sum_{m=1}^{N} M_{km} i_m(0) = \Phi, \qquad m \neq k, \qquad \sum_{k=1}^{N} i_m(0) = I_M \qquad (6.33)$$

where L_k is inductance, M_{km} is the mutual inductance of the conductors k and m, and I_M is the lightning input current. Now, in contrast to the steady-state mode, the currents will be different even in identical conductors if they are located at different sites of the sheath. We shall illustrate this with reference to three parallel conductors of length l and radius r located in the same plane so that the distances between the adjacent conductors are identical and equal to D. If i_1 is the current in the central conductor and i_2 is that in the end conductors, with M_{12} as the mutual inductance of the adjacent conductors and M_{23} of remote ones, we shall have

$$Li_1 + 2M_{12}i_2 = Li_2 + M_{12}i_1 + M_{23}i_2, \qquad i_1 + 2i_2 = I_M$$

$$i_1 = \frac{L - M_{12} - (M_{12} - M_{23})}{3L - 4M_{12} + M_{23}} I_M, \qquad i_2 = \frac{L - M_{12}}{3L - 4M_{12} + M_{23}} I_M$$

$$L \approx \frac{\mu_0 l}{2\pi} \ln\left(\frac{l}{r}\right), \qquad M_{12} \approx \frac{\mu_0 l}{2\pi} \ln\left(\frac{l}{D}\right), \qquad M_{23} \approx \frac{\mu_0 l}{2\pi} \ln\left(\frac{l}{2D}\right).$$

The current in the central conductor is lower than in the end conductors because of $M_{12} > M_{23}$.

It is easy to solve a set of equations of the type (6.33) even for a large number of conductors simulating a sheath. Only the calculation of inter-conductor distances is somewhat cumbersome, requiring knowledge of the cross section profile coordinates. We shall leave this problem to the reader and illustrate, instead, the analytical solution for the current distribution in a long cylindrical sheath with an elliptical cross section [9]. This solution is useful for the evaluation of many real profiles and for testing computation programmes:

$$J(x) = \frac{I_M}{2\pi[a^2 - x^2(1 - b^2/a^2)]^{1/2}}. \tag{6.34}$$

Here, a is the large and b the small semiaxis of the ellipse and x is the distance between the ellipse centre and the calculation point projection on the large axis. The ratio of the minimum linear current density (on the plane part of the ellipse) to the maximum one (on its tip) is $J_{max}/J_{min} = a/b$. The current non-uniformity may be great in real structures, such as the aircraft wing, $a/b > 100$.

There is no magnetic field in the sheath at the moment of time $t = 0$. This is the result of the initial current distribution among the conductors owing to the magnetic induction emf. With the redistribution of the currents under the action of ohmic resistance, a magnetic field will gradually arise in a non-circular sheath. The field becomes the source of overvoltages in the inner circuits of the object. By integrating numerically the set of equations

$$U = R_k i_k + L_k \frac{di_k}{dt} + \sum_{m=1}^{N} M_{mk} \frac{di_m}{dt}, \quad m \neq k, \quad \sum_{k=1}^{N} i_k = i(t) \tag{6.35}$$

where $U(t)$ is also the unknown voltage drop along the length of the sheath 'made up' of conductors, one can find the variation in the current distribution along the sheath perimeter. The initial condition for the integration is the solution to (6.33). The calculation accuracy increases with the number N of simulating conductors. But the limiting case of $N = 1$ is also suitable for the evaluation of the time constant of a transient process: $T_{tr} = L_1/R_1$, where L_1 and R_1 are the linear sheath inductance and resistance. The current is redistributed slowly, $T_{tr} \sim 0.1$ s, in the sheaths of large objects with radius $r \sim 1$ m and thickness $d \sim 1$ mm. During the action of a common lightning current impulse with $t_p \sim 100$ µs, the current distribution along the sheath perimeter differs but little from the initial distribution profile. The results of a computer simulation support this conclusion. The computation for a sheath of complex geometry (figure 6.8) with $L_1 = 0.57$ µH/m and $R_1 = 1.05 \times 10^{-5}$ Ω/m ($T_{tz} = 54$ ms) has shown that the linear current density at all characteristic points of the sheath takes the steady-state value for a time about 20 ms. During the first 200 µs typical of lightning current, the density cannot change appreciably.

Let us consider overvoltages across the insulation between an inner conductor and the sheath. Suppose the conductor is placed very close to the inner sheath surface. The contour area between the conductor and the wall will be very small, and the internal magnetic flux will be unable to create an appreciable induction emf. The voltage between the conductor and the sheath will be equal to the integral of the ohmic component of the

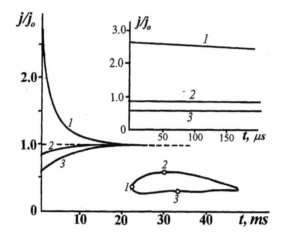

Figure 6.8. Evaluated evolution of a linear current density at indicated points of the wing-like sheath shown. $J_\infty \equiv J(t \to \infty)$.

longitudinal electric field $E_{in}(x)$ at the site of the conductor location x. But now, the evaluation of E_{in} should not be based on the average current density in the sheath, using the total current and linear resistance R_1. For a sheath of thickness d, we have

$$E_{in}(x) = J(x)\rho/d. \tag{6.36}$$

The nearer the current line with the maximum linear density, the higher the overvoltage across the conductor insulation relative to the object's shell. One practical conclusion is quite evident. To reduce manifold the overvoltage in the electrical line inside an aircraft wing and along its thinnest back end, where the current density is maximal, it is sufficient to shift the wire closer to the upper wing plane or, better, to the lower one, where the current density is minimal due to the wing curvature (figure 6.8). Laboratory measurements have confirmed this suggestion [10].

Note the seemingly ambiguous character of the evaluations. The sheath cross section is practically equipotential, so the inner conductor must be under the same voltage with respect to any point of the sheath in a particular cross section. However, the ohmic overvoltage component for a conductor inside an elliptical cylinder (figure 6.9) with respect to points 1 and 2 of the large and small semiaxes differ by a factor of $J_{max}/J_{min} = a/b$, in agreement with (6.34). This contradiction is superficial. In the presence of a magnetic field, there is the magnetic component, in addition to the electrical one, $U = U_e + U_M$. The distance between the conductor and current line 1 is practically zero, and the magnetic flux induces nothing in such a narrow circuit, $U_M \approx 0$. On the contrary, a wide circuit, made up of a conductor and remote current line 2, is affected by most of the internal magnetic flux.

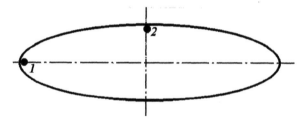

Figure 6.9. The conductors inside an elliptic cylinder.

The emf induced by the flux adds the ohmic voltage to the necessary value U. The evaluation of the magnetic flux direction will show that the signs of U_e and U_M coincide if the circuit, in which U_M is induced, is composed by the current line with a linear density less than the average value; otherwise, U_e and U_M have opposite directions. Therefore, the values of U_e and U_M vary with the design circuit chosen, but the sum remains the same.

We shall make use of this circumstance to find the time variation of the magnetic field inside the sheath. It has been pointed out above that lightning current $i(t)$ acts for such a short time that it cannot be redistributed radically along the sheath perimeter; therefore, we have $J(t) \sim i(t)$ at any point. Hence, we get $E_{in}(t) \sim i(t)$ for a thin sheath where the skin effect is inessential. Choosing a design circuit with $U_M = 0$, we find $U(t) = U_e(t) \sim E_{in}(t) \sim i(t)$. But in the general case, this is $U(t) = U_e(t) + U_M(t)$, with the values of U_e and U_M being comparable; hence, $U_M(t) \sim i(t)$.

Thus, the induction component of overvoltages in a sheath is proportional to lightning current rather than to its derivative! Therefore, the magnetic flux penetrating into the sheath varies in time as the integral of current $i(t)$. This remarkable result has been confirmed by experiments. The oscillograms in figure 6.10 illustrate a test current impulse, similar in shape to a lightning current impulse, and a magnetic pulse $H(t)$ inside a

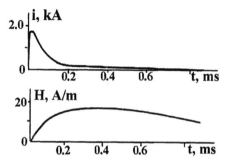

Figure 6.10. Oscillograms of the test current and magnetic field inside the wing-like sheath.

sheath simulating an aircraft wing [11]. The response time of the magnetic field detector did not exceed $0.5\,\mu s$, so that the $H(t)$ pulse front close to $300\,\mu s$ and an order of magnitude higher than the current impulse front causes no doubt.

6.2.6 Overvoltage in a double wire circuit

Although the use of a metallic sheath as a reverse wire saves on metal, most internal circuits of objects to be protected consist of two wires, because they are better screened from noises. When the magnetic field inside an object is zero, as is the case with a perfectly circular sheath, the lightning current raises the potential of each wire relative to the shell but no overvoltage arises between the wires. This is important because the electromagnetic induction can damage the insulation and produce noises in information transmission systems. The consequences of an information line disorder are often as hazardous as a failure in an electronic unit.

It follows from the previous section that the magnetic induction emf inside the sheath strongly depends on the wire location. The emf value is maximal when one of the wires goes near the inner sheath surface along the current line of maximum linear density and the other wire is immediately adjacent to the line with $J_{\min}$. The overvoltages U_1 and U_2 of the two wires relative to the shell are determined only by the ohmic components, since the wires immediately adjacent to the inner sheath produce with it zero area circuits: $U_1 = J_{\max}\rho l/d$ and $U_2 = J_{\min}\rho l/d$. The voltage between the wires $\Delta U_{\max} = U_1 - U_2 = (J_{\max} - J_{\min})\rho l/d$ is due to the internal magnetic field, so the variation rate of the magnetic flux penetrating through the circuit composed of the wires is $d\Phi_{\mathrm{in}}/dt = (J_{\max} - J_{\min})\rho l/d$. We can again conclude that the magnetic field pulse in the sheath is not similar to the lightning current but to its time integral. During a current impulse of a negative lightning $t_p \approx 100\,\mu s$ (on the 0.5 level) with $J_{\max} \approx$ const, the magnetic field within a non-circular sheath rises as $H(t) \sim t$ (for a circular sheath, $J_{\max} = J_{\min}$ and $H = 0$). At this lightning current, the higher the conductive sheath resistivity and the greater the non-uniformity of the initial current distribution along the sheath perimeter, the higher is the internal magnetic field.

To illustrate, an estimation will be made for an elliptical cylinder of length $l = 100\,\mathrm{m}$. The following parameters will be used: $a = 1\,\mathrm{m}$, $b = 1\,\mathrm{cm}$, the aluminium sheath thickness $d = 1\,\mathrm{mm}$, and $\rho = 3 \times 10^{-8}\,\Omega \cdot \mathrm{m}$. The lightning current amplitude will be taken to be $I_M = 200\,\mathrm{kA}$, a value used in aircraft tests for lightning resistance. Using formula (6.34), we obtain $J_{\max} \approx 3200\,\mathrm{kA/m}$, $J_{\min} \approx 32\,\mathrm{kA/m}$, and $\Delta U_{\max} \approx 9.5\,\mathrm{kV}$. In a real construction, such a great overvoltage could have resulted from a poor design of the internal electrical network. Wires running to the same electronic unit must not be separated so much from each other, nor should they be placed on the inner side of a metallic shell at places differing much in the surface curvature and, hence, in the linear current density. A compact packing of cable assemblies at sites of minimum surface curvature is a good and nearly free means of limiting overvoltages in internal circuits of objects with metallic shells.

Overvoltages rise considerably if the shell is made from a plastic and if its electric circuits are located in a special outer metallic jacket extending from the head to the tail. The linear resistance of the jacket may be 1–2 orders higher than that of the totally metallic fuselage. The ohmic component of overvoltage will increase respectively. To eliminate the magnetic component, associated with the penetration of the magnetic field into the jacket, it is very desirable to make it as a pipe with a circular cross section.

6.2.7 Laboratory tests of objects with metallic sheaths

The lightning protection practice involves a great many technical problems associated with the formation of test current and the measurement of all parameters of interest. Here, we shall be concerned with the physical aspects of testing, which could allow prediction of the object's response to lightning current in a real situation, generally different from laboratory conditions.

Let us begin with a laboratory current simulating lightning current. The best thing to do would be to make a laboratory generator reproduce lightning current exactly. The high requirements on the protection reliability make one apply maximum currents with an amplitude up to 200 kA, especially for testing aircraft. Tests on the 1:1 scale are attractive because they do not require overvoltage measurements. It is sufficient to examine the object's equipment after the tests to see that there is no damage. However, the generation of a high current creates problems when the object has a long length or when the current impulse front to be reproduced must be short. For example, the maximum steepness for the impulse $i(t) = I_0[1 - \exp(-\beta t)]$ with an exponential front is $(di/dt)_{max} = \beta I_0$. To generate such impulses, the source must develop the voltage $U_{max} = \beta I_0 L$, where L is the circuit inductance close to that of the test object; $L \approx L_1 l$ for an object of length l. The maximum voltage is $U_{max} \approx 12\,\text{MV}$ for $L_1 \approx 1\,\mu\text{H/m}$, $l \approx 100\,\text{m}$, $I_0 = 200\,\text{kA}$, and $\beta \approx 0.6\,\mu\text{s}^{-1}$, corresponding to the front $t_f \approx 5\,\mu\text{s}$ average for the current of the first negative lightning component. A generator with such parameters would have an enormous size and great cost.

The intuitive desire to elongate the current impulse front rather than to reduce its amplitude in the testing of objects with a solid metallic sheath has a reasonable physical basis. Due to the longer front duration t_f, the ohmic overvoltage in the internal circuits of the object to be designed could change only when there is an appreciable current redistribution along its cross section perimeter during the time $t \approx t_f$. This would require the time $t_f > 100\,\mu\text{s}$. Therefore, the application of impulse fronts with a duration of dozens of microseconds, instead of typical lightning impulses, cannot affect the test results. The same is true of overvoltages in a double wire circuit, induced by an internal magnetic flux. Consequently, the increase of the impulse front duration by one order of magnitude will be unable to affect appreciably the overvoltage in internal electrical circuits. This considerably reduces the requirements on the laboratory source of impulse current, because its operating voltage decreases in proportion with the increase in t_f. The decrease in U by an order of magnitude reduces the costs, because the costs of high-voltage technologies rise faster than the actual voltage.

Much attention should be given to the simulation of the lightning impulse duration in laboratory conditions. Anyway, the test impulse should not be shorter than the real one, for the overvoltage amplitude may thus be underestimated because of the skin-effect. It would be unreasonable to reproduce on the test bed the actual amplitude of the lightning current impulse if there are no non-linear elements in the test object's circuit and the overvoltages can be registered by detectors. Since the electrical and magnetic components of overvoltage are similar in shape and equally depend on the applied current amplitude, one can recalculate the measurements in proportion with higher currents and select the test impulse amplitude in terms of the highest possible accuracy and registration convenience.

Quite another matter is the situation when the test object's sheath is not solid but has slits or technological windows. The 'external' magnetic field of the lightning partly penetrates through the sheath; the field is proportional to the current and the induced overvoltages are proportional to the current impulse steepness. The total overvoltage now depends on both the current rise time and amplitude, so the engineer has no chance to select a convenient test impulse shape. In principle, the recalculation of measured pulses to real ones is also possible, but this requires a detailed analysis of the overvoltage mechanism and the responses of the object's circuits, which does not raise the testing reliability.

Another problem is to connect the test object to the laboratory generator. It is obvious that a conductor with the generated current should be connected to the site of a possible lightning stroke. In the case of terrestrial objects in natural conditions and on a test bed, the problem of current output is solved in a simple way – by using a grounding bus. The situation for aircraft and spacecraft is more complicated. In real conditions, the lightning current first flows through a metallic sheath (say, the fuselage) and then enters the ascending leader channel, whose length is much greater than that of the object. It is difficult to reproduce the real current path in laboratory conditions – this would require a very high voltage to make the impulse current run through the long conductor simulating a lightning channel. Besides, the test object and the numerous detectors would be under a very high potential relative to the earth.

The return current wire is normally located close to the test object. Its magnetic field interacts with the object's metallic sheath, through which forward current flows. As a result, the current distribution along the sheath

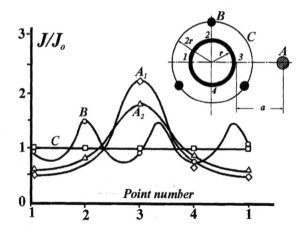

Figure 6.11. Measured angular distributions of the linear current density along the circular pipe perimeter at various locations of the reverse current conductor. Marked points (on the pipe scheme) are presented on the abscissa axis. The curve A_1 corresponds to a single reverse wire A for $a = 2r$, curve A_2 is that for $a = 4r$, curve B depicts three reverse wires B placed as shown in the scheme. Uniform distribution C corresponds to the coaxial reverse current cylinder of radius $2r$.

perimeter changes, the redistribution being considerable if the return current wire is close to the object. Inducing the emf of the opposite sign, the reverse current increases the current load in the nearby parts of the metallic sheath but decreases it in the remote parts.

For a particular geometry, the current distribution should be found numerically from the set of equations (6.33) by adding, to each equation, a term for the magnetic flux from the reverse current wire, $-I_M M_{k0}$, where M_{k0} is the mutual inductance between the return current wire and the kth conductor simulating the sheath. Under conditions typical of test beds, the distortions due to the return current path may be appreciable. The results presented in figure 6.11 have been obtained from the tests of a sheath shaped as a circular pipe. In order to avoid the effects of currents induced in the conductive soil, the sheath was raised above the earth at a height $H = 7r$, where r is the pipe radius. The role of the return current wire was performed by a thin conductor running parallel to the pipe at the distances $a = 2r$ and $a = 4r$ from it, three conductors located at 120° at the same distance, and a coaxial cylinder of radius $2r$. The latter design provides a perfectly uniform current distribution along the perimeter of the sheath cross section. The return current of a single wire distorts the current distribution to the greatest extent: its minimum linear density drops to $0.5j_{av}$ and the maximum density rises to $2.3j_{av}$ ($j_{av} = I_0/2\pi r$). The current distribution becomes more uniform when the number of reverse conductors is increased.

When a sheath has a complex geometry, it is hard to predict the reverse current effect on the test results. The current redistribution in the sheath may lead to both the overestimation and underestimation of overvoltages in the internal circuits. Much depends on the arrangement of the internal conductors and return current wire.

6.2.8 Overvoltage in a screened multilayer cable

Overvoltages in screened multilayer cables are due to the skin-effect. As a result the cable wire screens in the layers are loaded differently by the lightning current. Every layer is formed by wires arranged in a circle and having their own screens (figure 6.12(*a*)). Depending on the reliability requirements, a cable may have an outer metallic sheath or a dielectric coating protecting it from mechanical damage. However, a direct lightning stroke produces a breakdown of dielectric material, and the lightning current is distributed among the screens. The adjacent screens in a layer contact each other along the whole cable length. It can be assumed in a first approximation that they form a solid sheath of circular cross section with resistance $R_k = R/n_k$ and inductance L_k, where R is the resistance of an individual wire screen and n_k is the number of screened wires in the kth layer (figure 6.12(*b*)). For simplicity, we shall consider a double layer cable, marking the inner layer with $k = 1$ and the outer layer with $k = 2$. The adjacent screens of wires from the adjacent layers are also in contact with one another.

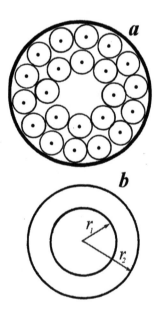

Figure 6.12. Multilayer cable (*a*) and solid sheath model (*b*).

For this reason, a set of circular sheaths can be regarded as a solid conductor, and the current penetration along its radius (from the second layer to the first one) can be considered as a skin effect. Such a system can also be treated as a set of discrete circular layers.

In the latter case, the current distribution among the layers at the initial moment of time $t = 0$ can be found from the condition of magnetic flux coupling equality (6.33). Equations (6.35) are valid at $t > 0$ and have the following solution for two layers at constant current $I_M = i_1 + i_2 = \text{const}$:

$$i_1(t) = \frac{R_2 I_M}{R_1 + R_2} [1 - \exp(-\lambda t)], \qquad i_2(t) = \frac{I_M}{R_1 + R_2} [R_1 + R_2 \exp(-\lambda t)]$$

$$(6.37)$$

where $\lambda = (R_1 + R_2)/(L_1 - L_2)$. Equations (6.35) allowed for the mutual inductance of the layers, $M_{12} = L_2$, as in the treatment of the screen-wire system in section 6.2. In accordance with the skin-effect law, the lightning current first loads the outer sheath and then gradually penetrates into the inner sheath. The current is distributed uniformly between the individual screens in each circular layer, $i_{s1} = i_1/n_1$ and $i_{s2} = i_2/n_2$. The overvoltage across the insulation between a wire and its own screen (providing that the skin-effect in an individual screen is neglected) is similar to the current in the layer, $U_1(t) = R_1 i_1(t)$ and $U_2(t) = R_2 i_2(t)$, but not to the lightning current.

If a double wire circuit uses the cores of one layer, there is no over-voltage in the instruments connected to it, because the potentials of the layer cores are identical. If the instruments are connected to the cores of different layers, the voltage between them is

$$U_{12} = U_2 - U_1 = I_M R_2 \exp(-\lambda t). \qquad (6.38)$$

At $I_M = 1$, expression (6.38) is a unit step function for the set of equations providing the solution for the lightning current impulse of an arbitrary shape. In particular, at $i(t) = I_M[\exp(-\alpha t) - \exp(-\beta t)]$, we have

$$U_{12} = I_M R_2 [B \exp(-\beta t) - A \exp(-\alpha t) - (B - A) \exp(-\lambda t)]$$

$$A = \alpha/(\lambda - \alpha), \qquad B = \beta/(\lambda - \beta). \qquad (6.39)$$

Owing to the relatively small value of $L_1 - L_2 \approx (\mu_0/2\pi) \ln (r_2/r_1)$ at close layer radii r_2 and r_1, the layer current ratio is redistributed rapidly, for $T = \lambda^{-1} \approx 10\,\mu s$. This is the reason for a fast damping of the overvoltage pulse U_{12} (figure 6.13), which may be remarkably shorter than the current impulse. It follows from (6.39) that the pulse U_{12} reverses the sign; its opposite tail is damped approximately at the rate of lightning current reduction. The overvoltage amplitude in a double wire cable is close to that in a wire-shell system, exactly as in a sheath with a sharply non-uniform current distribution. If the screens are thin and have a high resistance, the hazard of damaging the connected measuring instruments is fairly great.

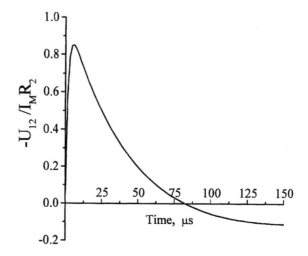

Figure 6.13. Overvoltage pulse on a two-layer cable for the bi-exponential current impulse with $\alpha = 0.007\,\mu s$, $\beta = 0.6\,\mu s$ and the redistribution time constant $T = 50\,\mu s$.

The problem for a multilayer cable can be solved in a similar way. The overvoltages between the cable cores grow with distance between the respective layers. Other conditions being equal, the overvoltages drop with the layer depth in the cable. The use of cores of one cable layer reduces considerably the overvoltage in a double wire system but does not eliminate it entirely. There are no perfectly circular cables – the cable is pressed under its own weight and becomes deformed during its winding on a drum. The result is that the current distribution along the sheath cross section perimeter becomes non-uniform, producing additional overvoltages between the cores of the same layer. To minimize these overvoltages, it is desirable to connect the equipment to the adjacent cores of the same layer. High precision equipment should be connected to the cores of deeper layers. Overvoltages arising in a multilayer cable can be evaluated from the same set of equations (6.35).

6.3   Metallic pipes as a high potential pathway

Modern constructions have an abundance of underground metallic pipes, and the lightning protection engineer must take them into account as a possible pathway for currents from remote lightning strokes. This actually happens when a pipe lies close to a high lightning rod or another object preferable to lightnings. Spreading through the earth away from the grounding electrode in a way described in section 6.2.2, some of the current enters a metallic pipe and runs along its length. A pipe is sometimes

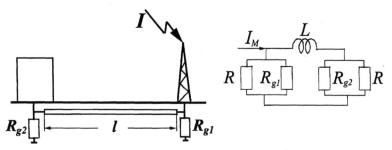

Figure 6.14. Underground pipe as the pathway for a lightning current and the design circuit for a simple evaluation of the object potential.

connected directly to the object grounding electrode. Figure 6.14 illustrates the typical situation when a metallic pipe line connects the grounding electrode (with grounding resistance R_{g1}) of an object, struck by lightning, to the grounding (with resistance R_{g2}) of a well-protected object. Although the lightning is unable to reach the latter directly, some of the current finds its way to its grounding electrode – the pipe. For applications, it is important to know the dependence of this current on the line length l and on the soil conductivity.

Section 6.2.3 considered the problem of current distribution for an underground pipe of infinite length. The limited line length in the present case is an important parameter, especially because it has the grounding resistances at its ends. Generally, this problem can be solved analytically using the Laplace transformation. But the final result is represented as a functional series too complex for a treatment, so numerical computations are necessary. It is, therefore, more expedient to solve this problem numerically from the very beginning. Before presenting the results of a computer simulation, we shall make a simple evaluation. Let us replace an underground pipe by the lumped inductance $L = L_1 l$ and its intrinsic grounding resistance $R_g = (G_1 l)^{-1}$. The latter will be represented as two identical resistors $R = 2R_g$ by connecting them to the ends of the line in parallel to the grounding resistances R_{g1} and R_{g2} of the objects it connects (figure 6.14). This rough approximation makes sense, since we are interested in the value of current i_2 at the far end connected to the grounding mat, rather than in its distribution along the line. In this approximation, we have

$$L\frac{di_2}{dt} + R_{e2}i_2 = (i - i_2)R_{e1}, \qquad R_{ej} = \frac{RR_{gj}}{R + R_{gj}}, \ j = 1, 2. \qquad (6.40)$$

Putting the lightning current to be $i = I_M \exp(-\alpha t)$ and $i_2(0) = 0$, we find

$$i_2(t) = \frac{R_{e1}\lambda I_M}{(R_{e1} + R_{e2})(\lambda - \alpha)}[\exp(-\alpha t) - \exp(-\lambda t)], \qquad \lambda = \frac{R_{e1} + R_{e2}}{L}.$$
$$(6.41)$$

At the beginning, while the effect of self-induction emf is still noticeable, the current largely flows through the equivalent resistance R_{e1} at the front end of the line. After time $T = \lambda^{-1}$, the current gradually penetrates to the far end of pipe. Some of it, $i_{g2} = i_2 R/(R + R_{e2})$, finds its way to the grounding electrode of the object of interest, raising its potential to the value $U_2 = i_{g2} R_{g2}$ relative to a remote point on the earth. For a longer line, the values of i_{g2} and U_2 decrease for two reasons. An increase in $L = L_1 l$ and $G = G_1 l$ raises the time constant T, and by the time the current has reached the far end of the pipe, the initial lightning current is considerably damped. Besides, a smaller portion of the current i_2 that has reached the far end enters the object's grounding electrode because of the greater pipe leakage. The dependence of i_{g2} and U_2 on l proves to be rather strong, especially when the effective duration of the lightning current, $t_p \approx \alpha^{-1}$, is comparable with $T = \lambda^{-1}$. Suppose we take $t_p = 100\,\mu s$ on the 0.5 level ($\alpha = 0.007\,\mu s^{-1}$), the grounding resistances $R_{g1} = R_{g2} = 10\,\Omega$, and $L_1 = 2.5\,\mu H/m$. A metallic pipe with a 10 cm diameter and 100 m in length, lying at the surface of the soil with $\rho = 200\,\Omega/m$ $(G_1 = 2.1 \times 10^{-3}\,(\Omega/m)^{-1}$, $R = 9.7\,\Omega)$, will deliver the current $i_{g2} \approx 0.171 I_M$ to the ground of the object located at its far end. The object's potential will be raised to $U_2 \approx 50\,kV$ at $I_M = 30\,kA$. At $l = 200\,m$, we have $i_{g2} \approx 0.0861 I_M$ and, at the same lightning current, $U_2 \approx 25\,kV$. But even this voltage is quite sufficient for a spark to be ignited between closely located elements of two metallic structures, provided that one of them is connected to the grounding electrode and the other is not. Such a spark can induce an explosion or fire in explosible premises.

In low conductivity soils, current can be transported through metallic pipes for many kilometres. This refers, to a still greater extent, to external pipes and rails mounted on a trestle which are grounded only locally, through the supports separated by dozens of metres. Here, evaluations can also be made with expression (6.42), putting $R = 2R'_g/n$, where R'_g is an average resistance of the support grounding and n is the number of supports.

A comparison of the estimates and computations is shown in figure 6.15 for the above example with $l = 200\,m$. The estimates for the current amplitude at the far end of the pipe and for the moment of maximum current show a satisfactory agreement with the numerical computations. The computations will be unnecessary if one finds it possible to ignore the initial portion of the pulse front and can put up with a 20–25% error.

Let us calculate the potential at the far end of the pipe unconnected to the grounding electrode at either end. This may happen due to careless design or poor maintenance of communications lines. The soil will be considered to have a low conductivity, $\rho = 1000\,\Omega/m$; $L_1 = 2.5\,\mu H/m$. The curves in figure 6.16 show the variation in the voltage and current amplitude ratio U_{max}/I_M for impulses of negative lightnings with $t_p = 100\,\mu s$ and for those of 'anomalous' positive lightnings, which are an order of magnitude longer. The pipe is capable of delivering a potential of dozens of kilovolts

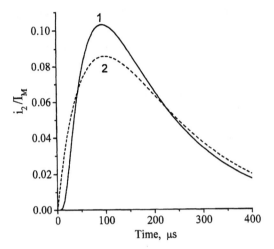

Figure 6.15. Portion of a lightning current passed to the object through the communication pipe of 200 m length. Curve 1: numerical computation, 2: simple evaluation.

for a distance of 1 km to the object even at a moderate lightning current of 30 kA. Damage of the contact between the pipe and the object's grounding electrode may be fatal if a spark arising in the air gap encounters an inflammable substance.

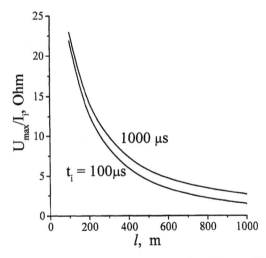

Figure 6.16. Computed maximum overvoltages transferred to an object at the far end of the underground pipe of 10 cm diameter and of length l. The pipe is not connected with the grounding of both an object and a lightning rod. Computations were made for the usual lightning current impulse of 100 μs duration, and for an 'anomalous' impulse of 1000 μs. Lightning stroke to the other end of the pipe.

The delivery of high potential can be controlled in a simple way – all communications lines must be connected to the same grounding mat. In that case, the voltage of all mat components will be raised equally by the brought current of a remote lightning stroke. It should be noted that this is a reliable means to cope with the overvoltage of kilovolt values. A simple connection of metallic sheaths to the grounding mat cannot remove pulse noises of tens or hundreds of volts having a short rise time. Steep current impulses spreading across the buses and components of the grounding mat always create an induction emf, producing abrupt voltage changes even in conductors of about 1 m in length. Electrical circuits must be mounted in such a way as to avoid the appearance of closed contours or joints of the conductor screens to points remote from each other in the grounding mat. This sometimes becomes such a delicate matter that the result depends on the engineer's intuition rather than on exact knowledge.

6.4 Direct stroke overvoltage

We described the manifestations of overvoltage due to a direct lightning stroke when discussing the lightning current propagation across a metallic sheath. The highest current enters the sheath when a lightning discharge strikes an object directly (section 6.2.1). This happens, for example, when an aircraft is affected by the return stroke current recharging the descending leader which has connected the aircraft to the earth. Below, we discuss a direct lightning stroke at a grounded terrestrial object. Specifically, we shall be interested in the voltage applied to the insulation of the object relative to the earth or another construction located nearby. The classical situation is that a voltage arises between the lightning rod that has intercepted the lightning and the nearby object being protected. A rough treatment of this situation was made in section 1.5.1. The fast variation of a high lightning current i along the metallic parts of a construction raises its potential by $U = R_g i + L\, di/dt$ relative to a remote point on the earth. Much depends on what is understood by the grounding resistance R_g and inductance L. These issues are discussed in much detail in the books on direct stroke overvoltages (e.g., [6]). Here, we outline the most important physical aspects of the problem.

6.4.1 The behaviour of a grounding electrode at high current impulses

An important parameter of a grounding electrode is the stationary grounding resistance usually measured during the spread of direct or low frequency alternative current of several amperes. The value of R_{g0} found from the measurements may be several times larger or smaller than $R_g = U_e/I_M$ corresponding to a rapidly varying kiloampere lightning current (here, U_e is the potential at the current input into the protector). We have discussed, at several points in the book, the two physical mechanisms affecting differently the ability of a metallic conductor to tap off the lightning current to the earth: the self-inductance and ionization expansion of the surface contacting the soil.

The voltage drop across the inductance prevents current flow into the conductor. A long conductor has to be treated as a line with distributed parameters. The input resistance of the line, $R_{in} = U(0, t)/i(0, t)$ varies in time, since the current diffuses along the line, and it takes some time for the whole conductor to be loaded by current more or less uniformly. As the limiting case, consider an infinite conductor in a soil with resistivity ρ. From formulae (6.21) and (6.22), the voltage at the conductor input is $U_e(t) \equiv U(0, t) \sim t^{-1/2}$ for the current $i(0, t) = \text{const} = I_0$ and $t > 0$. At $I_0 = 1$, formula (6.21) can be treated as a unit step function of the system, $y(t)$. This allows us to follow the input voltage of a horizontal grounding conductor at the lightning current $i(t)$ with a real impulse front by using the Duhamel–Carson integral:

$$U(0, t) = y(t)i(0) + \int_0^t y(\tau)i'(t - \tau)\, d\tau. \tag{6.42}$$

For a impulse with an exponential front $i(t) = I_0[1 - \exp(-\beta t)]$ we have

$$U(0, t) = 2I_0\left(\frac{\beta L_1}{\pi G_1}\right)^{1/2} h(\beta t) \tag{6.43}$$

where $h(\beta t)$ is a function given by the last integral in (6.29) and figure 6.5. Its maximum h_{max} at $\beta t_m \approx 0.9$ permits the calculation of the maximum voltage drop across the grounding electrode:

$$U_{max} \approx 1.08 I_0(\beta L_1/\pi G_1)^{1/2}. \tag{6.44}$$

The effective input resistance of an extended horizontal grounding electrode, corresponding to U_{max}, is expressed as

$$R_{g_{eff}} = \frac{U_{max}}{i(t_{max})} \approx 1.82\left(\frac{\beta L_1}{\pi G_1}\right)^{1/2}. \tag{6.45}$$

In contrast to a lumped grounding electrode with $R_g \approx \rho$, the input resistance of an extended one varies much less with the soil resistivity, $R_{g_{eff}} \sim \rho^{1/2}$. Extended grounding electrodes are ineffective, because only a short initial portion of their length , $l_{eff} \approx (R_{g_{eff}}G_1)^{-1}$, is actually operative during the impulse front time. For example, the effective resistance is $R_{g_{eff}} \approx 13\,\Omega$ and the effective length of a long grounding pipe with $L_1 = 2.5\,\mu\text{H/m}$ at the earth's surface is $l_{eff} \approx 22\,\text{m}$ in the case of the first component current of a negative lightning with the rise time $t_f \approx 5\,\mu\text{s}$ $(\beta \approx 0.6\,\mu\text{s}^{-1})$ and $\rho = 100\,\Omega \cdot \text{m}$. In a soil with an order of magnitude lower conductivity, the respective values are $R_{g_{eff}} \approx 42\,\Omega$ and $l_{eff} \approx 75\,\text{m}$.

Extending the grounding bus beyond the limit l_{eff}, we are still unable to reduce appreciably the maximum voltage drop across the bus. For this reason, it is better to introduce current at the centre of a long bus rather than at its end, such that two current waves would run in opposite directions along the half-length conductors. Still more effective are three conductors arranged at an angle of 120°, and so on. When a grounding mat with the lowest possible value of $R_{g_{eff}}$ is desired, it is preferable to load, more or less uniformly, the whole of the adjacent soil volume. For this aim, a set of horizontal conductors or a conductor network is combined with vertical rod electrodes. To avoid the interaction effect of the grounding elements and to achieve the maximum loading of them by current, the distance between the elements should be made comparable with their length (or with the height, for vertical rods). But even in that case, only part of the grounding mat, within the radius of l_{eff} from the current input, will operate effectively at the impulse front.

Thus, the resistance of a grounding electrode for rapidly varying currents is much higher than for direct current. A grounding mat network with numerous horizontal buses and vertical rods is able to reduce the effective resistance to the value of $R_{g_{eff}} \approx 1\,\Omega$. But when a large number of objects is being constructed, for example, the towers of a power transmission line, one has to deal with resistances as high as $R_{g_{eff}} \approx 10\,\Omega$ and more.

Laboratory experiments show that the grounding resistance of an electrode delivering to the earth very high currents is lower than for low currents. The grounding resistance decreases with the current rise. The grounding resistance ratio of a high impulsed current and low direct current, $\alpha_i = R_g/R_{g0}$, is often called the impulse coefficient of a grounding. The coefficients α_i used in the literature are sometimes as small as $\alpha_i \approx 0.1$. To illustrate, we shall cite the generalized function $\alpha_i = f(\rho I_M)$ which has been suggested for a vertical rod of 2.5 m in length from the results of small-scale laboratory experiments [7] (figure 6.17). The grounding resistance is reduced by a factor of four at $\rho = 1000\,\Omega \cdot m$ and $I_M = 30\,kA$.

In principle, this reduction in resistance might be due to a larger effective radius of the grounding electrode because of the soil air ionization. In section 6.2.2, we gave formula (6.15) for the linear conductivity of a long rod lying on the earth with one half of its surface contacting the soil. If the rod is fixed in the vertical position, the whole of its surface contacts the soil but its leakage conductivity is lower by a little less than a factor of 2 at the same length (due to the poorer operation of the upper end of the rod located at the earth's surface, because current cannot flow upward into the air). The linear conductivity G_1 and the grounding resistance R_g of a rod of radius r_0, fixed vertically into the earth for a length l, are

$$G_1 = \frac{2\pi}{\rho \ln (2l/r_0)}, \qquad R_g = \frac{\rho \ln (2l/r_0)}{2\pi l}. \qquad (6.46)$$

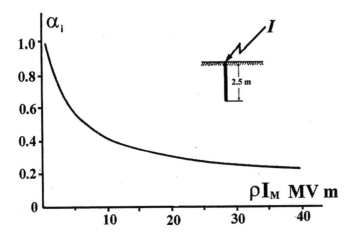

Figure 6.17. Impulse coefficient for the grounding rod of 2.5 m length.

To reduce R_g by a factor of 4 at the initial rod radius $r_0 = 1$ cm and $l = 2.5$ m, the radius must be increased to $r_1 = 105$ cm. The field at the ionized volume boundary must exceed the ionization threshold in the soil, $E_{ig} \approx 10$ kV/cm, and the current density at $\rho = 1000\,\Omega/$m must be $j = E_{ig}/\rho \approx 1$ kA/m^2. For the surface area of the ionized volume $S \approx 2\pi r_1 l + \pi r_1^2 \approx 24$ m^2, the total leakage current would be $I = jS = 24$ kA, corresponding to the current of a moderate lightning power.

However, the uniform radial ionization expansion of the initial grounding volume at a rate $r_1/t_f \approx 2 \times 10^5$ m/s (this process must be completed within the rise time of the current impulse, $t_f \approx 5\,\mu$s) can hardly occur in reality. Anyway, there is no experimental indication for this. More probable would be the rod 'elongation' owing to the leader development into the soil, because the current density and the field at the rod end are higher than at its lateral surface. The elongation of a grounding electrode is a more effective means of reducing the grounding resistance R_g, because of $R_g \sim 1/l$, since the resistance decreases only logarithmically with increasing radius (but only at $r \ll l$). However, even this process seems unlikely. There is no experimental evidence for the existence of long leaders in the soil bulk.

The leakage area of a grounding electrode is likely to increase due to the elongation of the leader creeping along the soil surface from an element of the grounding mat. The grounding resistance will then decrease, as $1/L$ at a long leader length L. This mechanism, observed in model laboratory experiments, seems optimal for a natural reduction in R_g due to the lightning current. To change appreciably the grounding resistance of a typical lightning protector, the leader must grow to $L \approx 10$ m in length (the total length of the protector electrodes) for the time $t_f \approx 5\,\mu$s of the lightning current rise. For this to happen, the leader must elongate at a rate of 2×10^6 m/s. A creeping leader develops in the air adjacent to the surface of a conductive soil and is shown by laboratory experiments to be devoid of a streamer zone and a charge sheath (section 6.2.2). In this respect, it is similar to a dart leader which develops a velocity of about 10^7 m/s at current $i \sim 1$ kA. Assuming that the development of a fast leader along the soil surface requires this current in the leader tip, i_t, let us estimate the tip radius, at which this appears possible.

If the grounding resistance is R_g and the lightning current is I_M, the leader is supported by the voltage $U \approx R_g I_M$ applied to its base. The tip possesses approximately the same potential, because the leader channel is a good conductor and not a large part of the voltage drops across it. The resistance of the leakage current from a hemispherical tip is equal, from formula (6.14), to $R_t \approx \rho/2\pi r_t$. The tip current is $i_t \sim U/R_t$. Only part of the lightning current enters the channel, I_0. This current mostly leaks into the soil through the lateral channel surface possessing, according to (6.15), a leakage resistance $R_{le} = (G_l L)^{-1} = \rho \ln(L/r_0)/\pi L$. Keeping in mind $U \sim R_{le} I_0$, we obtain the formula to be used for the estimation of the tip radius:

$$\frac{I_0}{i_t} = \frac{R_t}{R_{le}} = \frac{L}{2r_t \ln(2L/r_0)}. \tag{6.47}$$

The 6 cm radius obtained at $I_0 = I_M/2 = 15$ kA, $i_t = 1$ kA, $L = 10$ m, and $r_0 \approx r_t$, appears to be quite reasonable. The field at the channel lateral surface behind the tip, $E \sim \rho I_0/\pi r_t L \approx 80$ kV/cm, is high enough for the ionization expansion of the leader channel to occur there. Radii much larger than those of a conventional leader in air under similar conditions have been registered for laboratory leaders creeping along the soil. The photographs in figure 6.18 illustrate this quite clearly.

To conclude, the spread of high lightning currents reduces the grounding resistance, probably due to the excitation of one or several leaders creeping along the soil surface, thereby increasing the length of the grounding electrodes. But for a fast leader growth (otherwise, the leakage surface has no time to become larger for the short lightning rise time), a high current of about 1 kA must be delivered to the leader tip. This restricts the process of grounding resistance reduction by the condition under which the electrodes are arranged in compact groups. A fast reduction is hardly possible for a modern substation having an extended grounding network. The reduction is, however, quite possible in the case of a concentrated protector consisting of 2–3 horizontal conductors or several vertical rods.

It is worth saying a few words about the testing of lightning grounding. The great complexity of a large-scale simulation of lightning current makes one turn to model tests, in which the surface current density of small electrodes is preserved while the total current is reduced manifold. The laboratory studies indicates that the similarity laws are invalid for the leader process. The questions of how to interpret the small-scale simulation

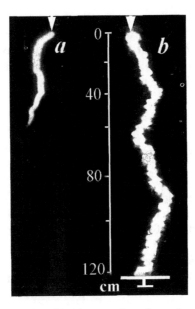

Figure 6.18. Still photographs of leader creeping along the soil during its development (*a*) and at the moment of gap bridging (*b*).

results and how well they reproduce the real process of lightning current spread are open to speculation.

6.4.2 Induction emf in an affected object

Let us consider a descending lightning stroke at an object of height h. The induction emf for the object is proportional to its inductance, $L = L_1 h$. The linear inductance can be estimated in a simple way, assuming that the current fills up a conductor composed of the object and the lightning channel of length $l \gg h$. If we assign to the conductor a constant radius $r_0 \ll l$ and assume a perfectly conducting soil, we shall obtain $L_1 \approx (\mu_0/2\pi) \ln (2l/r_0)$. This value will be $L_1 \approx 2.3\,\mu\mathrm{H/m}$ at $l \approx H = 3\,\mathrm{km}$ (H is the altitude of the negative cloud charge centre) and $r_0 = 5\,\mathrm{cm}$. Actually, the return stroke wave covers a much smaller distance during the time of the current impulse rise, when $\mathrm{d}i/\mathrm{d}t$ and the induction emf have maximum values. But owing to the logarithmic dependence of L_1 on the geometrical size of a long conductor, the change in the length of the lightning channel filled by current will affect but little the value of L_1. For example, we obtain $l = v_\mathrm{r} t_\mathrm{f} = 500\,\mathrm{m}$ and $L_1 \approx 2\,\mu\mathrm{H/m}$ for the return stroke velocity $v_\mathrm{r} = 10^8\,\mathrm{m/s}$ and $t_\mathrm{f} = 5\,\mu\mathrm{s}$, corresponding to the first component of a negative lightning. At $t_\mathrm{f} = 1\,\mu\mathrm{s}$ (the rise time of the subsequent component), L_1 will be only 20% lower. The same result is obtained when one uses the vector potential $\mathbf{A}(t)$ and vortex electric field $E_M = -\partial \mathbf{A}/\partial t$ for the calculations. Suppose the current rises linearly with a distance from the current wave front, $i(x) = b(x_f - x)$, where $x_f = v_r t$ and $b = \text{const}$. The current at the point x of the channel rises linearly with time, $A_i = \partial i/\partial t = b v_r$. Neglecting the delay time, as was done in (6.9) and (6.11), we have

$$\mathbf{A}(t) = \frac{\mu_0 A_i}{2\pi v_r} \int_0^{v_r t} \frac{(v_r t - x)\,dx}{(x^2 + r_0^2)^{1/2}}$$

$$E_M = -\frac{\mu_0 A_i}{2\pi} \ln \frac{v_r t + (v_r^2 t^2 + r_0^2)^{1/2}}{r_0} \approx -\left(\frac{\mu_0}{2\pi} \ln \frac{2v_r t}{r_0}\right) \frac{\partial i}{\partial t} \equiv -L_1 \frac{\partial i}{\partial t}$$

(6.48)

where r_0 is the average radius of the object affected by lightning. One can see that the formula for L_1 in (6.48) coincides with the one above, provided l is understood as the length of the channel loaded by current by the moment of time t.

If the finite velocity of an electromagnetic signal is taken into account and the object is located directly under the lightning channel, which happens in the case of a direct stroke, the evaluations made in section 6.1.1 give

$$E_M = -\frac{\mu_0 A_i}{2\pi} \ln \frac{v_r t + [v_r^2 t^2 + (1 + \beta)^2 r_0^2]^{1/2}}{(1 + \beta) r_0}, \qquad \beta = \frac{v_r}{c}.$$ (6.49)

Once again, we should like to emphasize the small contribution of the delay: the logarithms in formulae (6.48) and (6.49) differ less than by 3% at $\beta = 0.3$, $t_f = 5\,\mu s$, and $r_0 \approx 1\,m$.

Expressions (6.48) and (6.49) define rigorously the vortex electric field E_M at the earth's surface. When the object's height is $h \ll v_r t_f$, which is valid for many practical situations, the variation in E_M along the object can be ignored and the induction emf is $U_M \approx E_M h$. The emf rises linearly with increasing h. In particular, if we have $h = 30\,m$, $v_r = 0.3c$ and the lightning current rises to the amplitude $I_M = 100\,kA$ for the time $t_f = 5\,\mu s$, the maximum value of the induction component of the voltage at $r_0 = 1\,m$ is $U_{M_{max}} = 780\,kV$, a value comparable with the electrical component $U_{e_{max}} = R_g I_M \approx 1000\,kV$ at $R_g \approx 10\,\Omega$. Of course, the effects of the electrical component may be more serious because of its longer action. Indeed, in the first approximation, the pulse $U_e(t)$ is similar in shape to the lightning current impulse and $U_M(t)$ to its time derivative.

Formula (6.49) can also be used to evaluate the magnetic component after the current impulse maximum. For this, the real current entering the lightning channel should be represented as a sum of the two components: $i_1 = A_i t$, $i_2 = -A_i(t - t_f)$, and $i_2 = 0$ at $t \leqslant t_f$. It is not surprising that U_M is non-zero behind the impulse front, since the magnetic field continues to grow, as the lightning channel is filled by current. The total overvoltage pulse of a direct stroke, $U_d(t) = U_e(t) + U_M(t)$, is very different from the

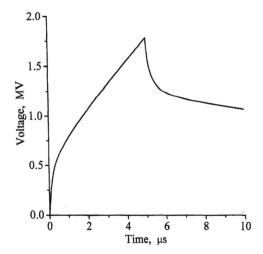

Figure 6.19. Computed overvoltage of the direct stroke at the transmission line tower with the grounding resistance $R_g = 10\,\Omega$ for the lightning current with $t_f = 5\,\mu s$ and amplitude of 100 kA; $v_r = 0.3c$.

lightning current impulse because of an abrupt rise and an equally abrupt fall of U_M with time (figure 6.19).

6.4.3 Voltage between the affected and neighbouring objects

It is important for many applications to know the voltage affecting the insulation gap between an object of height h, affected by a direct lightning stroke, and another object of height $h_1 < h$, located nearby. For this, one should find the difference between the evaluated overvoltage of the direct stroke, $U_{d_{max}}$, and the maximum overvoltage $U_{in_{max}}$, induced on the neighbouring object. The latter value strongly depends on the object's construction. So, let us analyse two extremal situations.

Suppose a lightning strikes a lightning rod located near the mast it protects (figure 6.20). The magnetic components of the overvoltages are determined by the vortex field strengths E_M from formula (6.49). For the rod, r_0 can be taken to be equal to its average radius r_1. For the mast, r_0 can be assumed to be equal to the distance d between the rod and the mast. The maximum time-dependent difference between the magnetic components of the voltage at the height h_1, where the distance between the constructions is minimal, is equal to

$$\Delta U_{M_{max}} \approx \frac{\mu_0 A_i h_1}{2\pi} \ln \frac{d}{r_1}. \tag{6.50}$$

Expression (6.50) allows for the value $v_r t \gg d$ at the moment this maximum occurs. The magnetic component of the overvoltage across the insulation

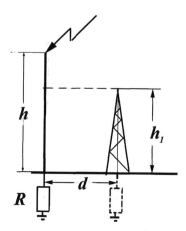

Figure 6.20. Estimation the voltage between a lightning rod struck and an object.

gap $\Delta U_{M_{max}} = U_{M_{rod}} - U_{M_{obj}}$ increases with distance d, because $U_{M_{rod}}$ is independent of d and $U_{M_{obj}}$ drops as the distance between the object and the current increases. The return stroke velocity v_r has practically no effect on $\Delta U_{M_{max}}$. Its upper limit is the value of $U_{M_{max}}$ for the affected lightning rod ($\Delta U_{M_{max}} \approx U_{M_{max}}/2$ at $d/r_1 \approx 100$).

The situation with the electrical component of overvoltage is less definite. The overvoltage is also determined by the difference between the two values, $\Delta U_e = U_{e_{rod}} - U_{e_{obj}}$, but $U_{e_{rod}} = -R_{gi}I_M$ is a definite quantity and $U_{e_{obj}}$ varies with the design of the object's grounding mat. The latter may be common with the lightning rod grounding grid and quite compact; in that case, we have $\Delta U_e = 0$ because the bases of the rod and the object are interconnected. There may be another extreme situation: the grounding mat of the object may be so far from that of the lightning rod that it may be unaffected by the electric field of the lightning current spreading through the soil. In that case, we shall not have $U_{e_{obj}} \equiv 0$ and $\Delta U_{e_{max}} = R_{gi}I_M$, because this would be possible only in the absence of current through the object's grounding mat. In reality, there is an electric charge induced on the object's surface due to the electrostatic induction (section 6.1.1), so a current flows across the object, creating the electrical component of the overvoltage. Its value can be found from formula (6.5) and the maximum value from (6.6), provided that the return stroke is simulated by a rectangular current wave in a vertical lightning channel. Let us evaluate the possible voltage from formula (6.6).

To go beyond the zone of the current spread away from the lighting rod grounding, it is necessary to move away at a distance ~20 m from it. The radius of the grounding grid, within which the electrodes are located, is hardly larger than 5 m, so that the distance between the rod and the object in formula (6.6) can be taken to be $r \approx 25 \, \mathrm{m}$. Assuming that the height of a typical object is $h = 30 \, \mathrm{m}$, its linear capacitance $C_1 \approx 10 \, \mathrm{pF/m}$, $C = C_1 h = 300 \, \mathrm{pF}$, and $R_\mathrm{g} = 10 \, \Omega$, we shall have $U_{e_{obj}} \approx 30 \, \mathrm{kV}$ at the lightning current $I_M = 30 \, \mathrm{kA}$. Although $U_{e_{rod}}$ and $U_{e_{obj}}$ have different signs and $|\Delta U_e| > |U_{e_{rod}}|$, the additional value is not essential because it is an order of magnitude less than $R_\mathrm{g} I_M$ in the above example. The situation when a lightning rod is put up at a distance sufficient for the separation of its own grounding grid and that of the object is quite realistic. This is done for the protection of especially important constructions to avoid pulse noises or sparking due to the induction emf, when some of the current finds its way to the object's grounding through the soil.

Another extreme case, in which the electrical component of the object overvoltage is dominant, is a lightning stroke at a metallic grounded tower of a power transmission line. The direct stroke overvoltage affects an insulator string, to which a power wire is suspended. Consider first a simple and frequent variant (in lines with an operation voltage below 110 kV) when the line has no protecting wire. In that case, we do not have to solve the difficult problem of lightning current distribution between the affected tower and the wire, repeatedly grounded by the adjacent towers. Nor should we bother about the electromagnetic effect of the protecting wire on the power wire (section 6.4.4). As in the previous situation, the insulator string is affected by the overvoltage ΔU equal to the potential difference of the tower at the point of the string suspension and the power wire. The calculation of the tower overvoltage is similar to that for a lightning rod, just described. A specific feature of this problem is the existence of the wire. Being suspended horizontally, it does not respond to the magnetic field of the current in the lightning vertical channel. The power wire is well insulated from the tower grounding by the insulator string. Owing to its far end being grounded, it would be able to maintain zero potential, but for the current created by the redistribution of the charge induced on the wire. The induced charge is very high because some of the wire length is located close to the lightning channel, and the total capacitance of a long wire is very large. Naturally, the small distance between the wire and the lightning channel does not mean the existence of a direct contact between them, so we can speak only of the effect of electrical induction on the wire.

Even though the wire is connected to the earth at zero resistance, the induced charge cannot respond immediately to the lightning charge variation and the wire potential cannot remain at zero. The grounding point is located far away, at the end of the wire, so the charge liberated by the induction cannot be delivered to it faster than with light velocity c. For the induced charge q_{in} to appear at the point x, a current wave must be excited at this point, which will eventually transport the charge $-q_{in}$ out of the wire to the earth. This wave will propagate at light velocity. During its motion, the potential at the wave front will rise due the voltage drop on the wave resistance of the line with distributed parameters, i.e., along a long wire. Elementary current and potential waves arise at any point on the wire, where the induced charge is changed by the lightning field. Propagating with light velocity to the left and to the right of the origin, the currents of elementary waves are summed, raising the voltage between the wire and the earth. After the waves are damped, this voltage, naturally, drops to zero, because the wire is grounded. The response of a long line to the external field $E_{0x}(x, t)$ acting along a horizontally suspended wire is described by the equations

$$-\frac{\partial U_e}{\partial x} = R_1 i + L_1 \frac{\partial i}{\partial t} - E_{0x}(x, t), \qquad -\frac{\partial i}{\partial x} = C_1 \frac{\partial U_e}{\partial t} \qquad (6.51)$$

where the potential $U_e(x, t)$ is due exclusively to the line response to the field $E_{0x}(x, t)$. The total potential of the wire relative to the earth, $U_{ge}(x, t) = U_0(x, t) + U_e(x, t)$, contains another component, $U_0(x, t)$, defined by the charges of the lightning return stroke. Neglecting the ohmic voltage drop relative to the induction term and taking $\partial U_0 / \partial x = -E_{0x}$ into account, we arrive at the wave equation with a distributed driving force and containing no damping term:

$$\frac{\partial^2 U_{ge}}{\partial x^2} - \frac{1}{c^2} \frac{\partial^2 U_{ge}}{\partial t^2} = -\frac{1}{c^2} \frac{\partial^2 U_0}{\partial t^2}, \qquad c = (L_1 C_1)^{-1/2}. \qquad (6.52)$$

The solution to this equation represents a general solution to a homogeneous equation and a particular solution to an inhomogeneous one, corresponding to the two identical waves propagating in opposite directions along the line:

$$U_{ge}(x, t) = \frac{1}{2} \int_0^t \frac{\partial}{\partial \Theta} U_0(X_1, \Theta) \, d\Theta + \frac{1}{2} \int_0^t \frac{\partial}{\partial \Theta} U_0(X_2, \Theta) \, d\Theta \qquad (6.53)$$

$$X_1 = x - c(t - \Theta), \qquad X_2 = x + c(t - \Theta).$$

The integrals give the sum of the above elementary waves moving at light velocity. The waves are excited by the time variation of the external field potential U_0. For the elementary wave to arrive at the point x at the moment of time t, the causative variation in U_0 must occur at the points $x \pm \Delta x$ earlier, by the time $\Theta = \Delta x / c$. If the time is counted from the moment of the lightning contact with the line tower, the lower, zero limit of the integrals of (6.53) should be replaced by the time-of-flight of light for a minimum distance between the lightning channel and the wire.

In the general case, the difficulties that arise in the calculation of the integrals depend on how one approximates the lightning current related to the linear charge in the return stroke wave inducing the field E_0, as well as on the lightning channel position relative to the wire. Of significance are the following factors: what object the lightning strikes (the earth or an element of a power transmission line raised above the ground), the channel deviation from the normal, as well as its bendings and branching. It is impossible to solve this problem without numerical computations. The question then arises as to the stage in the study, at which a computer simulation is most helpful. One should not ignore a numerical integration of initial equations (6.51), allowing the control of the effect of active resistance R_1 which sometimes has a large value. The effective value of R_1 may be much higher than the resistance of the line wire to direct current because of the skin-effect, the soil resistance, used by the wave as a 'return wire', and due to the energy consumed by the impulse corona. The corona is excited in the wire by overvoltages and absorbs some of the propagating wave energy, contributing to its damping. The impulse corona also increases the effective linear capacitance of the wire, since the electrical charge is localized not only on the wire surface but in the adjacent air. The charge is delivered there by streamers of metre lengths. The capacitance $C_{1\mathrm{eff}}$ depending on the local wire overvoltage varies together with the velocity of perturbations in the wire, $v = (C_{1\mathrm{eff}}L_1)^{-1/2}$. This greatly distorts the wave front, since different sections of the wave front have different velocities. The problem becomes greatly non-linear and definitely requires a numerical solution.

The calculation formulas given below describe simple situations neglecting the wave damping in the wire. They have been derived by direct integration of (6.53) and borrowed from [3]. The lightning channel is considered to be vertical; the return stroke wave moves towards the cloud at constant velocity v_r. For a rectangular charge wave in the channel of lightning that has struck the earth (but not a line tower) at a horizontal distance r from the wire, we obtain for the wire point nearest to the lightning channel

$$U_{\mathrm{ge}}(0, t) = -\frac{I_M h}{2\pi\varepsilon_0 c r} \frac{1}{\kappa^2 + 1}\left[\kappa - \frac{\beta}{(\kappa^2 + 1 - \beta^2)^{1/2}}\right], \qquad \kappa = \frac{v_r t}{r} \quad (6.54)$$

where $\beta = v_r/c$, v_r is the return stroke velocity, and h is the wire height above the earth. The time in (6.54) is counted from the moment of the lightning channel contact with the earth. This formula can be used at $t > r/c$, i.e., after the electromagnetic signal has covered the distance between the channel and the wire. The overvoltage is still active at a large distance from the stroke point ($x \to \infty$), where the lightning field effect is negligible, $E_{0x} \approx 0$. The wave reaches that point through the wire, as in the case of a communications line (this occurs without damping at $R_1 = 0$). Such overvoltage waves are known as wandering waves. For these waves, we have

$$U_{\mathrm{ge}}(\infty, t) = -\frac{I_M h v_r t}{2\pi\varepsilon_0 c(v_r^2 t^2 + r^2)} \quad (6.55)$$

where the time is counted from the moment of the wave front arrival at the 'infinitely' remote point of interest. At $t_m = r/v_r$, the function $-U_{\mathrm{ge}}(\infty, t)$ has

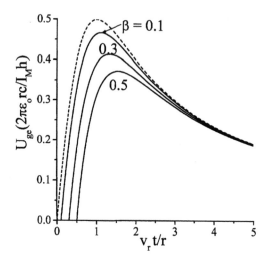

Figure 6.21. Evolution of an overvoltage at the wire point nearest to the lightning stroke point (solid curves) and a wandering wave voltage (dashed curve). For a return stroke, a rectangular current wave model is used.

a maximum:

$$-U_{ge}(\infty, t_m) = \frac{I_M h}{4\pi\varepsilon_0 rc} \tag{6.56}$$

which is independent of v_r. At $t \gg t_m$, the overvoltage is damped as t^{-1}.

Typically, the amplitude of a wandering wave is somewhat higher than the voltage amplitude relative to the earth at the site where the wire is closest to the stroke point (figure 6.21). The reason for this is the opposite signs of U_0 and U_e, causing a reduction in the value of $U_{ge} = U_e + U_0$ in the close vicinity of the wire, where $U_0 \neq 0$. Far from the stroke point, we have $U_0 \approx 0$, and the overvoltage is totally defined by the wire response. Although the overvoltage maximum at the closest point does vary with v_r, this variation is not appreciable. This is good because there are few measurements of the return stroke velocity and practically no synchronized measurements of the lightning current.

- If the lightning current is supposed to rise at the impulse front as $i(t) = A_i t$ with $A_i = \text{const}$, the overvoltage $U_{ge}(0, t)$, for the same conditions and designations as in (6.54), is

$$U_{ge}(0, t) = -\frac{A_i h}{2\pi\varepsilon_0 c v_r} \ln \frac{(\kappa^2 + 1 - \beta^2)^{1/2} - \beta\kappa}{1 - \beta^2}. \tag{6.57}$$

Formula (6.57) has a sense at $r/c \leqslant t \leqslant t_f + r/c$ ($\beta \leqslant \kappa \leqslant (v_r t_f/r) + \beta$). After the current impulse maximum, $t > t_f$, the calculation can be made using this formula and a superposition, by representing a real current wave as two waves of different signs shifted in time, as was done in the comments on formula (6.49). As the time increases within the lightning current rise time, the value of $U_{ge}(0, t)$ rises monotonically. The calculated pulses $U_{ge}(0, t)$ for a rectangular current wave and for a wave with a linearly rising current have something in common at $r/v_r \approx t_f$, which is valid for remote lightning strokes with $r \geqslant 100$ m. Both approximations lead to an increase in the over-voltage during the current front. But for close strokes, especially for a direct stroke at a line tower, the discrepancy between the calculated pulses becomes remarkable. This is the reason for the sceptical attitude to analytical solutions, which we showed at the beginning of the discussion. No doubt, a linearly rising current is closer to reality than a rectangular impulse, but it cannot simulate the actual current rise accurately. The same is true of the impulse amplitude. The discrepancy in the calculations made within the models considered grows with decreasing distance r.

Nevertheless, another analytical solution [3] may be useful for the estimations. It concerns the case of a direct lightning stroke at a transmission line tower, when the shortest distance between the lightning channel and the wire is determined only by the height difference between the tower, h_s, and the wire, h. If a charge wave corresponding to the lightning current front $i(t) = A_i t$ moves up along the vertical lightning channel from the tower top to the cloud with constant velocity v_r, we shall have at the point of the wire suspension

$$U_{ge}(0, t) \approx -\frac{A_i h}{4\pi\varepsilon_0 c v_r} \ln \frac{(v_r t + h_s)[(v_r t + h_s - h)(v_r t + h_s + h)]^{1/2}}{(1 + \beta)h_s[(h_s - h)(h_s + h)]^{1/2}}. \quad (6.58)$$

Let us calculate the overvoltage due to the first component of a negative lightning with the average parameters $I_M = 30$ kA, $t_f = 5\,\mu s$, $A_i = 6$ kA/μs, and $v_r = 0.3c = 90$ m/μs. We shall have $U_{ge}(0, t_f) \approx 320$ kV for $h_0 = 30$ m and $h = 20$ m at $t = t_f$. Similar, but of the opposite sign, is the potential rise on the tower grounding resistance $R_g \approx 10\,\Omega$ due to the lightning current, $R_g I_M = 300$ kV. This doubles the electrical component of the overvoltage across the insulator string.

It is worth noting the specific effect of overvoltage on the line insulation. Overvoltage is not strictly related to any point on the line, as is the case with the voltage drop across the tower grounding. It has been mentioned that the charge liberated by electrical induction moves along the wire, creating a wandering overvoltage wave. With a negligible damping, it can cover a distance of several kilometres, affecting, on its way, all the insulator strings it encounters. An insulation breakdown may occur even far from the lightning stroke, where the line insulation is poor for some reason. Really hazardous is the encounter of the wandering wave with a high-voltage substation, because the overvoltage wave penetrates to the internal insulation of its transformers and generators, which is always poorer than the external insulation.

A wandering wave also arises when a lightning strikes a power wire. The lightning current spreads along the wire from the stroke point in both directions, producing very strong overvoltage waves, $U(x, t) = Zi(x, t)/2$. Since the wave resistance is $Z \approx 250–400\,\Omega$ (the smaller value is typical of ultrahigh voltage lines with split wires of a large equivalent radius), the current $I_M = 30\,kA$ would produce an overvoltage with an amplitude of 3.5–6 MV. In reality, the overvoltage is limited to the value of breakdown voltage for the tower insulator string closest to the lightning stroke, where the flashover does occur. A wave with an amplitude equal to the string break-down voltage is a wandering wave in this case. Of course, the overvoltage may rise again, after the string flashover, due to the self-induction emf of the tower and to the voltage drop across its grounding, to which the lightning current runs after the string flashover. Wandering waves are damped by the same processes that determine the resistance R_1 in (6.51).

It is important for lightning protection practice to compare the over-voltages due to direct strokes at a line tower and a wire. In the former case, the voltage drop across the insulator string is the sum of three components. The voltage drop across the tower grounding and the induced voltage of the wire are approximately the same quantitatively but have opposite signs. This totally gives about $\sim 2R_gI_M$ over the string. The mag-netic component L_sA_i has a real effect only on the current impulse front, and its average value is equal to L_sI_M/t_f (L_s is the tower self-inductance). The magnetic component for the first leader of a moderate negative lightning ($I_M = 30\,kA$, $t_f = 5\,\mu s$) and for a tower of standard size ($h_0 \approx 30\,m$) does not exceed 200 kV but $2R_gI_M > 600\,kV$ because of $R_g \geqslant 10\,\Omega$. It appears that overvoltages due to a direct moderate stroke at a tower can flashover the insulation only in lines with voltages less than 110 kV, which have strings less than 1 m in length. For a 220 kV transmission line, a hazard may arise when the currents are twice as high as the average value, but such lightnings occur only with a 10% frequency. The hazard of a lightning stroke at a tower is not high for 500–750 kV transmission lines, since they have long strings. A reverse flashover may arise from a lightning with 100 kA currents and more, but their number is less than 1% of the total. If the lightning current strikes the wire, the current spreads in both directions along it. With the wave resis-tance $Z > 200\,\Omega$, we get $ZI_M/2 > 3\,MV$ even for a moderate lightning. This is sufficient to flashover the insulation of any of the currently operating lines. A lightning stroke at a wire should always be considered to be hazardous.

6.4.4   Lines with overhead ground-wires

When a lightning strikes a tower of a power transmission line protected by a grounded wire, the current is split between the tower and the grounded wire, due to which the current load on the tower is reduced. However, the engineer is then faced with a complex problem of calculating the current distribution.

Another aspect of this problem is the account of the screening effect of a protecting wire. Since the wire is connected to the tower, it acquires, in a first approximation, the tower potential, thus creating a voltage wave of the same sign. Owing to the electromagnetic induction, a similar wave of a lower amplitude is excited in the power wire. As a result, the voltage in the insulator string, equal to the potential difference between the tower and the power wire, drops. These additional problems complicate the calculation of direct stroke overvoltage for a line with an overhead ground-wire. The problems that arise here relate to electric circuit theory rather than to physics, so we shall discuss them only briefly.

Many engineers try to calculate the current distribution between a tower and an overhead wire within the model of an equivalent circuit with concentrated parameters. The lightning channel is regarded as a source of current $i(t)$. The tower is replaced by its inductance L_s and grounding resistance R_g, the two grounding wire branches (on the left and on the right of the stroke point) are represented by the branch inductances $L_c/2$ and their grounding resistances in the adjacent towers, $R_g/2$, connected in parallel. One also introduces the mutual-induction emf $M_c\, di/dt$, induced by the lightning current in the wire–towers–earth circuit (figure 6.22). This circuit can be simplified further by putting $R_g = 0$, because the principal interest is focused on the current front of $t_f \approx 1\text{--}10\,\mu s$ and because the cable inductance along the many hundreds of metres of its length is as high as hundreds of microhenries and the time constant is usually taken to be $L_s/R_g > 100\,\mu s$. This model circuit then presents no calculation problems, provided that the mutual inductance of the vertical lightning channel and the circuit including the ground-wire is known. As the return stroke wave moves up, the channel is filled by current so that the value of M_c rises in time. The calculations similar to those for the derivation of formula (6.49) and allowing for the time delay yield [3]

$$U_c(t) = \frac{\mu_0 h_0}{2\pi}\left[\ln\frac{v_r t + 2h_0}{2(1+\beta)h_0} + 1\right]. \tag{6.59}$$

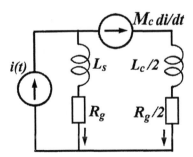

Figure 6.22. The design circuit for a current a tower of the line with the protective cable.

The considerable simplification of the real process can be justified only at low grounding resistances of the towers, when the current in the wire circuit is limited mostly by its inductance, and one can neglect the current branching off to the grounding resistances of all other towers except the one nearest to the affected tower.

The value of R_g in a real transmission line in areas with low conductivity soils may be several times higher than the normal value, reaching $100 \, \Omega$. Then the current distribution problem must also take into account the removal of some of the lightning current to 2–5 towers away from the stroke point. The equivalent circuit becomes more complicated (chain-like), representing a series of link circuits identical to the first one. For a more rigorous solution, the ground-wire is to be considered as a long line with a wave resistance Z_c and many local non-uniformities produced where the ground-wire contacts a tower. Each tower is then represented as a chain of L_s and R_g connected in series. Figure 6.23 illustrates the variation of the current impulse in the tower with the design circuit. For a circuit with lumped parameters, the neglect of the tower grounding resistance $R_g \approx 10 \, \Omega$ does not affect the result much, while at $R_g \approx 100 \, \Omega$, the tower current impulse shape changes radically. A circuit with distributed parameters permits one to follow the effect of consecutive wave reflections at the contacts between the ground-wire and the towers. The current impulse distortion by the reflected waves is especially

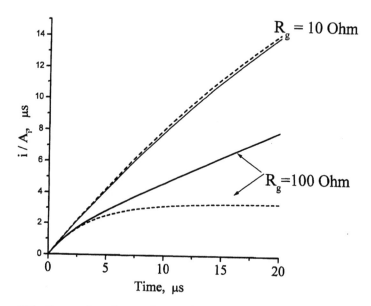

Figure 6.23. Current impulse on the struck tower with (solid curves) and without (dashed curves) allowance for a grounding resistance of the nearest tower. The model with a linearly raising current front is used for return stroke.

appreciable for a short impulse front t_f characteristic of the subsequent lightning components. As the linear resistance of the line, R_1, rises, the effect of
the reflections becomes less pronounced because the reflected waves are
damped more strongly. In any case, the overhead wire takes some of the
lightning current away, thereby unloading the affected tower; this current
fraction cannot be less than $2R_g/Z_c$.

Let us evaluate the screening effect of the protecting wire, which also
reduces the direct stroke overvoltage. Engineers had become aware of this
effect long before overhead wires were used as lightning protectors. Some
even supposed that a wire could reduce the voltage of an insulator string
to a value lower than the flashover voltage. The wire acquires the electrical
potential of the tower, which has increased by the value of the voltage
drop across the grounding resistance. As a result, a high-voltage wave runs
along the wire. The nearby power wire finds itself in its electromagnetic
field inducing a similar wave. If U_c is the voltage wave amplitude in the
ground-wire, the voltage produced in the power wire is $U_{coup} = k_{coup}U_c$,
where $k_{coup} = Z_{cw}/Z_c$ is a coupling coefficient and Z_{cw} is the wave resistance
of the grounded wire–power wire system which can also be regarded as a long
line. We have $Z_{cw} = (L_{1cw}/C_{1cw})^{1/2}$ by definition. The linear inductance L_{1cw}
and the capacitance C_{1cw} between this two wires are calculated in a
conventional way, with the allowance for the earth's effect. With $L_{1cw} \sim$
$\ln[(h_0 + h)/(h_0 - h)]$ and $C_{1cw} \sim \{\ln[(h_0 + h)/(h_0 - h)]\}^{-1}$, the coupling
coefficient is

$$k_{coup} \approx \frac{\ln[(h_0 + h)/(h_0 - h)]}{\ln(2h_0/r_c)} \qquad (6.60)$$

where r_c is the ground-wire radius. For a rigorous calculation, the geometrical radius in (6.60) should be replaced by an equivalent radius of the space
charge region at the wire, (the space charge is incorporated by a impulse
corona under the action of high voltage). This somewhat increases the
value of k_{coup}. Measurements give approximately $k_{coup} \approx 0.25$, instead of
the calculated 'geometrical' value of $k_{coup} \approx 0.2$. Therefore, the electrical
component of power wire overvoltage is reduced once more, this time by
the value $U_{coup} = k_{coup}U_c$. The total overvoltage reduction owing to the
ground-wire makes up several dozens percent, decreasing the tower-stroke
effect on the transmission line insulation.

We should like to mention a certain relationship between the type of
lightning action and the transmission line cut-off. Even induced overvoltages
are hazardous for low voltage lines (primarily those of 0.4–10 kV). Induced
overvoltages are much more frequent than direct strokes and are the main
reason for the line cut-offs. A protecting wire is useless in this case, so low
voltage transmission lines do not have it at all. For a line of 35 kV or
more, induced overvoltages are practically harmless and direct lightning
strokes are dominant. The favourable effect of an overhead grounded wire

becomes apparent at an operating voltage $U \geqslant 110\,\mathrm{kV}$, when the lightning current leading to the insulation flashover after a stroke at a line tower exceeds an average value $\sim 30\,\mathrm{kA}$. About 50% of cut-offs for 110–$220\,\mathrm{kV}$ lines equipped with a grounded wire are due to strokes at towers and 50% of cut-offs occur when lightning breaks through to get to the power wire. Beginning with $500\,\mathrm{kV}$, an increasing number of cut-offs are due to lightning breakthroughs to the power wire.

6.5   Concluding remarks

We finish this chapter and the book by describing the lightning effect on power transmission lines. Scientists are still unable to offer a clear mathematical description of its complicated mechanism. Modern computer simulations can infinitely specify and refine a mathematical model of the lightning effect, with respect to both the electromagnetic field and the object's response to a stroke. This is, to some extent, interesting, useful and makes sense. The process of refining computations has no limit. To illustrate, a detailed analytical treatment of long line parameters with the account of the earth's effect has taken several hundreds of pages in the work by Sunde [6]. Suppose a superprogramme has been created for the solution of the lightning protection problem; its application will immediately show that the great efforts it has required can change but little the existing low predictability of lightning-induced cut-offs. The key problem today is not a rigorous mathematical solution of the available equations but an adequate physical description of the principal physical processes producing a lightning discharge, its electromagnetic field, and the object's response to it. For this reason, we have tried to present simple qualitative models rather than stringent solutions to the equations. On the other hand, many aspects of this problem have been omitted, partly because they are not directly related to lightning as a physical phenomenon and partly due to the lack of space or to the limited knowledge about the key physical phenomena.

Let us look back at the material presented in this book in order to emphasize the points of primary importance. After the numerical value of an overvoltage has been calculated, it is necessary to compare the result obtained with the flashover voltage of the insulation in order to identify its possible flashover. Most of the voltage–time characteristics of insulation strings have been found from tests by standard $1.2/50\,\mu\mathrm{s}$ impulses (here, the first value is the front duration and the second is the impulse duration on the 0.5 level). Such a refined impulse has little to do with lightning over-voltages, and this is clear from figure 6.24. A lightning-induced overvoltage has necessarily a short-term overshoot arising not only from the current wave reflection by the grounding of the neighbouring towers but also from the magnetic induction emf. It is not quite clear how this rapidly damping

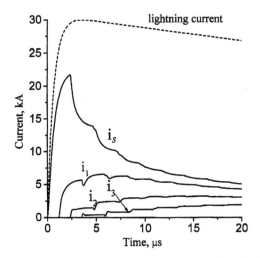

Figure 6.24. Current impulses in a struck tower (i_s) and in the few neighbouring towers (i_1-i_3). The wave problem was solved allowing for wave reflection from the places where grounded wire is connected with the towers.

overshoot affects the electrical strength of an insulation string. Under certain conditions, a powerful corona flash saturates the gap with a large space charge and can 'lock up' the leader process, increasing the strength [12]. As for 'anomalously' long overvoltages induced by positive lightnings, the electrical strength of air may, on the contrary, be several dozens percent smaller than what standard tests give [5, 13]. The question of the real electrical strength of the UHV transmission line insulation is still to be answered.

The return stroke models discussed above ignore the lightning channel branching and bending, whereas an actual discharge channel is far from being a straight vertical conductor. The channel can deviate from the normal by dozens of degrees, especially when it approaches high constructions. Another complicating point in a return stroke model is the counterleader. The assumption that the return wave starts directly from the top of an affected construction, say, from a line tower, is far from the reality. The length of a counterleader is comparable with the height of the construction it starts from. Together with the total length of the streamer zones of the descending and ascending leaders, this will give a value 1.5–3 times greater than the object's height. Such a high altitude of the return stroke origin and its propagation in both directions from the point of contact, not only towards the cloud, may have a considerable effect on the electromagnetic field of the lightning. The available theoretical models do not take these facts into account. There are no data on counterleaders related to the subsequent lightning components. Today, it is even impossible to confirm, or to disprove, the mere existence of a leader travelling to meet a dart leader.

Another weak point of the models is the set of statistical data on the amplitude and time characteristics of the lightning current impulse. There is some information on medium current lightnings, because these are numerous, whereas lightnings of extremal parameters are poorly understood. The consequences of this are quite serious. The choice of protection means and measures depends, to a large extent, on what has actually caused the storm cut-off of a particular transmission line – a reverse flashover of the insulation string, when the lightning strikes a tower, or the lightning breakthrough to the power wire bypassing the overhead protecting wire. Underestimating or, on the contrary, overestimating the high current probability by ignorance, one may arrive at the wrong conclusion concerning the contribution of reverse flashover in UHV transmission lines, which may cause great losses. The determination of extremal lightning parameters is one of the key problems in natural investigations. We should like to emphasize again that the exceptions are more important than the rules to lightning protection practice.

References

[1] Wagner C F 1956 *Trans. AIII* **75** (Pt 3) 1233
[2] Lundholm R, Finn R B and Price W S 1958 *Power Apparatus and Systems* **34** 1271
[3] Razevig D V 1959 *Thunderstorm Overvoltage on Transmission Lines* (Moscow: Gosrenrgoizdat) p 216 (in Russian)
[4] Golde R H (ed) *Lightning* 1977 vol. 2 (London, New York: Academic Press)
[5] Bazelyan E M and Raizer Yu P 1997 *Spark Discharge* (Boca Raton: CRC Press) p 294
[6] Sunde E D 1949 *Earth Conduction Effects in Transmission Systems* (Toronto: Van Nostrand) p 373
[7] Ryabkova E Ya 1978 *Grounding in High-Voltage Installations* (Moscow: Energiya) (in Russian)
[8] Bazelyan E M, Chlapov A V and Shkilev A V 1992 *Elektrichesrvo* **9** 19
[9] Kaden H 1934 *Archiv fur Electrotechnik* **12** 818
[10] Babinov M B and Bazelyan E M 1983 *Elektrichesrvo* **6** 44
[11] Babinov M B, Bazelyan E M and Goryunov A Yu 1991 *Elektrichesrvo* **1** 29
[12] Bazelyan E M and Stekolnikov I S 1964 *Dokl. Akad. Nauk SSSR* **155** 784
[13] Burmistrov M V 1982 *Elektrot. promyshlennost'; Ser. Appar. wysokogo napryazheniya* **1** 123

Index

Transmission line
  effect of operating voltage 253, 254
  protected by overhead wire 25, 314
  wandering wave 312, 314
  without protection 309, 314

Vector potential 272, 306
Velocity
  of dart leader 98
  of leader, empiric formula 82, 140,
    141
  of alpha and beta leader type 97
  of return stroke 116, 117, 173, 175,
    190
  of streamer 35, 37

Vortex electric field 272, 306

Wave
  in cable sheath 279
  in lightning channel 186
  ionization 32, 212, 214
  of potential, reflection from
    cloud 191
  resistance 175
  wandering in transmission line 312,
    314
Wire
  circle protector 248
  protector 250–252, 314

Milton Keynes UK
Ingram Content Group UK Ltd.
UKHW021626071024
449327UK00020BA/1203